AF316751

Glycans in Cell Interaction and Recognition

Glycans in Cell Interaction and Recognition
Therapeutic Aspects

Edited by

Michèle Aubery

Laboratoire de Glycobiologie
et Reconnaisance Cellulaire
Université René Descartes – Paris V
France

harwood academic publishers
Australia • Canada • France • Germany • India • Japan
Luxembourg • Malaysia • The Netherlands • Russia • Singapore
Switzerland

Amsteldijk 166
1st Floor
1079 LH Amsterdam
The Netherlands

British Library Cataloguing in Publication Data

A catalogue record for this book is available from the British Library.

ISBN: 90-5823-052-X

This book is dedicated to Professor André Verbert, in memory of a friendly glycomaniac

Contents

Preface

Cellular interactions are involved in many biological events, including regulation of gene expression, cell growth, cell migration, cell differentiation, and apoptosis, and might be modulated in several physiological and pathological processes, e.g., inflammation, arthritis, metastasis. During these interactions, cell–extracellular matrix and cell–cell adhesions occur through specific surface receptors, cadherins, integrins, selectins and/or proteoglycans. The last decade has witnessed major discoveries of protein moieties of the adhesive molecules and their membrane receptors, and significant advances have been made in the functional sequences, the nature of the promoter and the design of therapeutic techniques using adhesive molecules. It must be kept in mind that most adhesion molecules and adhesion receptors are glycoproteins, and advances in glycobiology have revealed the possibility that the carbohydrate chains of adhesion molecules might play important roles in cell–extracellular matrix and cell–cell recognition. It is necessary to consider the contribution of glycans borne by adhesion molecules in the function and regulation of cell recognition in terms of two complementary aspects: 1) basic investigations to elucidate mechanism(s) and regulation of glycans, and 2) medical, pharmaceutical and therapeutic aspects and diagnostic techniques.

The book has been organized into two parts. The first deals with general aspects of structure and biosynthesis of glycoconjugates. In addition, a review of the structure and metabolism of glycosaminoglycans and glycolipids is given, since some of these molecules also play a role in cellular interactions. A general account of cell adhesion and the extracellular matrix is also included in the first part. The second part addresses the pathology aspects of glycoconjugates as regards to cancer progression, neurophysiopathologies and Carbohydrate Deficient Glycoprotein Syndromes, and their therapeutic aspects. The editor would like to express her gratitude to P. Méhul for the book cover illustration.

Contributors

Valérie Altemayer
Glycobiologie
Vectorologie et Trafic Intracellulaire
CBM-CNRS
rue Charles-Sadron
45071 Orléans-Cedex 2
France

Michèle Aubery
Laboratoire de Glycobiologie et
Reconnaissance Cellulaire
Université René-Descartes–Paris V
UFR Biomédicale des Saints-Pères
45 rue des Saints-Pères
75006 Paris
France

Monique Aumailley
Institute for Biochemistry II
University of Cologne
Joseph-Stelzmann-Str. 52
50931 Cologne
Germany

Sylvain Bourgerie
Glycobiologie
Vectorologie et Trafic Intracellulaire
CBM-CNRS
rue Charles-Sadron
45071 Orléans-Cedex 2
France

Violaine Carriere
Glycobiologie
Vectorologie et Trafic Intracellulaire
CBM-CNRS
rue Charles-Sadron
45071 Orléans-Cedex 2
France

James W. Dennis
Samuel Lunenfeld Research Institute
Mount Sinai Hospital
600 University Ave, Rm. 876
Toronto, Ontario, M5G 7X5
Canada

Christian Derappe
Laboratoire de Glycobiologie et
Reconnaissance Cellulaire
Université René-Descartes–Paris V
UFR Biomédicale des Saints-Pères
45 rue des Saints-Pères
75006 Paris
France

Eric Duverger
Glycobiologie
Vectorologie et Trafic Intracellulaire
CBM-CNRS
rue Charles-Sadron
45071 Orléans-Cedex 2
France

Hudson H. Freeze
The Burnham Institute
10901N. Torrey Pines Rd.
La Jolla, CA 92037
USA

Con-Xiao Gao
Department of Biochemistry
Osaka University Medical School
2-2 Yamadaoka Suita
Osaka 565-0871
Japan

Martine Gilleron
Institut de Pharmacologie et Biologie
Structurale
CNRS, 205 route de Narbonne
31077 Toulouse Cedex
France

Hinrich K. Harms
Department of Pediatric
Gastroenterology
Dr. Von Haunersches Kinderspital
Ludwig Maximilian University
80337 München
Germany

Yoshita Ihara
Department of Biochemistry
Osaka University Medical School
2-2 Yamadaoka Suita
Osaka 565-0871
Japan

Yoshitaka Ikeda
Department of Biochemistry
Osaka University Medical School
2-2 Yamadaoka Suita
Osaka 565-0871
Japan

Thorsten Marquardt
Klinik und Poliklinik fur
Kinderheilkunde
Albert-Schweitzer-Str. 33
48149 Munster
Germany

Roger Mayer
Glycobiologie
Vectorologie et Trafic Intracellulaire
CBM-CNRS
rue Charles-Sadron
45071 Orléans-Cedex 2
France

Jean-Claude Michalski
Laboratoire de Glycobiologie
Structurale et Fonctionnelle
Université des Sciences & Technologies de Lille
Bâtiment C9-59655 Villeneuve d'Ascq Cedex
France

Patrick Midoux
Glycobiologie
Vectorologie et Trafic Intracellulaire
CBM-CNRS
rue Charles-Sadron
45071 Orléans-Cedex 2
France

Eiji Miyoshi
Department of Biochemistry
Osaka University Medical School
2-2 Yamadaoka Suita
Osaka 565-0871
Japan

Michel Monsigny
Glycobiologie
Vectorologie et Trafic Intracellulaire
CBM-CNRS
rue Charles-Sadron
45071 Orléans-Cedex 2
France

Chantal Pichon
Glycobiologie
Vectorologie et Trafic Intracellulaire
CBM-CNRS
rue Charles-Sadron
45071 Orléans-Cedex 2
France

Germain Puzo
Institut de Pharmacologie et Biologie
Structurale
CNRS, 205 route de Narbonne
31077 Toulouse Cedex
France

Christophe Quetard
Glycobiologie
Vectorologie et Trafic Intracellulaire
CBM-CNRS
rue Charles-Sadron
45071 Orléans-Cedex 2
France

Michel Riviere
Institut de Pharmacologie et Biologie
Structurale
CNRS, 205 route de Narbonne
31077 Toulouse Cedex
France

Annie-Claude Roche
Glycobiologie
Vectorologie et Trafic Intracellulaire
CBM-CNRS
rue Charles-Sadron
45071 Orléans-Cedex 2
France

Yin Sheng
Department of Biochemistry
Osaka University Medical School
2-2 Yamadaoka Suita
Osaka 565-0871
Japan

Ahmed S. Sultan
Department of Biochemistry
Osaka University Medical School
2-2 Yamadaoka Suita
Osaka 565-0871
Japan

Naoyuki Taniguchi
Department of Biochemistry
Osaka University Medical School
2-2 Yamadaoka Suita
Osaka 565-0871
Japan

Marafumi Yoshimura
Department of Biochemistry
Osaka University Medical School
2-2 Yamadaoka Suita
Osaka 565-0871
Japan

Jean-Pierre Zanetta
Laboratoire de Glycobiologie
Structurale et Fonctionnelle
Université des Sciences & Technologies de Lille
Bâtiment C9-59655 Villeneuve d'Ascq Cedex
France

Part A

General Aspects

A1. Biosynthesis of Membrane Glycoproteins

† André Verbert

1. INTRODUCTION

To become a membrane glycoprotein, a protein, after its birth in the ribosomal machinery, will have to fulfill three different requirements:

– to be integrated in the membrane bilayer,
– to be segregated, at least partially, in the lumen of the endoplamic reticulum where the whole co- and post-translational processes will take place,
– to be properly folded to reach the plasma membrane as a biologically active component.

The purpose of this chapter is to give the basic knowledge on biosynthesis of glycoproteins in order to appreciate how the glycoproteins are correctly synthesized in order to play their roles at the surface in cell-cell and cell-matrix adhesion, and to understand their relevance to pathology and therapeutical aspects.

2. SEGREGATION IN THE LUMEN OF ENDOPLASMIC RETICULUM AND MEMBRANE INSERTION

The destiny of a protein to be inserted in the membrane bilayer is written in its own protein sequence. A ''signal peptide'' occuring around the thirtieth amino acid allows the formation of a cytosolic tertiary complex between the nascent protein, the ribosome and a signal recognition particle (SRP). This complex associates to a docking protein of the ER membrane allowing the growth of the protein chain through the lipid bilayer. The protein chain is anchored in the membrane via an hydrophobic sequence known as the ''anchor sequence'' and the rest of the protein passes into the lumen. This membrane anchored protein will stay membrane bound if it does not possess any specific cleavable sequence susceptible to rough ER specific proteases. Quite a number of membrane proteins are also multispanned through the membrane, it is mainly the result of the occurence of hydrophobic sequences of around twenty amino-acid long which anchor the protein at various places.

It is out of the scope of this review to detail the complete translocation process of a protein within the rough ER lumen (for reviews, Walter and Lingappa, 1986; Kalies and

Hartmann, 1998) however it must be kept in mind that a protein crosses the lipid bilayer via a large assembly of proteins known as the "translocon". This polypeptide translocation complex contains the set of ribosomal docking proteins, the translocator proteins themselves but also the oligosaccharidyltransferase complex responsible for the initial step of the N-glycosylation (Gilmore, 1993). More surprising, it will be mentionned later in this chapter that a nascent protein can be retrotranslocated in the cytoplasm via a similar transmembrane complex.

Another way to afford the stable association of glycoproteins with the plasma membrane is the covalent linkage via a glycosyl phosphatidyl inositol anchor (GPI, see Figure 1e). GPIs are ubiquitous in eucaryotes. These structures exist covalently linked to specific proteins or glycoproteins and as free glycolipids representing biosynthetic intermediates. GPI-linked proteins are confined to the outer leaflet of the cell membranes (Ferguson, 1991).

3. THE GLYCOSYLATION STATUS OF MEMBRANE GLYCOPROTEINS

The cellular machinery for glycosylation appears quite complex. Could it be different when considering the multiplicity of glycan structures borne by glycoproteins? It is always surprising to realize that 3 identical amino-acids can only make 1 tripeptide although 3 identical monosaccharides can be combined in 176 different manners, and for larger oligosaccharides the possible combinations reach millions of possibilities. Quite complex glycosylation, since each linkage requires a specific glycosyltransferase; quite complex glycosylation, since pathways are controlled by an ordered action of glycosyltransferases and glycosidases. The synthesis of a simple biantennary glycans of a dozen sugars requires more than 40 step-after-step reactions. Finally, control must be very tight because the tridimensionnal structure- thus the function- of a glycan may be highly affected by a simple residue missing or in addition.

The glycoproteins can have their glycan moieties linked either to an asparaginyl residue (N-glycoproteins) or to a seryl or threonyl residue (O-glycoproteins) (for review, Vliegenthart and Montreuil, 1995). However, all amino acyl residues of this type are not glycosylated and consensus neighbouring sequences are required for a residue to be glycosylated. Such a consensus sequence is well known for N-glycan i.e Asn-Xaa-Thr/Ser (Xaa being any amino-acid residue but proline). But this is not sufficient and additional requirements must exist such as, presumably, an adequate folding of the protein around the glycosylation sites.

The two types of linkage can be found on the same protein. N-linked glycans are characterized by possessing an identical pentasaccharide core ($Man_3GlcNAc_2$), on which are attached a variety of antennae having at their non reducing end peripherals sugars, (quite often sialic acids and fucose). Three main types of N-glycans are described: the oligomannosidic type (also called high-mannose type), the N-acetyllactosaminic type (also called complex type) and the hybrid type. These three types are illustrated on Figure 1a, b, and c. It has to be noted that the complex type has often more than two antennae and multi antennae (tri-, tetra and penta-) are currently reported (for details, see the chapter on glycan structures by J.C. Michalski in this book).

This arrangement of sugar residues around a common inner core is also recovered for O-linked glycans but in this case 7 different inner cores have been reported as well as a wide variety of antennae (an example is given on Figure 1d).

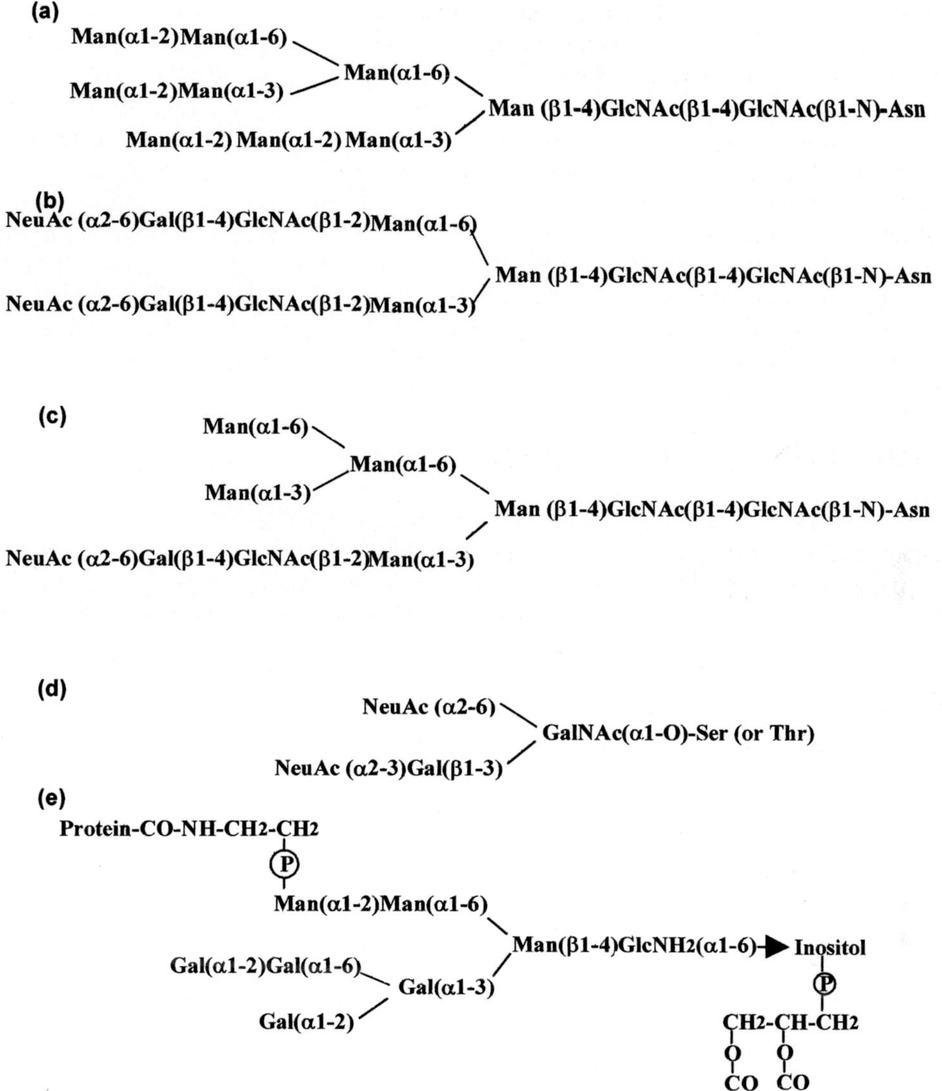

Figure 1: Some examples of structures of glycans of membrane glycoproteins.
(a) Oligomannoside type, (b) biantennary complex type, (c) Hybrid type, (d) O-glycan core as found in human glycophorin, (e) GPI type found in variant surface glycoprotein from *Trypanosoma brucei.*

Thus, the same general scheme is adopted for any glycoprotein glycans: a common core, various branches or antennae and few terminal sugars. From what has been recently described for N-glycans, it can be pointed out that concomitantly to the formation of the common oligomannosidic core, the correct folding of the glycoproteins is controlled. The terminal sugars, more involved in the function or half life of the protein, are added when

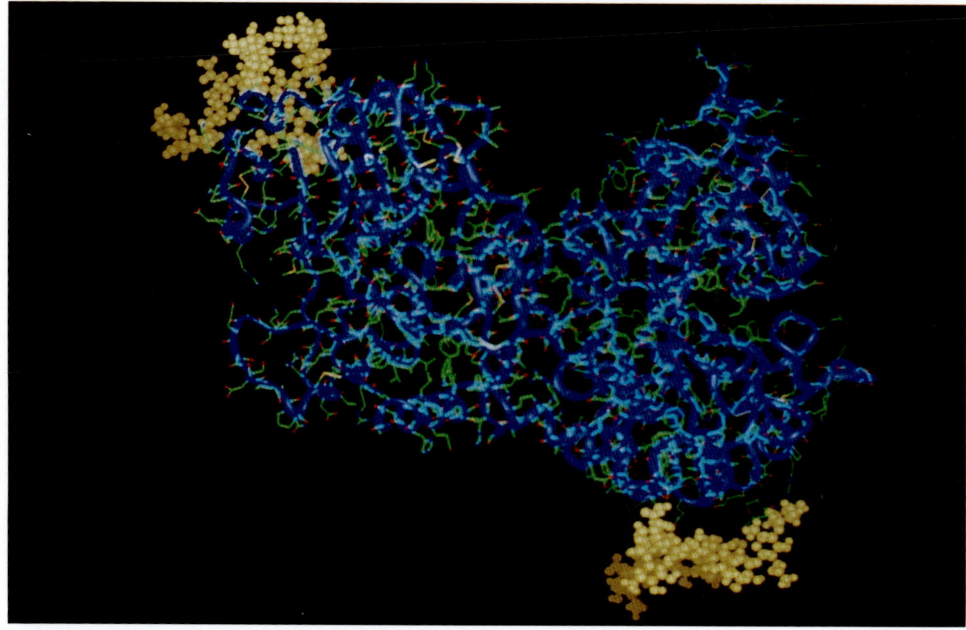

Figure 2: Three dimensional model of a glycoprotein.
Molecular modelling of human lactoferrin. The protein is in blue and the glycans are in yellow. Both glycans are biantennary complex types.

the protein has practically reached its final three dimensional structure. Thus, the common core would be more involved in the quality control of glycoproteins although terminal sugars would be more related to their final biological roles.

Last, but not least, it must be kept in mind that glycan moieties cover a large of the protein outer surface as illustrated in Figure 2. This is of importance for understanding the biological roles of glycans regarding solubility and folding of proteins in aqueous environment, protection of the protein against proteolytic attacks, epitope masking and all the recognition phenomena.

4. THE COMPLEXITY OF THE GLYCOSYLATION PROCESSES

This multiplicity of structures will imply the complexity of the various pathways of glycosylation and multiple controls of each step.

The type and structure of glycans will be mainly under the dependence of the enzymatic cell machinery and under the coordinated expression of glycosyltransferases. This will be also controlled by a precise subcellular topography of the enzymes and by the trafficking of the nascent glycoprotein within the different sites of glycosylation from the rough endoplasmic reticulum, through the Golgi vesicles, to the trans-Golgi network and the plasma membrane.

The general reaction for glycosylation consists in the transfer of a sugar residue from its activated donor form to a specific acceptor as shown here under:

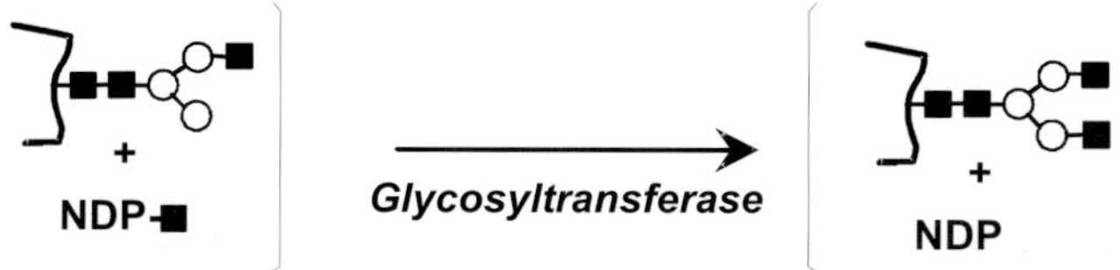

These reactions are very specific regarding the enzyme, the donor, the acceptor and the type of linkage. This already indicates that the glycosylation process is under general control of the gene expression of specific glycosyltransferases. This is true for any kind of glycosylation.

In all these reactions, the primary donor is a sugar-nucleotide as UDP-GlcNAc, UDP-GalNAc, GDP-Man, UDP-Glc, UDP-Gal, GDP-Fuc and CMP-NeuAc. It is worthwhile to note that these hydrophilic molecules are located in the cytoplasm although most of the glycosylation reactions occur in the lumen of ER or Golgi vesicles. The transport of these donors through the intracellular membranes to reach their reaction sites offer another regulation point based on compartmentalization and carrier dependent control of their transport. This transport is achieved by two means:

1 An hydrophobic membrane molecule will be charged with the sugar moiety from a nucleotide sugar by a transfer reaction at the cytoplasmic face of the membrane. Then, via a still non explained mechanism this carrier molecule "flip-flops" the sugar at the lumenal side of the vesicle and will act now as donor for the lumenal transfer reactions. This will be the case for the core formation of N-glycoprotein and GPIs using phospho-dolichol as lipid carrier, this will be detailed underneath.

2 The nucleotide sugar can be transported as such via a carrier protein. This is now well documented (Hirschberg, 1987) and some transporter has been cloned and expressed (Eckhardt, 1996). So far these carriers have been characterized as antiport proteins which exchange nucleotide sugars against nucleoside monophosphates produced themselves by the action of lumenal phosphatases on the residual nucleoside diphosphates released by the transfer reaction (see scheme above). This transport mechanism reaches a double goal: to allow and control the entry of sugar donors to the glycosylation sites and to exit the nucleoside phosphates, produced during the glycosylation, which are known to be inhibitors of the glycosyltransferase reactions.

5. GPIs BIOSYNTHESIS

GPIs biosynthesis (Menon *et al.*, 1997) is illustrated in Figure 3. It is initiated in the endoplasmic reticulum by the transfer of GlcNAc from UDP-GlcNAc to phosphatidyli-nositol (PI) to yield GlcNAc-PI. In the second step, GlcNAc-PI is de-N-acetylated to yield GlcN-PI which is further elaborated via the transfer of at least three mannose residues (from mannose-phospho-dolichol) and one or more phosphoethanolamine groups linked to the third mannose from the reducing terminus of the glycan is derived from phosphatidylethanolamine (PE). Side-chain modifications of the core GPI structure may

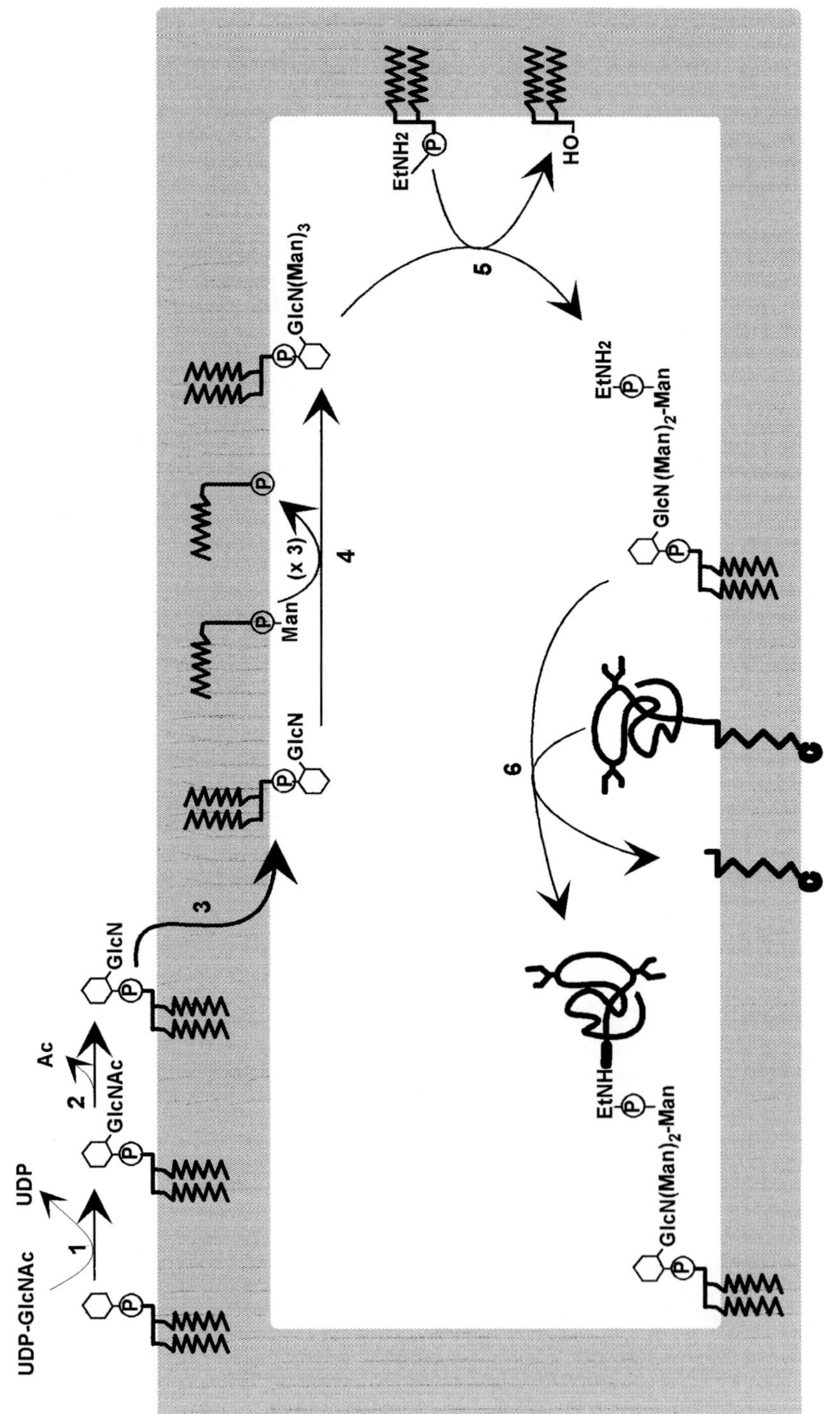

Figure 3: Pathway for biosynthesis of GPIs.

The topography of the pathway within the ER membrane is not yet fully determined. Step 1: Addition of GlcNAc on phosphatidyl inositol; Step 2: Deacetylation of GlcNAc Step 3: Presumed flip-flop to the ER lumen; Step 4: Addition of mannose residues from Man-P-Dol; Step 5: Addition of phosphoethanolamine on mannose residue from phosphatidylethanolamine; Step 6: Transamidation reaction linking the protein onto the GPI anchor.

occur during assembly or after completion of the phosphoethanolamine-containing structure. The completed GPI moiety is then attached to specific proteins via a transamidation reaction in which a C-terminal GPI-directing signal sequence in the acceptor protein is removed and replaced with the GPI moiety linked, via ethanolamine, to the newly exposed C-terminal amino acid protein. This last step of transamidation occurs in the lumen of rough ER. This implies that GPI has to be translocated from the cytosolic face to the lumenal face of rough ER membrane. So far, this flip-flop has been suggested to occur at several steps of the biosynthetic pathways.

6. O-GLYCOSYLATION

The formation of O-linked glycans occurs by a stepwise addition of each monosaccharide from the corresponding sugar nucleotide donor. The first monosaccharide is transfered on a serine or threonine residue presumably in the Golgi or earlier compartment. In the case of mucin type glycoprotein, this first sugar is an N-acetylgalactosamine residue and so far no consensus sequences have been identified. It is feasible that a large spectra of GalNActransferases are necessary (Clausen and Bennett, 1996).

Further sequential additions of monosaccharides occur while the protein is trafficking within the different Golgi stacks and trans Golgi network. As a general rule for the O-glycosylation pathway, a 3-linked residue has to be introduced before a 6-linked branch can be added. Thus the multiplicity of structures is generated by the enzymatic equipment of the cell and its subcellular location. The sequential addition of sugars is thus depending on the traffic of the glycoproteins within the different stacks of the Golgi apparatus (for review, Brockhausen, 1995)

7. N-GLYCOSYLATION

The N-glycosylation is a rather more complex process starting in the rough endoplasmic reticulum as a cotranslational process and continues in every Golgi stacks and trans Golgi network (for basic review, Kornfeld and Kornfeld, 1985). In the ER, a 14-mer oligosaccharide will be preassembled on a lipid intermediate (a pyrophospho-dolichol) and this unique oligosaccharide will be transferred ''en bloc'' on an asparagine residue of the nascent protein. In the Golgi apparatus, the glycan will be further processed by trimming and additions to give the various structures: oligomannoside type, N-acetyllactosamine type, poly N-acetyllactosamine type or hybride type.

Besides this segregation of the substrates of the glycosylation reaction, there is one more regulation point regarding the intracellular trafficking of glycoproteins. This regulation relies on the sequential distribution of the various glycosyltransferases along the cytological compartments: the rough endoplasmic reticulum where the common oligosaccharide precursor is built up, the ''cis'' and the ''medium'' Golgi where antennae are synthesized, and the ''trans Golgi'' and ''trans Golgi network'' when the terminal sugars are added. There again, regulation may act by controlling the shuttle vesicle system driving the glycoprotein content from one compartment to the other.

7.1. The N-glycosylation Process in the Rough Endoplasmic Reticulum

The princeps reaction of N-glycosylation is the transfer "en bloc" of a tetradecasaccharide on the acceptor asparagine residue of a nascent glycoprotein. This tetradecasaccharide is preassembled on a specific polyisoprenol, a dolichol, *via* a pyrophosphate bond to form the $Glc_3Man_9GlcNAc_2$-PP-Dol (for review, Hemming, 1995). This compound is assembled, step by step, in the rough endoplasmic reticulum. The $GlcNAc_2$-PP-Dol is first synthesized, on the cytoplasmic face of the ER membrane, by the addition of GlcNAc-1-P on the phospho-dolichol, then of GlcNAc on the GlcNAc-PP-Dol previously formed. In both cases, the donor is UDP-GlcNAc.

The next steps will consist in the addition of nine mannose residues. The first five being added from GDP-Man by a serie of five sequential reactions at the cytoplasmic face of the ER membrane. The next four mannoses being added, in contrast, on the lumenal face of the ER membrane. This process raises two questions: in the first place, how is the cytoplasmic $Man_5GlcNAc_2$-PP-Dol translocated at the lumenal face? In the second place, how do the last four activated mannoses (GDP-Man) cross the membrane? The flip-flop mechanism of the $Man_5GlcNAc_2$-PP-Dol is not yet understood and is thought to be mediated by a vectorial mannosylation reaction. The second question has its proper answer through the involvement of a secondary donor, Man-P-Dol, synthesized at the cytoplasmic face from GDP-Man and which is, due to its hydrophobic characteristic and possible involvement of a protein, able to cross the membrane and reach a lumenal mannosylation site (Rush and Waechter, 1995). The same mechanism is proposed for the addition of the last three glucose residues using Glc-P-Dol as donor.

This ultimate step of glycosylation confers to the lipid intermediate a higher affinity for the oligosaccharidyltransferase and the $Glc_3Man_9GlcNAc_2$ species is the favourite substrate to be transferred onto an asparagine residue of the acceptor protein. As mentionned above, the asparagine must be in a specific Asn-Xaa-Thr/Ser sequence, yet, since all the asparagine in such triplets are not always glycosylated, it is assumed that other controls may exist. They are presumably related to the neighbouring tridimensional structure of the protein and to the involvement of chaperone molecules.

Once transferred onto protein, the glycan moiety will be stepwisely trimmed to reach the $Man_5GlcNAc_2$ structure (for review, Verbert, 1995). The first event will be the removal of the three glucose residues by the action of two different glucosidases. Glucosidase I will remove the α-1,2-linked glucose residue and glucosidase II will cleave the next two α-1,3-linked glucose residues. Both enzymes can be inhibited (both in cell-free system and whole cell) by castasnospermine and deoxynojirimycin and related derivatives.

The question of correct folding of the protein is important and we still have to learn a lot about rough ER or related early Golgi compartment where degradation of proteins has recently been observed. It has been shown that the misfolded or misassembled (glyco)proteins are retrotranslocated to the cytoplasm by the "reverse use" of the translocon machinery (Wiertz *et al.*, 1996; Bonifacino, 1996). Once in the cytoplasm the protein is digested by the proteolytic complex formed by the proteasome. The cleavage and fate of the glycan moiety is still under investigation (Cacan and Verbert, 1997; Cacan *et al.*, 1998).

The glycoprotein will be retained in the rough ER by $Glc_1Man_9GlcNAc_2$ recognizing lectins, the calnexin or calreticulin (Hammond and Helenius, 1993). A specific glycosyltransferase recognizing unfolded or misfolded glycoprotein will be in charge to

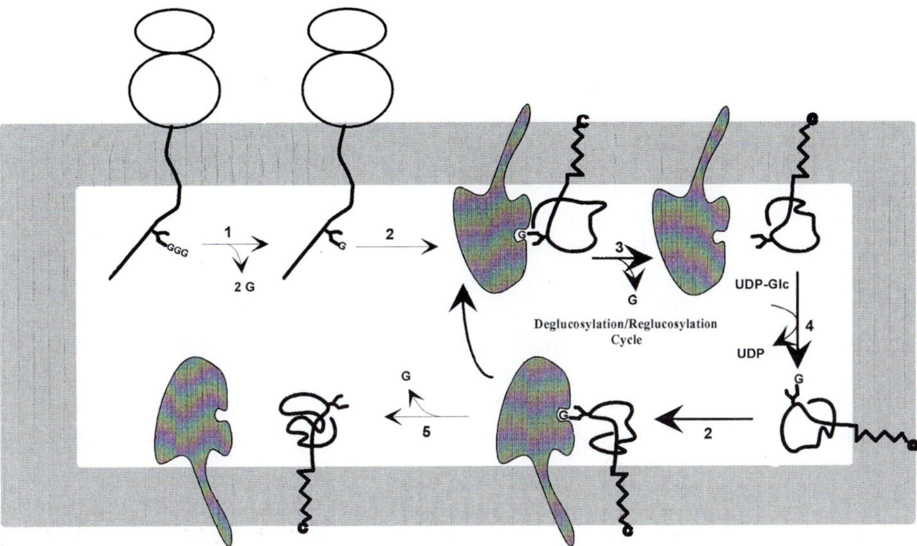

Figure 4: Quality control of nascent glycoproteins.
Step 1: Deglucosylation of the first two mannose residues of nascent glycoproteins; Step 2: Retention by calnexin of the monoglucosylated species; Step 3: deglucosylation; Step 4: Re-glucosylation of the non-yet folded glycoprotein; Step 5: release of the correctly folded glycoprotein.

keep the glycan glucosylated as long as it has to be retained in the rough ER to acquire its correct spatial conformation (Parodi, 1996). After this quality control process (illustrated in Figure 4), the glycan can be further processed to $Man_8GlcNAc_2$ structure which is the key structure to leave the rough ER and to be shuttled to cis-Golgi by β-COP (coat-protein) covered vesicles.

Although α-mannosidases were described as Golgi enzymes since 1979, several lines of evidence suggested that some mannose residues could be removed from certain glycoproteins before they reach the Golgi compartment. Indeed, Bischoff and Kornfeld (1986) have shown the occurence of an ER mannosidase and its activity is highly specific and insensitive to the mannose analogue 1-deoxymannojirimycin, an inhibitor of certain mannosidases. This α-mannosidase could be similar to yeast enzyme which converts Man_9 species to the Man_8 species which can then be elongated to the typical yeast mannan structures.

This high specifity of the ER α-mannosidase could be related to certain specific conformation of the glycan in its protein environment. This could explain why some glycoprotein $Man_9GlcNAc_2$ structures (or $Glc_1Man_9GlcNAc_2$ structures) might escape the action of this enzyme and reach the Golgi without being processed to Man_8 structures.

Finally, it has to be mentioned that an alternate pathway may occur when glucosidase activity is low or deficient. In this case $Glc_1Man_9GlcNAc_2$ protein may enter the Golgi where deglucosylation is achieved via the action of an endo-α-mannosidase by removing a $Glc(\alpha$-1,3)Man disaccharide and yield to only one $Man_8GlcNAc_2$ isomer. This salvage pathway indicates how crucial is the removal of glucose for further glycoprotein maturation.

7.2. The N-glycosylation Process in the Golgi-apparatus

In the *cis-Golgi*, the glycans will be further trimmed by specific mannosidases to reach the $Man_5GlcNAc_2$ structure. This requires the cleavage of the three remaining α-1,2-linked mannose residues. This will be achieved in an ordered way, going through the formation of $Man_7GlcNAc_2$, $Man_6GlcNAc_2$ isomers and then the $Man_5GlcNAc_2$ isomer which will be the substrate for the GlcNAc transferase I (see Figure 5)

The mannosidases involved in this trimming are named mannosidases IA and/or IB. They are located in the cis-Golgi cisternae and, in contrast to the previous ER mannosidase, are very sensitive to deoxymannojirimycin.

The key enzyme for the conversion of the $Man_5GlcNAc_2$-protein species to complex type glycan is the N-acetylglucosaminyl transferase (GnT-I). As described in Figure 5, the newly formed $GlcNAcMan_5GlcNAc_2$-protein structure is the required substrate for the Golgi mannosidase II which removes the two mannose residues linked to the α-6 mannose. The resulting $GlcNAcMan_5GlcNAc_2$-protein is now the acceptor for the N-acetylglucosaminyl transferase (GnT-II) channeling the way to multiantennary structures. In fact, the presence of a β-1,2-linked GlcNAc residue at the non reducing end of the $Man(\alpha1-3)Man(\beta1-4)GlcNAc$ arm is essential for subsequent action of several enzymes in the processing pathway: α-3/6-mannosidase II, GlcNAc transferase II, III and IV, and the α-1,6-fucosyltransferase which adds fucose in α-1,6 linkage to the Asn-linked GlcNAc. This type of control of glycosylation, illustrated in Figure 6 and known as ''go or no-go'' pathways by H. Schachter, is based on enzyme specificity and allows the chanelling toward different glycosylation pathways leading to the different types of glycans.

Then, the stepwise action of a series of different glycosyltransferases (galactosyltransferases, fucosyltransferases and sialyltransferases mainly, for review Schachter, 1995) will lead to the building of hundreds of N-glycan structures recovered on glycoproteins (steps 8 to 10, on Figure 5). As mentionned earlier, different glycans will be obtained according to the enzymatic equipment of cells and will also depend on many physiological factors.

To make a long story short, the way a protein is glycosylated depends on many factors: at the gene level, according to the active glycosyltransferase genes of the cell, at the enzyme expression level, according to the physiological state of the cell, and at the intracellular trafficking level, since, all along the path through the ER and Golgi vesicles, compartmentalization may interfere to add complexity in the ordered specificity of glycosyltransferases.

At this stage it is worthy to mention in this book devoted to cell adhesion and pathological implications that the impairment of glycosylation might be caused by a defect at any of these level of regulation. This will be illustrated by three examples.

8. THREE EXAMPLES OF PATHOLOGIES RELATED TO GLYCOPROTEINS BIOSYNTHESIS DISORDERS

1) An example can be taken when the gene for an enzyme of the glycosylation machinery is missing or inactive. This is the case for the different pathological disorders known as ''Carbohydrate Deficient Glycoprotein'' syndrome. From a biochemical point of view the CDG syndrome has been characterized as a lack of glycosylation or an hyposialylation revealed with transferrin glycoforms recovered in serum of the patients (Jaeken and

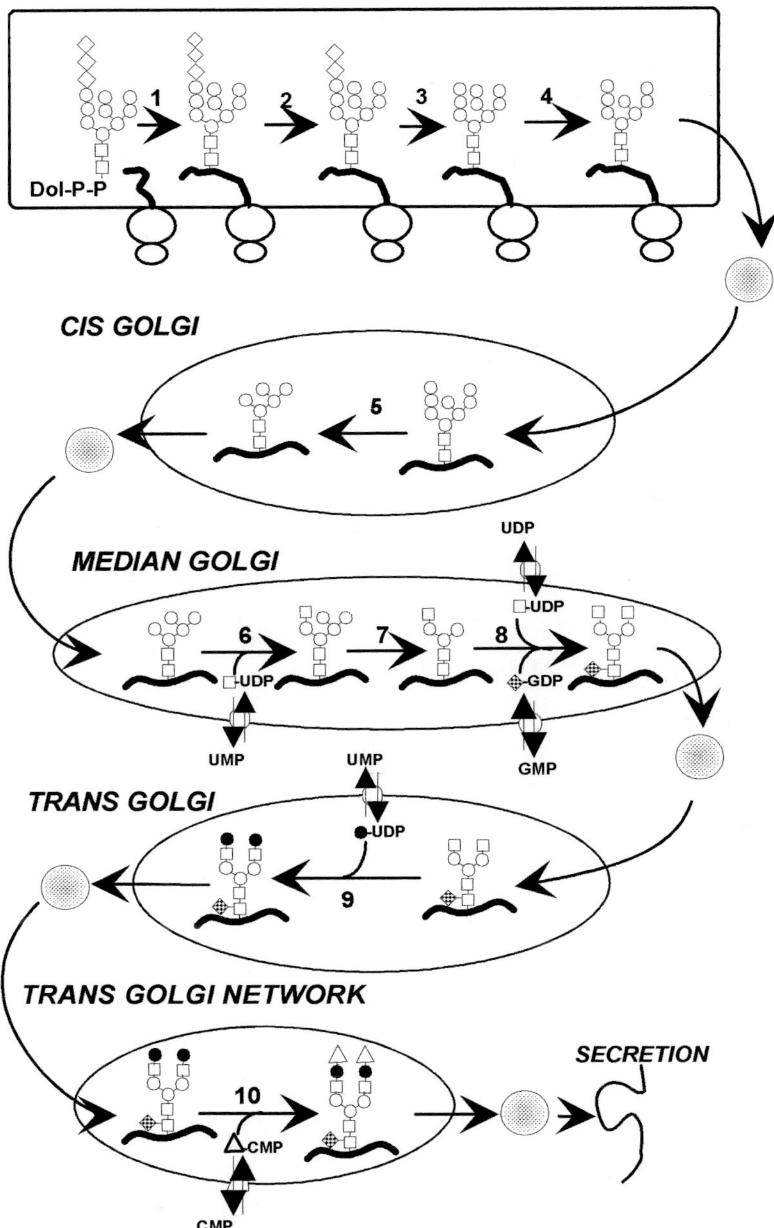

Figure 5: Biosynthetic pathway of N-glycosylation of proteins.
The main reactions and their intracellular locations are given for the biosynthesis of a biantennary complex type glycan, scheme adapted from Kornfeld and Kornfeld (1985).
Step 1: Transfer "en bloc"; Steps 2 and 3: deglucosylation; Step 4: ER mannosidase; Step 5: Golgi mannosidase I; Step 6: GlcNActransferase I; Step 7: Golgi mannosidase II; Step 8: GlcNActransferase II and fucosyltransferase; Step 9: galactosyltransferase; Step 10: sialyltransferases.

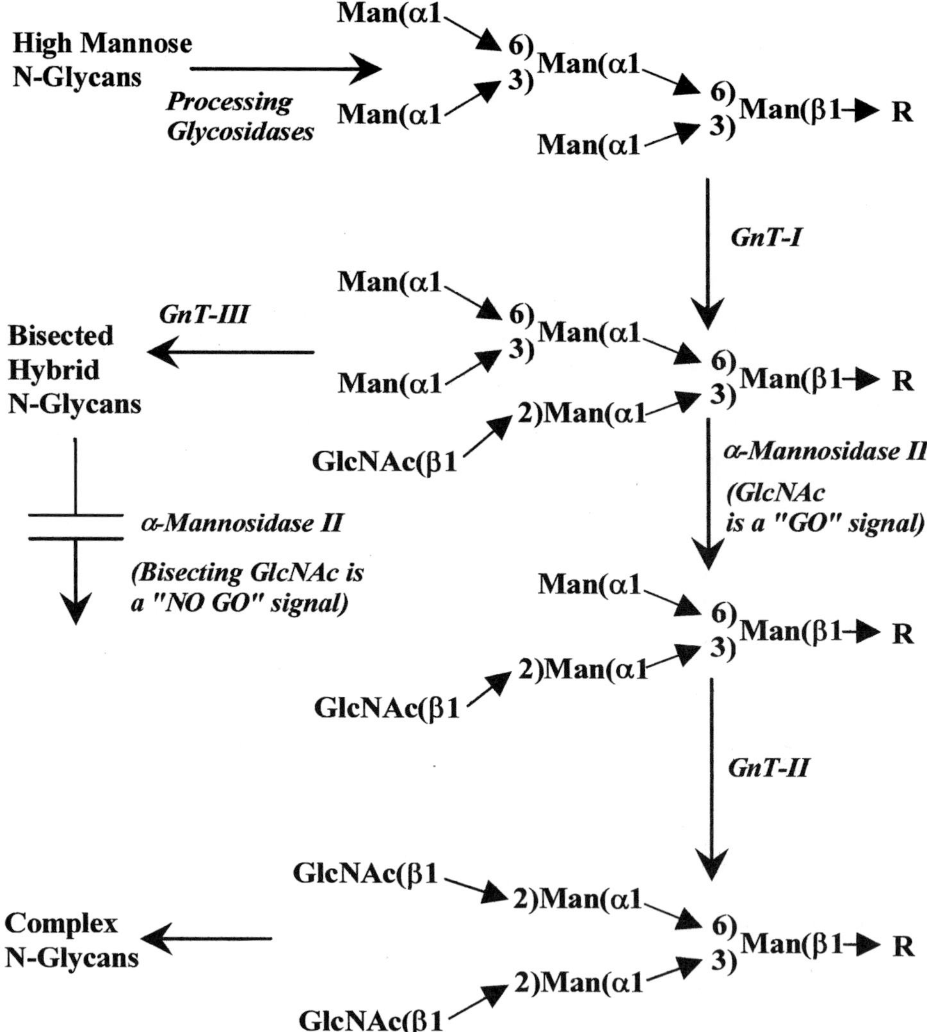

Figure 6: Processing of N-glycans.
''GO'' and ''NO GO'' signals are illustrated in the pathway from high mannose type to complex type glycan (GO signal) and the formation of bisected hybrid type glycans (NO GO signal).

Carchon, 1993). In type-I CDG, one or two of the normally glycosylated sites have no more linked oligosaccharides indicating a failure in the steps synthesizing the lipid intermediates or at the transfer ''en bloc'' reaction. For quite a few cases, it has been shown that it was due to the very low capacity of cells to synthesize the GDP-mannose required to make the lipid intermediate precursors. The deficiency is either due to a lack of active phosphomannose mutase (type-Ia) or of active phosphomannose isomerase (type-Ib). In type-II CDG, the two sites of glycosylation of serotransferrin are occupied but they miss the α-6 branch of their normally biantennary glycans. It has been clearly demonstrated that it was due to a lack of GlcNActransferase II.

2) Regarding differential expression of glycosyltransferase activities, the consequence can be the channeling to different glycosylation pathways. Indeed, the activity of a glycosyltransferase can lead to make a product which is no more a substrate for the subsequent or further located glycosyltransferases ("go/no-go" competition). A clear example is the formation of poly-.N-acetyllactosamine branches in cancer cells. Poly-N-acetyllactosamine antennae are predominantly attached to GlcNAc residue transferred by the action of the GlcNActransferase V. The competition between glycosyltransferases as a modulator of the glycosylation status of cells is illustrated by the work of Taniguchi and collaborators (Yoshimura *et al.*, 1995) who have shown by recombination procedure that overexpression of GlcNActransferase III restores the normal phenotype. Indeed GlcNActransferase III introduces the bisecting GlcNAc in most of the processed glycoproteins and renders their glycans no more acceptors for the GlcNActransferase V.A lot of similar examples have been exemplifyied by H. Schachter to explain the different pathways leadind to the diversity of glycoforms (Schachter, 1991).

3) Finally in terms of trafficking it is well documented that lysosomal storage diseases are a consequence of a default in the mannose-6-phosphate signal borne by lysosomal enzymes. To stay in the scope of plasma membrane glycoproteins, an example will be taken in the therapeutic trials to prevent correct folding of viral protein and as such disturbing the biological activity of a membrane located viral protein. Many viruses contain an outer envelope which is composed of one or more viral glycoproteins. One example will illustrate how N-linked glycan processing can be used as a target for anti-viral therapy. HIV-I, the causative agent of AIDS encodes two heavily N-glycosylated envelope glycoproteins (gp120 and gp41). During infection, gp120 binds to the cellular receptor (CD4 antigen) and undergoes a conformational change that exposes gp41 and triggers the fusion with the cell membrane thus mediating the entry of the virus. It has been shown that treatment of HIV-1 infected cells with N-butyl deoxynojirimycin (NB-DNJ), an inhibitor of rough ER α-glucosidases inhibits syncytium formation and the formation of infective virus (Mehta *et al.*, 1995). Indeed, NB-DNJ prevents ER processing of glucose residue leading to a misfolding of the gp120 due to its inability to be retained by the calnexin and/or calreticulin system. These misfolding is sufficient to inhibit the viral fusion process.

ACKNOWLEDGEMENT

I am very thankful to my colleague and friend, Professor René Cacan, Université des Sciences et Technologies de Lille, who has patiently read the manuscript and greatly contribute to improve it by stimulative and questions and discussions.

REFERENCES

Bischoff, J., and Kornfeld, R. (1986) The soluble form of rat liver alpha-mannosidase is immunologically related to the endoplasmic reticulum membrane alpha-mannosidase. *J. Biol. Chem.*, **261**, 4758–4765

Bonifacino, J.S. (1996) Reversal of fortune for nascent proteins. *Nature,* **384**, 405–406.

Brockausen, I. (1995) Biosynthesis of O-glycans of the N-acetylgalactosamine-α-Ser/Thr linkage type. In J. Montreuil, J.F.G. Vliegenthart, and H. Schaehter, (eds.), *Glycoproteins*, Elsevier Science 3.V., pp. 201–260.

Cacan, R., and Verbert, A. (1997) Free oligomannosides produced during the N-glycosylation process: origin, intracellular trafficking and putative roles. *TIGG*, **9**, 365–377.

Cacan, R., Duvet, S., Kmiecik, D., Labiau, O., Mir, A.M., and Verbert, A. (1998) "Glyco-deglyco" processes during the synthesis of N-glycoproteins. *Biochimie*, **80**, 59–68.

Clausen, A., and Bennett, E.P. (1996) A family of polypeptide N-acetylgalactogaminyl-transferase control the initiation of mucin-type O-linked glycosylation. *Glycobiology*, **6**, 635–646.

Eckhardt, M., Ahlenhoff, M., Bethe, A., and Gerardy-Schahn, R. (1996) Expression cloning of the Golgi CMP-sialic acid transporter. *Proc. Nath. Acad. Sci.*, USA, **93**, 7572–7576.

Ferguson, M.A.J. (1991) Lipid anchors on membrane proteins. *Current Biology*, **1**, 522–529.

Gilmore, R. (1993) Protein Translocation across the endoplasmic reticulum: a tunnel with toll booths at entry and exit. *Cell*, **75**, 589–592.

Hammond, G., and Helenius, A. (1993) A chaperone with a sweet tooth. *Current Biology*, **3**, N12.

Hemming, F. (1995) The coenzymic roles of phosphodolichols. In J. Montreuil, J.F.G. Vliegenthart, and H. Schaehter, (eds.), *Glycoproteins*, Elsevier Science 3.V., pp. 125–143.

Hirschberg, C.B. (1987) Topography of Glycosylation in the rough endoplasmic reticulum and Golgi apparatus. *Ann. Rev. Biochem*, **56**, 63–67.

Jaeken, J., and Carchon, N. (1993) The carbohydrate-deficient glycoprotein syndrome: on overview. *J. Inher. Metab. Dis.*, **16**, 813–820.

Kalies, K.W., and Hartmann, E. (1998) Protein translocation into the endoplasmic reticulum (ER) Two similar routes with different modes. *Eur. J. Biochem.*, **254**, 1–5.

Kornfeld, R., and Kornfeld, S. (1985) Assembly of asparagine-linked oligosaccharides. *Ann. Rev. Biochem.*, **54**, 631–664.

Mehta, A., Rudd, P.M., Block, T.M., and Dwek, R.A. (1995) A strategy for antiviral intervention: the use of α-glucosidase inhibitors to prevent chaperone-mediated folding of viral envelope glycoproteins. *Biochem. Soc. Trans.*, **25**, 1188–1193.

Menon, A.K., Baumann, N.A., van't Hof, W., and Vidugiriene, J. (1997) Glycosylphosphatidylionositols: biosynthesis and intracellular transport. *Biochem. Soc. Trans.*, **25**, 861–865.

Parodi., A.J. (1996) The UDP-GLC: glycoprotein glucosyltransferase and the quality control of glycoprotein folding in the endoplasmic reticulum. *TIGG*, **8**, 1–12.

Rush, J.S., and Waechter, C.J. (1995) "Transmembrane movement of a water-soluble analogue of mannosylphosphoryldolichol is mediated by an endosplasmic reticulum protein". *J. Cell Biol.*, **130**, 529–536.

Schachter, H. (1991) The "Yellow brick road" to branch complex N-glycans. *Glycobiology*, **1**, 453–461.

Schachter, H. (1995) Glycosyltranferases involves in the synthesis of N-glycan antennae. In J. Montreuil, J.F.G. Vliegenthart, and H. Schaehter, (eds.), *Glycoproteins*, Elsevier Science 3.V., pp. 153–200.

Verbert, A. (1995) From Glc$_3$ Man$_9$ GlcNAc$_2$-protein to Man5GlcNAc$_2$-protein: transfer "en bloc" and processing. In J. Montreuil, J.F.G. Vliegenthart, and H. Schaehter, (eds.), *Glycoproteins*, Elsevier Science 3.V., pp. 145–152.

Vliegenthart, J.F.G. and Montreuil, J. (1995) Primary structure of glycoprotein glycans. In J. Montreuil, J.F.G. Vliegenthart, and H. Schaehter, (eds.), *Glycoproteins*, Elsevier Science 3.V., pp. 13–28.

Walter, P., Lingappa, V.R. (1986) Mechanism of protein translocation across the endoplasmic reticulum membrane. *Ann. Rev. Cell Biol.*, **2**, 499–516.

Wiertz, E.J.H.J., Tortorella, D., Boggo, M., Yu, J., Mothes, W., Jones, T.R., Rapoport, T.A., and Ploegh, H.L. (1996) Sec61-mediated transfer of a membrane protein from the endoplasmic retiuculum to the proteasome for destruction. *Nature*, **384**, 432–438.

Yoshimura, M., Nishikawa, A., Ihara, Y., Taniguchi, S., and Taniguchi N. (1995) Suppression of lung metastasis of B16 mouse melanoma by N-acetylglucosaminyltransferase III gene transfection. *Proc. Natl Acad Sci.*, USA, **92**, 8754–8758.

A2. Structure and Analysis of Glycoprotein-Associated Oligosaccharides

Jean-Claude Michalski

Glycosylation constitutes a complex and ubiquitous post-translational event. Most proteins within living organisms contain sugar chains or glycans. Glycans differ in their mode of attachment to the peptidic chains and also present a large heterogeneity directly related to the glycosylation-machinery of the cell or the producing organism. In most cases, this heterogeneity is directly related to the biological function of glycans. Moreover, glycoproteins generally exist as populations of glycosylated variants (glycoforms) of a single polypeptide. In this chapter, we present a review of the structures of the different glycan families, as well as principal analytical techniques and recent developments of structural glycobiology.

1. INTRODUCTION

Glycosylation constitutes the most common and diverse post-traductional event of proteins in living organisms (Montreuil, 1982; Kobata, 1992 and Lis & Sharon, 1993). Diversity comes both from the kinds of amino acid, which may be modified, and the structural heterogeneity of oligosaccharidic chains. Striking advances have been made during the last years in structural glycobiology or glycotechnologies with refinements of separation and analytical methods as well as introduction of new technologies (Dwek *et al.*, 1993; Rudd & Dwek, 1997). As a consequence the number of known structures of glycans has grown immensely. New types of carbohydrate–protein linkages have been discovered and unusual constituents identified (Vliegenthart & Casset, 1998). Increased knowledge of carbohydrate structures directly contributes to a better understanding of glycan functions. In the present chapter, we define the scope of glycans heterogeneity and the task and principal technologies involved in the analysis of protein glycosylation.

2. NATURE OF MONOSACCHARIDES

Four classes of monosaccharides are present in glycoprotein glycans the most common as listed first:

1. Neutral sugars: D-galactose, D-mannose, D-glucose, L-fucose, L-furanoarabinose, D-xylose

2. Amino sugars: N-acetyl-D-glucosamine, N-acetyl-D-actosamine
3. Uronic acids: D-glucuronic acid, L-iduronic acid
4. Sialic acids: N-acetyl-neuraminic acid, N-glycolyl-neuraminic acid

Sialic acids present a great heterogeneity (Schauer, 1991; Varki, 1992). They differ not only in the substituent on the amino group (acetyl or glycolyl) but also in the number (up to three), position (4, 7, 8 and 9) and nature (acetyl, lactoyl and methyl) of substituents on the hydroxyl groups of neuraminic acid.

Sulfated sugars occur both in N- and O-glycans. For example, 4-SO_3^--GalNAc is found in pituitary hormones (Baenziger & Green, 1988) and in Tamm Horsfall Glycoproteins (Van Rooijen *et al.*, 1998), 3-SO_3^- – Gal is found in thyroglobulin (De Waard *et al.*, 1991) and in human mucins (Lamblin *et al.*, 1991); 6-SO_3^--GlcNAc is found in Thyroglobulin; 3-SO_3^--GlcNAc in N-CAM_3 (Kudo *et al.*, 1996). 4 or 6 sulfated mannose occur in ovalbumin (Yamashita *et al.*, 1983) and 3 sulfated glucuronic acid (HNK-1 antigen) in bovine peripheral myelin glycoproteins PO (Voshol *et al.*, 1996). Many other rare constituents have also been described, for reviews see (Lis & Sharon, 1993; Montreuil, 1995).

3. CARBOHYDRATE-PEPTIDIC LINKAGES

Until recently the only known linking groups were N-acetylglucosamine–asparagine (GlcNAc-Asn) and O-acetylactosamine–serine/threonine (GalNAc-Ser/Thr). During recent years several new types of sugar–peptide linkages have been discovered and are listed in Table 1.

Tyrosine has been added to the list of O-linked amino acids with the identification of Glc α-Tyr in glycogenin (Smythe *et al.*, 1988). Glucose bound in a β-linkage to serine has been described in the bovine clotting factor IX (Hase *et al.*, 1990). It is now well established that O-glucosaminylation constitutes a widely distributed type of glycosylation among cytosolic and nuclear proteins (Hart *et al.*, 1989). O-linked fucose is an unusual form of glycosylation group of serine or threonine residues within epidermal growth factor-like domains of a number of serum proteins (Moloney *et al.*, 1997). A new O-mannosyl-type glycosylation has been described on bovine peripheral nerve α-dystroglycans. These O-glycans are generally sialylated (Chiba *et al.*, 1997). Finally, a new type of glycosylation namely C-mannosylation has been described in human RNase 2 (De Beer *et al.*, 1995), in which the anomeric carbon atom of an α-D-mannopyranose residue is directly linked to the C_2 atom of trytophan residue.

4. THE GLYCOSYL-PHOSPHATIDYL ANCHOR

Some membrane proteins are ''anchored'' with a glycosyl-phosphatidyl inositol directly attached to the C-terminal amino acid carboxyl group of the protein (Fergusson &

Table 1. Linkage between glycan and polypeptide chain.

Type	Monosaccharide	Amino acid	Occurrence
N-Glycoside	β-GlcNAc	Asn	Common
	β-GalNAc	Asn	Archaebacteria
	β-Glc	Asn	Archaebacteria
	α/β-Glc	Asn	Animals
	L-Rha	Asn	Eubacteria
O-Glycoside	α-GalNAc	Ser/Thr	Animals (mucin type)
	β-GalNAc	Ser/Thr	Animals
	β-GlcNAc	Ser/Thr	Animals (intracellular type)
	α-Xyl	Ser	Animals (collagen type)
	α-Fuc	Ser/Thr	Animals
	β-Gal	HyP	Animals
	α-Glc	Tyr	Animals (glycogenin type)
	β-Glc	Ser/Thr	Animals, Eubacteria
	α-Man	Ser/Thr	Animals, Yeasts
	α-Gal	HyP/Ser	Plants, Eubacteria
	β-L-Ara*f*	HyP	Plants
	β-Gal	Tyr	Eubacteria
C-Glycoside	α-Man	Trp	Human RNase

Williams, 1988). This type of anchorage has been described in organisms representing most stages of eukaryotic evolution, going from protozoa to man. GPI have been described in rat brain glycoprotein Thy-1, human erythrocyte acetylcholinesterase, the variant surface glycoprotein (VSG) and 1G7 antigen of the parasitic protozoan T-brucei. All these structures possess a common tetrasaccharidic core Man(α1-2) Man(α1-6) Man(α1-4)GlcNH$_2\alpha$. The tetrasaccharide is bound *via* the 6-hydroxyl group of mannose to ethanolamine phosphate, which is itself attached by an amide linkage to the α-carboxyl group of the C-terminal amino acid of the protein. The reducing end of the tetrasaccharide is glycosidically linked to an inositolphospholipid, which is embedded in the lipid layer GPI of the cell surface membrane. The GPI differ in the nature and number of additional carbohydrate and extra ethanolamine phosphate linked to the tetrasaccharide core (Figure 1).

5. STRUCTURE OF GLYCANS

5.1. The Sequon and the Glycoforms

As previously mentioned the N-linking group consists of the sequence (GlcNAc-Asn) (Figure 2), with the asparagine as part of consensus sequence Asn-Xaa-Ser/Thr (Osawa & Tsuji, 1987). Amino acid X is an important determinant of the efficiency of N-glycosylation. The presence of proline at the X position completely blocks glycosylation, whereas tryptophan, aspartic acid, glutamic acid or leucine are associated with inefficient N-glycosylation (Shakin–Eshleman *et al.*, 1996). Each glycosylation site

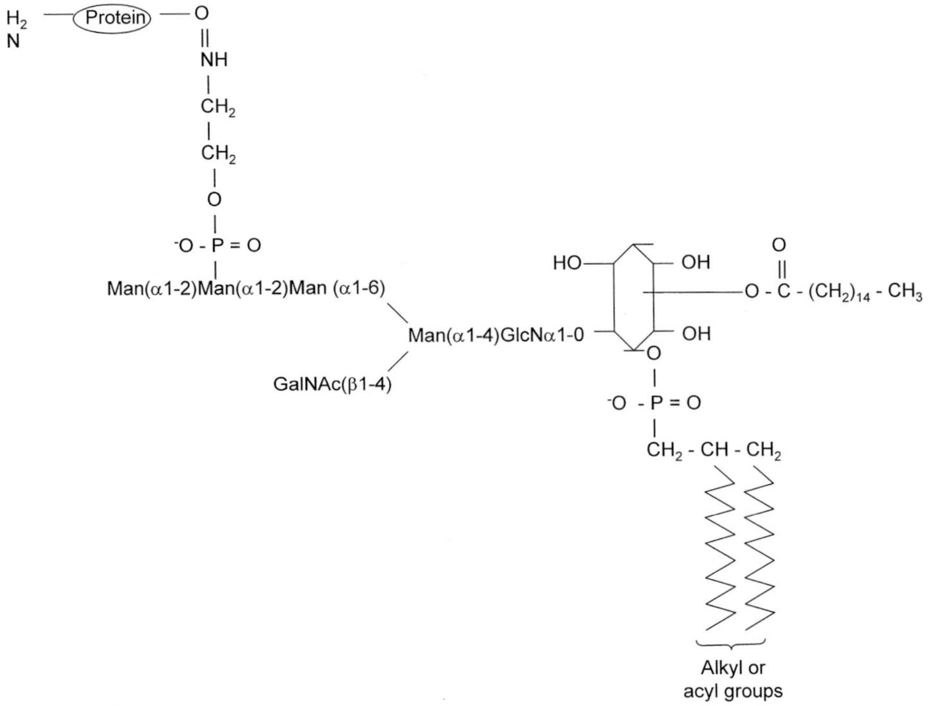

Figure 1: Structure of the Thy-1-glycoprotein glycosyl anchor.

may be occupied or not, moreover each attachment site frequently accommodates different glycans (site heterogeneity). For that reason, glycoproteins generally exist as populations variants of a single polypeptide called "glycoforms" (Rademacher *et al.*, 1988). Most glycoproteins emerge with characteristic glycosylation patterns and heterogeneous populations of glycans at each glycosylation site.

5.2. Subgroups of N-glycans

All N-glycans contain a common pentasaccharidic core (Figure 3) so-called the "di-N-acetyl-chitobiose trimannosyl core". According to the structures and the location of the extra sugar residues added to the core, N-glycans are further classified into three subgroups.

5.2.1. "N-acetyl-lactosaminic" or "complex" type N-glycans

The trimannosyl core is substituted with a different number of N-acetyl-lactosamine units which constitute the "antennae" (Montreuil, 1982). Glycans will be classified from bi- to penta-antennary. The presence or absence of an α-fucosyl residue linked to C-6 position of the proximal N-acetylglucosamine residue and β-N-acetylglucosamine linked to the C-4 position of the β-mannosyl residue-bisecting GlcNAc) contributes to increase the structural

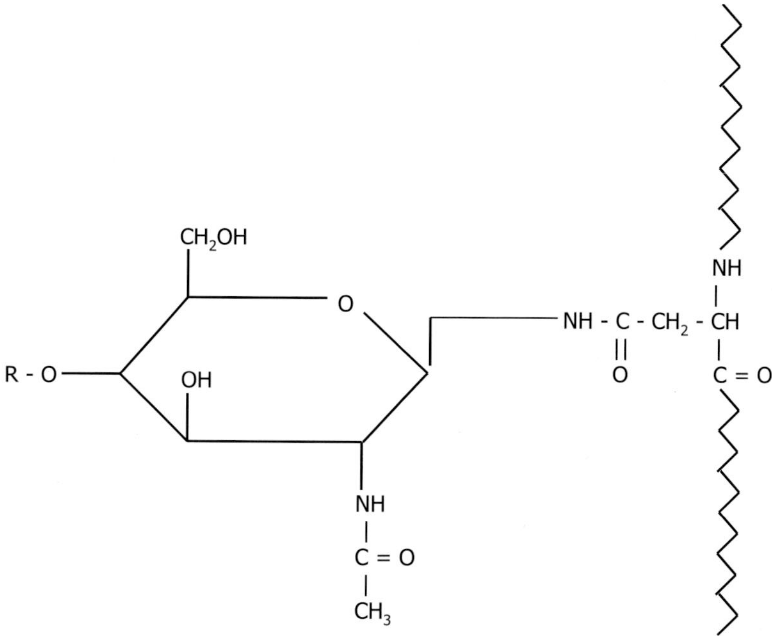

Figure 2: Structure of the GLcNAc-Asn linkage of N-glycosylproteins.

variation of N-acetyl-lactosaminic type glycans. The antennae terminate in a great diversity of outer-chain structures (Figure 4).

Plant glycoproteins. In most plant glycoproteins, a xylose residue is attached in β1,2-linkage to the β-linked-mannose of the core and L-fucose linked α-1,3 is found on GlcNAc-*1* attached to the asparagine residue. N-acetyl-lactosaminic outer-chain may be present; some of them express the Le^{a} antigen (Fichette-Laine 1997) (Figure 5).

Polylactosaminic structures. The branches of many N-oligosaccharides in animal cells (Fukuda, 1994) contain poly-N-acetyllactosamine chains, a polymer of β-1,3 N-acetyllactosamine (Figure 6).

This type of chains are predominant on the Man(α1-6)Man(β-) branch of the core. Chain extension with lactosamine is in competition with chain termination with sialic acid (Rudd & Dwek, 1996).

Polysialyl-glycans. Polysialyl-glycans are found in neural cell adhesion molecules (N-CAMs) (Troy, 1992). Polymers of α-2,8-linked sialic acid (up to 50 residues) are attached to multiantennary glycans, nevertheless polysialylation occurs assimetrically on the antennae and the presence of GlcNAc(β1-6) linked on the Man(α1-6) arm is required for the polysialylation of the core (Kudo *et al.*, 1996).

5.2.2. *Oligomannose or high-mannose type N-glycans*

These oligosaccharidic chains contain only α-mannose in addition to the trimannosyl core. In mammalian cells the limit structure consist of a Man$_9$ oligosaccharides. The

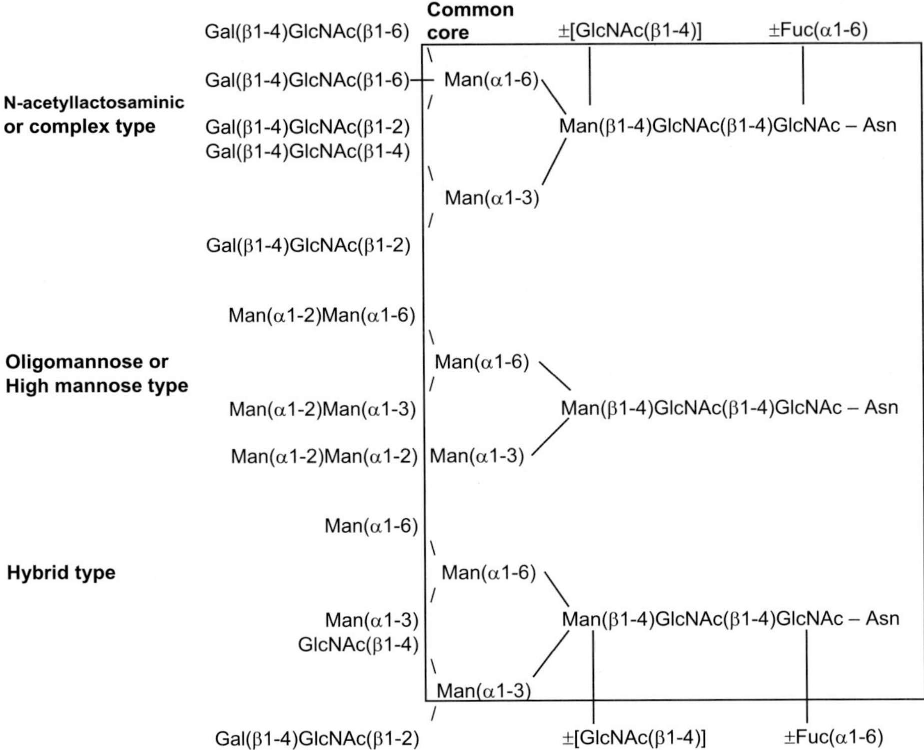

Figure 3: Different N-glycans families.

heptasaccharide Man(α1-6)[Man(α1-3)] Man(α1-6)[Man(α1-3)] Man(β1-4)GlcNAc (β1-4)GlcNAc referred as Man$_5$ is commonly found in numerous glycoproteins. Even if high-mannose type N-glycans do not present an heterogeneity comparable to that of complex type glycans, an important variation is formed in these sugar chains both by the numbers, the locations and the branching pattern of the mannose residues, creating a great number of position isomers. This heterogeneity may be the support for the recognition by different intra-cellular mannose type lectins.

5.2.3. Hybrid type N-glycans

The third group of N-glycans is called hybrid type because the oligosaccharides have the structural features of both high-mannose and complex type glycans. The Man(α1-6) branch accomodates one or two mannosyl residues as in the case of high-mannose type glycans and the Man(α1-3) arm bears lactosaminic outer-chains as in complex type glycans. The ''bisecting'' GlcNAc generally occurs on hybrid glycans the presence or absence of the α1-6 linked fucose produces additive structural variation. Hybrid glycans are commonly found in bird glycoproteins.

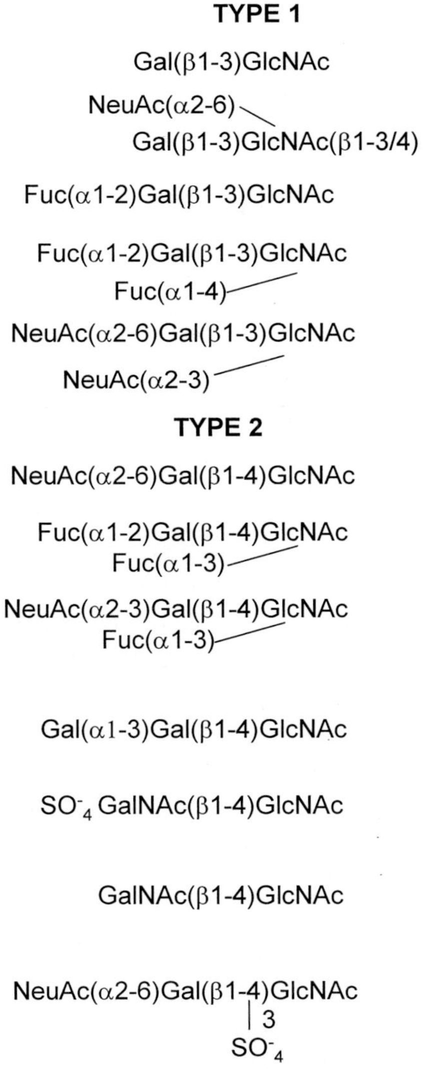

TYPE 1

Gal(β1-3)GlcNAc

NeuAc(α2-6)
　　　　　　　　Gal(β1-3)GlcNAc(β1-3/4)

Fuc(α1-2)Gal(β1-3)GlcNAc

Fuc(α1-2)Gal(β1-3)GlcNAc
　　　　Fuc(α1-4)

NeuAc(α2-6)Gal(β1-3)GlcNAc
　　　　NeuAc(α2-3)

TYPE 2

NeuAc(α2-6)Gal(β1-4)GlcNAc

Fuc(α1-2)Gal(β1-4)GlcNAc
　　　Fuc(α1-3)

NeuAc(α2-3)Gal(β1-4)GlcNAc
　　　Fuc(α1-3)

Gal(α1-3)Gal(β1-4)GlcNAc

SO_4^- GalNAc(β1-4)GlcNAc

GalNAc(β1-4)GlcNAc

NeuAc(α2-6)Gal(β1-4)GlcNAc
　　　　　　　　　| 3
　　　　　　　　SO_4^-

Figure 4: Examples of outer-chain structures of complex type N-glycans.

6. STRUCTURE OF O-GLYCANS

O-glycans are largely represented in mucins where they represent up to 80% weight of the molecule (Roussel *et al.*, 1988), they also occur in N- and O-glycosylproteins. These oligosaccharides are linked to the peptidic backbone though an O-glycosidic linkage between the hemi-acetal function of N-acetylactosamine and the alcohol group of a serine or threonine residue (Figure 7). This type of linkage is rather sensitive to alkaline reagents. O-glycans are generally released from the glycoprotein by reductive alkaline treatment in

Jean-Claude Michalski

```
Fuc(α1-4)
         \
Gal(β1-4)GlcNAc – Man(α1-6)
                          \
              Xyl(β1-2) – Man(β1-4)GlcNAc(β1-4)GlcNAc – Asn
                         /                          /
Gal(β1-4)GlcNAc-Man(α1-3)                    Fuc(α1-3)
              /
Fuc(α1-4)
```

Figure 5: Structure of plant complex-type-N-glycans.

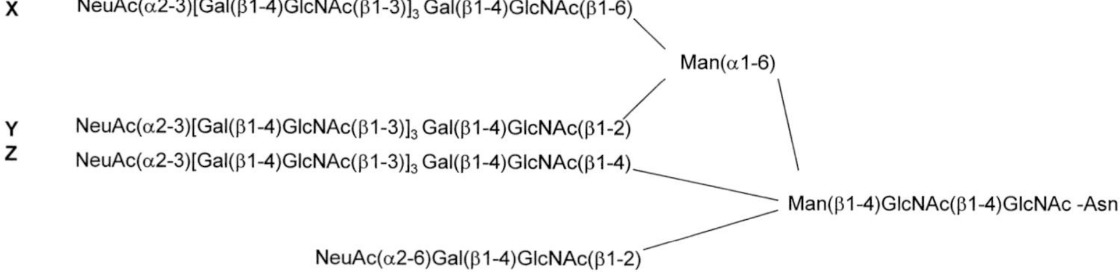

Figure 6: Complex antennary structure with poly-N-acetyl-lactosamine extensions on three branches (X > Y > Z).

the form of oligosaccharide-alditols (Carlson, 1968). O-glycans are characterized by a large heterogeneity found in their chain length which may be comprised between 1 to 20 monosaccharides units (Hounsell *et al.*, 1996). The main monosaccharides are N-acetylactosamine (GalNAc), Fucose (Fuc) and sialic acids (NeuAc). Galactose and N-acetylglucosamine may be sulfated.

Far from this extreme heterogeneity, O-glycans are all constructed on a same structural model made of three distinct regions (Figure 8) (Roussel & Lamblin, 1996):

- The "core" which includes the GalNAc linked to the peptidic and one or two other additive monosaccharides
- The "backbone" which constitutes a hinge region made of Gal and GalNAc sequences
- The "periphery" made with different monosaccharides generally assembled as antigenic determinants (Table 2).

6.1. The Cores

Up to date eight different O-glycannic cores have been described in Figure 9.

1. Core 1: (β1-3)GalNAc α Ser/Thr is the most common one initially described in pig submaxillar mucin (Carlson, 1968). This sequence is known as Thomsen-Friedenreich

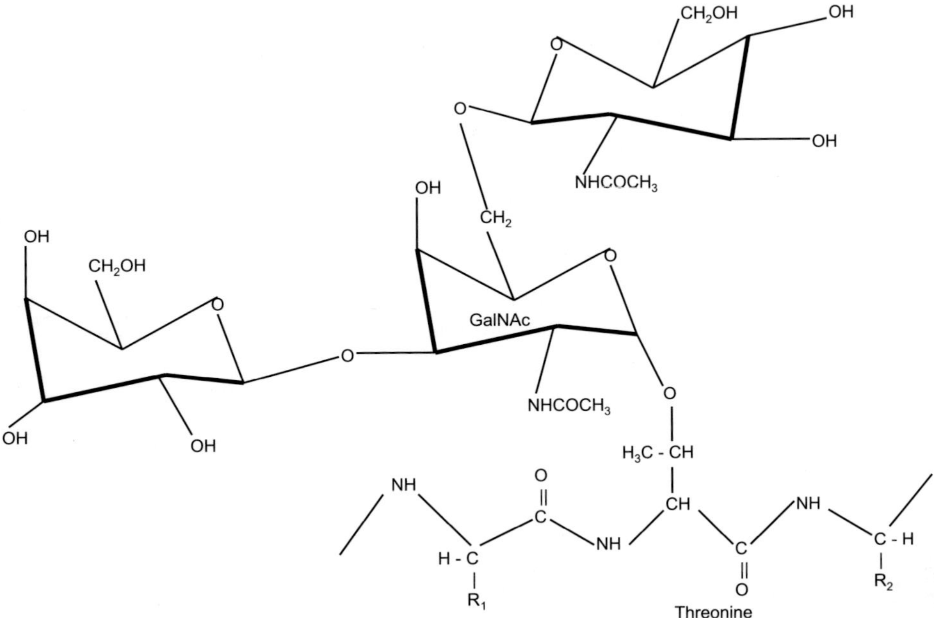

Figure 7: Structure of the ''mucin-like'' O-linkage.

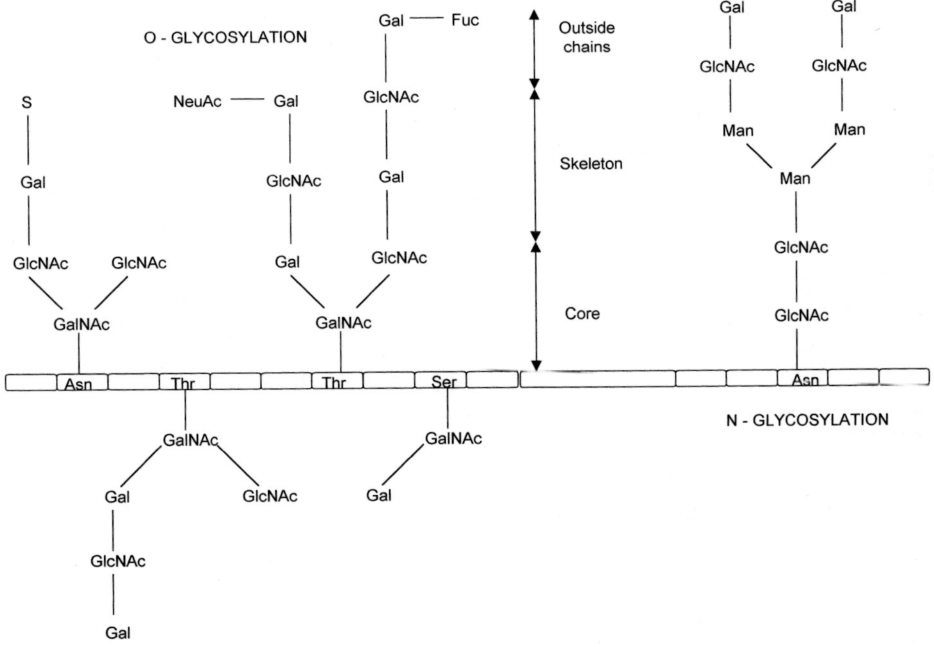

Figure 8: General structural scheme of O-glycans.

                    Jean-Claude Michalski

antigen recognized by the PNA lectin (Lotan *et al.*, 1975) and has been described in mucins of different species (Lamblin *et al.*, 1980, Maes *et al.*, 1997). Core 1 is generally sialylated in glycoproteins:

NeuAc(α2-3)Gal(β1-3)GalNAc(α1-0)Ser/Thr
NeuAc(α2-3)[Gal(β1-3)]GalNAc(α1-0)Ser/Thr
NeuAc(α2-3)[NeuAc(α2-3)Gal(β1-3)]GalNAc(α1-0)Ser/Thr

2. Core 2: (β1-3)[GlcNAc(β1-6)]GalNAc α Ser/Thr originates from core 1 by the addition of a β1-6 linked GlcNAc (Brockhausen *et al.*, 1985). It has been described in different secreted mucins such as human gastric mucin (Qates *et al.*, 1974) or swallow nest mucin (Wieruszeski *et al.*, 1987).

3. Core 3: GlcNAc(β1-3)GalNAc α Ser/Thr. First isolated from rat intestinal mucins (Carlson *et al.*, 1978), this core occurs in numerous mucins (Chai *et al.*, 1992a; Hanisch *et al.*, 1992, 1993; Strecker *et al.*, 1992).

4. Core 4: GlcNAc(β1-6)[GlcNAc(β1-3)]GalNAc α Ser/Thr initially described in ovine gastric mucin (Hounsell *et al.*, 1980), it has further been described in human mucins (Breg *et al.*, 1988; Klein *et al.*, 1992, 1993).

5. Core 5: GalNAc(α1-3)GalNAc α Ser/Thr constitutes a rare sequence initially described in human colon adenocarcinoma (Kurosawa *et al.*, 1983) and human meconium (Hounsell *et al.*, 1985).

Core Number	Sequence
1	Gal(β1-3)GalNAc

2	GlcNAc(β1-6)
	    \\
	      GalNAc
	    /
	Gal(β1-3)

3	GlcNAc(β1-3)GalNAc

4	GlcNAc(β1-6)
	    \\
	      GalNAc
	    /
	GlcNAc(β1-3)

5	GalNAc(α1-3)GalNAc
6	GlcNAc(β1-6)GalNAc
7	GalNAc(α1-6)GalNAc
8	Gal(α1-3)GalNAc

Figure 9: Principal O-glycans cores.

Table 2. Principal carbohydrate antigens found in outside chains of O-glycans

Antigen	*Structure*
A	GalNAc(α1-3)[Fuc(α1-2)]Galβ-
B	Gal(α1-3)[Fuc(α1-2)]Galβ-
B linear (Galili)	Gal(α1-3)Gal-
H	[Fuc(α1-2)]Galβ-
T	Gal(β1-3)GalNAcα-Ser/Thr
Tn	GAlNAcα-Ser/Thr
Sialyl Tn	NeuAc(α2-6)GalNAcα-Ser/Thr
S[d] (Cad)	GAlNAc(β1-4)[SA(α2-3)]Galβ-
I	Gal(β1-4)GlcNAc(β1-3)Galβ-
I	Gal(β1-4)GlcNAc | (β1-6) Gal(β1-4)GlcNAc(β1-3)Ga1β
Le[a]	Gal(β1-3)[Fuc(α1-4)]GlcNAc(β1-3)Gal-
Le[b]	[Fuc(α1-2)]Gal(β1-3)[Fuc(α1-4)]GlcNAc(β1-3)Gal-
Le[x]	Gal(β1-4)[Fuc(α1-3)]GlcNAc(β1-3)Gal-
Le[y]	[Fuc(α1-2)]Gal(β1-4)[Fuc(α1-3)]GlcNAc(β1-3)Gal-
Sialyl-Le[x]	NeuAc(α2-3)Gal(β1-4)[Fuc(α1-3)]GlcNAc(β1-3)Gal-
Forssman	GalNAc(α1-3)GalNAc((β1-3)Gal-

6. Core 6: GlcNAc(β1-6)GalNAc α Ser/Thr is exclusively found in human mucins (Hounsell *et al.*, 1985; Schachter & Brockhausen, 1992).
7. Core 7: GalNAc(α1-6)GalNAc α Ser/Thr was described in bovine submaxillary mucins (Chai *et al.*, 1992b).
8. Core 8: (α1-3)GalNAc α Ser/Thr was described in human respiratory mucins (Van Halbeek *et al.*, 1994).

6.2. O-Glycans Backbones

Addition of Gal and GalNAc residues on the different cores leads to the formation of different backbones (Figure 10).

6.3. O-Glycans Periphery

O-glycans are terminated by different monosaccharides or different oligosaccharide sequences, which may be related to the ABH or Lewis blood groups antigens (Schachter & Brockhausen, 1992; Van den Steen *et al.*, 1998). The different types of external substituents are summarised on Table 2. Commonly sialic acid is 5-N-acetyl-neuraminic acid (Neu5Ac) which may also occur in O-acetylated form (Varki, 1992; Reid *et al.*, 1984). N-glycolyl neuraminic acid (NeuGc) by opposition to other species is normally absent in human O-glycans. Nevertheless, it has been characterized in cancer cells as Hanganutziu-Deicher antigen (Devine *et al.*, 1991). Sulfated groups substitute galactose residues either at position **C3** (Capon *et al.*, 1989; Kuhns *et al.*, 1995; Lo Guidice *et al.*, 1997), or position **C6** and **C4** (Sangadala *et al.*, 1993; Mawhinney *et al.*, 1987), they are also found on N-acetylglucosamine (Mawhinney *et al.*, 1992; Lo Guidice *et al.*, 1994; Strecker *et al.*, 1987).

Backbone	**trivial name or Ref.**
Gal(β1-3)GlcNAc(β–)	Type 1 or Lacto
Gal(β1-4)GlcNAc(β–)	Type 2 or neo-lacto
[Gal(β1-4)GlcNAc(β1-3)]$_n$	i antigen
[Gal(β1-4)GlcNAc(β1-6)]$_n$	-

Gal(β1-4)GlcNAc(β1-6)   I antigen
                \
                    Gal(β-)
                /
Gal(β1-4)GlcNAc(β1-3)

GlcNAc(β1-6)Gal(β1-)

Gal(β1-3)GalNAc(α, β)   Type 3 and 4

Figure 10: Different types of O-glycans backbones.

7. ANALYSIS OF PROTEIN GLYCOSYLATION

Physical, chemical and enzymatic methods can be applied to the analysis of glycans. Each type of glycoconjugates presents its own unique analytical problems, and no general receipt can be proposed for the study of a peculiar glycan. Nevertheless, general strategies developed for the isolation and characterization of glycans, and a very wide range of approaches and techniques have been adopted for the study of protein glycosylation. Methodologies of structural glycobiology improve very quickly and are now very sensitive, allowing characterization of very small amounts of material below the nanomole level. In this chapter we will review general methods used in structural analysis and will focus on new strategies used in glycans separation and sequencing.

7.1. Glycans Release from Native Glycoprotein

Glycans may be released from glycoproteins by chemical or enzymatic methods.

7.1.1. N-glycans

N-glycans can be released by the hydrazinolysis procedure (Michalski *et al.*, 1984; Patel *et al.*, 1993). This method is quantitative and has been fully automatized (Oxford Glycosystem, Glycoprep®). It should be mentioned that de-N-acetylation occurs during the reaction and sialic acids substituents (such as N-glycolyl or O-acetyl-groups are lost).

   Release of glycan moieties may be easily carried out by enzymatic hydrolysis. Two kinds of enzyme namely endo-N-acetyl-β-D-glucosaminidase and peptid-N-glycosidase are used (Karamanos, 1997). Endo-N-acetyl-β-D-glucosaminidases form a homogenous family of enzymes from different sources that act on the β1-4-N-acetylglucosaminyl bond of the N-N'-diacetylchitobiose core present in all N-glycosylproteins.

Among "endo" enzymes, Endo-H is strictly specific of high-mannose type structures. These enzymes are convenient tool for discriminate the type of glycosylation.

Peptide N-glycanase [peptide–N^4–(N-acetyl-β-glucosaminyl) asparagine amidase] is able to split the monosaccharide-peptide linkage removing the completely glycan moiety from the glycoprotein. These enzymes may be use in preparative scale for releasing the pool of oligosaccharides before subsequent separation.

7.1.2. O-glycans

Classically O-glycans are released by alkaline reductive treatment (Carlson, 1968) as oligosaccharide-alditols. The use of O-glycanase (endo-α-N-acetyl-galactosaminidase) is limited due to the narrow specificity of the enzyme, which requires at least desialylation of the oligosaccharides (Endo & Kobata, 1976).

7.2. Glycan Fractionation

Various approaches have been developed to achieve separation of oligosaccharides with high resolution and great sensitivity of detection (Davies & Hounsell, 1996). The fractionation of oligosaccharides mixture is based on discrete parameters such as charge, size or hydrophobicity (Michalski & Alonso, 1998). The most common strategy is to first separate the different oligosaccharides into different classes on the basis of their charge, follow by further fractionation according to their size. In most cases, multidimensional separation protocols are required to obtain homogenous oligosaccharides. "Mapping" strategies using 2- and 3-dimensional glycan analysis in which glycan pools are resolved into individual species by successive passage though anion-exchange, hydrophobic and hydrophilic interaction HPLC have been developed (Tomiya *et al.*, 1988; Hase, 1994; Kopp *et al.*, 1997). Glycan structures are assigned by comparison of the 2- or 3-D coordinates of the elution positions of the samples with these of known structures. Oligosaccharides may be separated into the native form or fluorescently labelled with different dyes, such as 2-amino-pyridine, 2-amino-benzamide or anthranilic acid (Rudd & Dwek, 1997a; Anumula & Dhume, 1998). Among HPLC techniques, separation by high-pH anion exchange chromatography (HPAEC) coupled to pulse amperometric detection allows powerful separation of oligosaccharide isomers (Townsend, 1995). Assignments of glycan structures may be confirmed by sequential exoglycosidases digestion using highly specific enzymes (Edge *et al.*, 1992). Other separation techniques include capillary electrophoresis (Kakehi & Honda, 1996; Hermentin *et al.*, 1994) and polysaccharide gel electrophoresis (Jackson, 1991).

In addition gel permeation (Kobata, 1994) or serial lectin affinity chromatography (Cummings, 1994) can also fractionate glycan mixtures.

7.3. Glycan Structural Analysis

Complete structural determination of glycans requires a combination of different techniques. The choice of the investigation methods depends on the type of glycan and the amount available. Chemical, enzymatical and physico-chemical methods will generally be used. Complete structural analysis of purified oligosaccharides requires determination of 1) the type of number of constituent monosaccharides, their anomeric linkages (α and β) and eventually their ring conformation (furanose or pyranose) 2) nature and position of the linkages between monosaccharides 3) the type and position of aglycon substituents (acetyl, methyl, phosphate or sulfate groups).

7.3.1. Monosaccharide composition

Monosaccharide composition analysis is a prerequisite step for determining the approach to be used for detailed structural characterization, it also provides a first clue on the structure of the glycans (Montreuil *et al.*, 1994).

The analysis involves:

1. The release of individual monosaccharides by acid hydrolysis.
2. The fractionation and molar ratio quantification by gas liquid chromatography (GLC) or High Performance Liquid Chromatography (HPLC).
3. Identification of the individual monosaccharides by gas liquid chromatography/mass spectrometry (GLC–MS).

When sialic acids are present they have to be analyzed separately in the native form (Manzi & Varki, 1994) or after derivatization (Klein *et al.*, 1997).

7.3.2. Sequence determination

1. Nuclear Magnetic Resonance (NMR)

NMR spectroscopy provide complete information on the glycan's structure including 1) nature of the monosaccharides and ring size of sugars 2) anomericity and nature of the linkage 3) the nature, number and position of substituents (Vliegenthart *et al.*, 1983; Van Halbeek 1994). Limitation of the NMR study are the quantity and purity of the oligosaccharides (from 20 to 50 nmoles).

2. Mass spectrometry (MS)

Developments of MS techniques are proceeding very quickly. Due to the low quantity of material required (1 to 10 pmoles), MS becomes the method of choice for structural analysis (Dell *et al.*, 1994). Various techniques such as Fast Atom Bombardment (FAB), Electrospray ionization (ES) or Matrix Assisted Laser Desorption Ionization (MALDI), mass spectrometry can be used. The structural information obtained includes molecular mass, monosaccharide composition and presence of substituents. Specific fragment ions generated during MS analysis also provide sequence informations.

MS is also used for site heterogeneity study. Analysis of glycopeptides generated after partial proteolysis, provide information both on the nature of glycans and on the peptide sequence (Sasaki *et al.*, 1988).

3. Sequential exoglycosidases hydrolysis

The recurrent sequential degradation with highly specific exo-glycosidases gives information on the identity, the anomericity, the nature of linkage and the sequences of the oligosaccharide (Edge *et al.*, 1992). The complete sequence of an oligosaccharidic chain can be obtained by stepwise treatment with a panel of different exoglycosidases. The method can be associated with different chromatographic procedures such as gel-permeation, HPLC or Fluorophore Assisted Carbohydrates Electrophoresis (FACE) (Guile *et al.*, 1996; Jackson 1994; Küster *et al.*, 1997).

4. Chemical methods

Permethylation analysis combined with GLC/MS is useful to give an insight on the substitution pattern of a given oligosaccharide (i.e. position of glycosidic linkages, terminal sugars, branching points, sequence) (Levery & Hakomori, 1987; Geyer & Geyer 1994; Fournet *et al.*, 1981). The method can be applied to very low amount of material ranging from 5 to 30 pmoles.

8. CONCLUDING REMARKS

Structural glycobiology is a booming field, mainly due to the improvements of new technologies for characterizing glycans below the nanomole level. A large array of new structures has been established recently. Detailed structural analysis provide a comprehensive information for the study of both conformation and biological function of glycans. Due to the impressive number of glycans which may be present in glycoproteins, isolation and establishment of their structure remain a difficult challenge requiring a panel of different methods including sophisticated instruments. Finally, it is clear that primary structures of glycans represent just a part of our knowledge in understanding the function of glycosylation. The 3D structures of both the protein and the oligosaccharides are other features to consider.

ACKNOWLEDGEMENTS

We would like to acknowledge Mrs Catherine Alonso for her help in preparation of this manuscript.

REFERENCES

Anumula, K.R., Dhume, S.T. (1998) High resolution and high sensitivity methods for oligosaccharide mapping and characterization by normal phase high performance liquid chromatography following characterization with highly fluorescent anthranilic acid. *Glycobiology*, **8**, 685–694.

Baenziger, J.V., Green, E.D. (1988) Pituitary glycoprotein hormone oligosaccharides: structure, synthesis and function of the asparagine-linked oligosaccharides in lutropin, follitropin and thyrotropin. *Biochim. Biophys. Acta.*, **947**, 287–306.

Breg, J., Van Halbeek, H., Vliegenthart, J.F.G., Klein, A., Lamblin, G., Roussel, Ph. (1988) Primary structure of neutral oligosaccharides derived from respiratory mucus glycoproteins of a patient suffering from bronchiectasis, determined combination of 500 MHz ^{1}H NMR spectroscopy and quantitative analysis. *Eur. J. Biochem.*, **171**, 643–654.

Brockhausen, I., Matta, K.L., Orr, J., Schachter, H. (1985) Mucin synthesis UDPGlcNAc: NAc-R-β3-N-acetylglucosaminyl transferase and UDP-GlcNAc: GlcNAc β1-3 GalNAc-R (GlcNAc to GalNAc) β6-N-acetylglucosaminyl transferase from pig and rat colon mucosa. *Biochemistry*, **24**, 1866–1874.

Capon, C., Leroy, Y., Wieruszeski, J.M., Ricart, G., Strecker, G., Montreuil, J., Fournet, B. (1989) Structures of O-glycosidically linked oligosaccharides isolated from human meconium glycoproteins. *Eur. J. Biochem.*, **182**, 139–152.

Carlson, D.M. (1968) Structures and immunochemical properties of oligosaccharides isolated from pig submaxillary mucins. *J. Biol. Chem.*, **243**, 616–626.

Carlson, H.E., Sundblad, G., Hammarström, S., Lönngren, J. (1978) Structure of some oligosaccharides derived from rat intestinal glycoproteins. *Carbohydr. Res.*, **64**, 181–188.

Chai, W., Hounsell, E.F., Cashmore, G.C., Rosankiewicz, J.R. Bauer, C.J., Feeney, J., Feizi, T., Lawson, A.M. (1992b) Neutral oligosaccharides of bovine submaxillary mucin. A combined mass spectrometry and ^{1}H NMR study. *Eur. J. Biochem.*, **203**, 257–268.

Chai, W., Hounsell, E.F., Cashmore, G.C., Rosankiewicz, J.R. Feeney, J., Lawson, A.M. (1992a) Characterization by mass spectrometry and ^{1}H NMR of novel hexasaccharides among the acidic O-linked carbohydrate chains of bovine submaxillary mucin. *Eur. J. Biochem.*, **207**, 973–980.

Chiba, A., Matsumura, K., Yamada, H., Inazu, T., Shimizu, T., Kusunoki, S., Kanazawa, A., Kobata, A., Endo, T. (1997) Structures of sialylated O-linked oligosaccharides of bovine peripheral nerve α-dystroglycan. *J. Biol. Chem.*, **272**, 2156–2162.

Cummings, R.D. (1994) Use of lectins in analysis of glycoconjugates. *Methods Enzymol*, 230, 66–85.

De Beer, T., Vijegenthart, J.F.G., Cöffler, A., Hofsteenge, J. (1995) The hexopyryranosyl residue that is C-glycosidically linked to the side chain of tryptophan-7 in human RNase U_s is α-mannopyranose. *Biochemistry*, **34**, 11785–11789.

De Waard, P., Koorevaar, A., Kamerling, J.P., Vliegenthart, J.F.G. (1992) Structure determiantion by ^{1}H NMR spectrometry of (sulfated) sialylated N-linked carbohydrate chains released from porcine thyroglobulin by peptide-N4-(N-acetyl-β-glucosaminyl)asparagine amidase. *J. Biol. Chem.*, **266**, 4237–4243.

Dell, A., Reason, A.J., Khoo, K.H., Panico, M., Mc Dowell, R.A., Morris, H.R. (1994) Mass spectrometric approaches for glycobiology. *Methods Enzymol.*, **230**, 108–132.

Devine, P.L., Clark, P.A., Birell, G.W., Laytin, G.T., Ward, B.G., Alewood, P.F., Mc Kenzie, I.F.C. (1991) The breast tumor-associated epitope defined by monoclonal antibody 3^E-1.2 is an O-linked is an O-linked mucin carbohydrate containing N-glycolylneuraminic acid. *Cancer Res.*, **51**, 5826–5836.

Davies, M.J., Hounsell, E.F. (1996) Carbohydrate chromatography: towards yoctomole sensitivity. *Biomed. Chrom.*, **10**, 285–289.

Dwek, R.A., Edge, C.J., Harvey, D.J., Wormald, M.R. (1993) Analysis of glycoprotein associated oligosaccharides. *Ann. Rev. Biochem.*, **62**, 65–100.

Edge, C.J., Rademacher, T.W., Wormald, M.R., Parekh, R.B., Butters, T.D., Wing, D.R., Dwek, R.A. (1992) Fast sequencing of oligosaccharides: the reagent-array analysis method. *Proc. Natl. Acad. Sci. USA*, **89**, 6338–6342.

Endo, Y., Kobata, A. (1976) Partial purification and characterization of an endoalpha-N-acetylactosaminidase from the culture medium of *Diplococcus pneumoniae*. *J. Biochem.*, **80** (Tokyo), 1–8.

Ferguson, M.A.J., Williams, A.F. (1988) Cell surface anchoring of proteins via glycosylphosphatidylinositol structures. *Ann. Rev. Biochem.*, **57**, 285–320.

Fichette-Laine, A.C., Gomord, V., Cabanes, M., Michalski, J.C., Saint-Marcary, M., Foucher, B., Cavelier, B., Hawes, C., Lerouge, P., Faye, L. (1997) N-glycans harbouring the Lewis a epitope are expressed at the surface of plant cells. *Plant J.*, **12**, 1411–1417.

Fournet, B., Strecker, G., Leroy, Y., Montreuil, J. (1981) Gas-liquid chromatography and mass spectrometry of methylated and acetylated methyl-glycosides. Application to the structural analysis of glycoprotein glycans. *Anal. Biochem.*, **116**, 489–502.

Fukuda, M. (1994) Cell surface carbohydrate: cell type specific expression. In Molecular Glycobiology (Fukuda, M., Hindsgaul, O. eds) Oxford, IRL Press

Geyer, R., Gyer, H. (1994) Saccharide linkage analysis using methylation and other techniques. *Methods Enzymol.* **230**, 86–108.

Guile, G.R., Rudd, P.M., Wing, D.R., Prime, S.B., Dwek, R.A. (1996) A rapid high-resolution high-performance liquid chromatography method for separating glycan mixtures and analizing oligosaccharide profiles. *Anal. Biochem.*, **240**, 210–226.

Hanisch, F.G., Chai, W., Rosankiewicz, J.R., Lawson, A.M., Stoll, M.S., Feizi, T. (1993) Core typing of O-linked glycans from human gastric mucins lack of evidence for the occurrence of the core sequence 1-6GalNAc. *Eur. J. Biochem.*, **217**, 645–655.

Hanisch, F.G., Peter-Katalinic, J., (1992) Structural studies of fetal mucins from human amniotic fluid. Core typing of short chain O-linked glycans. *Eur. J. Biochem.*, **205**, 527–535.

Hart, G.W., Haltiwanger, R.S., Holt, G.D., Kelly, W.G. (1989) Glycosylation in the nucleus and cytoplasm. *Ann. Rev. Biochem.*, **58**, 871–874.

Hase, S. (1994) High performance liquid chromatography of pyridylaminated saccharides. *Methods Enzymol.*, **230**, 225–237.

Hase, S., Nishimura, H., Kawabata, S.I., Iwanaga, S., Ikenaka, T. (1990) The structure of (xylose)2Glucose-O-Ser S3 found in the first epidermal growth factor linked domain of bovine blood. *J. Biol. Chem.*, **265**, 1858–1861.

Hermentin, P., Doenges, R., Witzel, R., Hokke, C.H., Vliegenthart, J.F.G., Kamerling, J.P., Conradt, H.S., Nimtzi, M., Brazel, D. (1994) A strategy for the mapping of Nglycans by high performance capillary electrophoresis. *Anal. Biochem.*, **221**, 29–41.

Hounsell, E.F., Davies, M.J., Renouf, D.V. (1996) O-linked protein glycosylation structure and function. *Glycoconj. J.*, **13**, 19–25.

Hounsell, E.F., Fukuda, M., Powell, M.E., Feizi, T., Hakomori, S. (1980) A new O-glycosidically linked tri-hexosamine core structure in sheep gastric mucin: a preliminary note. *Biochem. Biophys. Res. Commun.*, **92**, 1143–1150.

Hounsell, E.F., Lawson, A.M., Feeney, J., Gooi, H.C., Pickering, N.J., Stoll, M.S., Lui, S.C., Feizi, T. (1985) Structural analysis of the O-glycosidically linked core region oligosaccharides of human meconium glycoproteins which express onco-foetal antigens. *Eur. J. Biochem.*, **148**, 367–377.

Jackson, P. (1991) Polyacrylamide gel electrophoresis of reducing saccharides labeled with the fluorophore 2-aminoacidone: subpicomolar detection using an imaging system based on cooled charge-coupled device. *Anal. Biochem.*, **196**, 238–244.

Jackson, P. (1994) High-resolution polyacrylamide gel electrophoresis of fluorophorelabeled reducing saccharides. *Methods Enzymol.*, **230**, 250–264.

Kakeki, K., Honda, S.J. (1996) Analysis of glycoproteins, glycopeptides and glycoprotein-derived oligosaccharides by high-performance capillary electrophoresis. *J. Chromatogr. A.*, **720**, 377–393.

Karamanos, Y. (1997) Endo-N-acetyl-beta-D-glucosaminidases and their potential substrates: structure I function relationships. *Res. Microbiol.*, **148**, 661–671.

Klein, A., Carnoy, C., Lamblin, G., Roussel, Ph., Van Kuik, J.A., Vliegenthart, J.F.G. (1993) Isolation and structural characterization of novel sialylated oligosaccharidealditols from respiratory mucus glycoproteins of a patient suffering from bronchiectasis. *Eur. J. Biochem.*, **211**, 491–500.

Klein, A., Carnoy, C., Wieruszeski, J.M., Strecker, G., Strong, A.M., Van Halbeek, H., Roussel, Ph., Lamblin, G. (1992) The broad diversity of neutral and sialylated oligosaccharides derived from human salivary mucins. *Biochemistry.*, **31**, 6152–6165.

Klein, A., Diaz, S., Ferreira, I., Lamblin, G., Rousel, Ph., Manzi, A.E. (1997) New sialic acids from biological sources identified by a comprehensive and sensitive approach: liquid chromatography–electrospray ionization–mass spectrometry (LC-ISI-MS) of SIA quinoxalinones. *Glycobiology*, **7**, 421–432.

Kobata, A. (1992) Structures and functions of the sugar chains of glycoproteins. *Eur. J. Biochem.*, **209**, 483–501.

Kobata, A. (1994) Size fractionation of oligosaccharides. *Methods Enzymol*, **230**, 200–208.

Kopp, K., Schlüter, M., Werner, R.G. (1997) 2-aminobenzamide labeling of desialylated oligosaccharides. A sensitive method for monitoring lot to lot consisting of recombinant glycoproteins: In Towsend & Hutckins (eds) *Techniques in Glycobiology*, Dekker, N.Y., pp. 475–489.

Kudo, M., Kitajima, K., Inoue, S., Shiokawa, K., Morris, H.R., Dell, A., Inoue, Y. (1996) Characterization of the major core structures of the α2-8-linked polysialic acid containing glycan chains present in neural cell adhesion molecule in embryonic chick brains. *J. Biol. Chem.*, **271**, 32667–32677.

Kuhns, W., Jam, R.K., Matta, K.L., Paulsen, H., Baker, M.A., Geyer, R., Brockhausen I. (1995) Characterization of a novel mucin sulphotransferase activity sulphated O-glycan core 1,3-sulphate Galbeta 1-3-GalNAc alpha—R. *Glycobiology*, **5**, 689–697.

Kurosawa, A., Nakajima, H., Funakashi, I, Matsuyama, M., Nagayo, T., Yamashina, I. (1983) Structures of the major oligosaccharides from a human rectal adenocarcinoma glycoprotein. *J. Biol. Chem.*, **258**, 11594–11598.

Küster, B., Wheeler, S.F., Hunter, A.P., Dwek, R.A., Harrey, D.J. (1997) Sequencing of N-linked oligosaccharides from protein gels: In gel deglycosylation followed by matrix-assisted laser desorption/ionization mass spectrometry and normal phase high performance liquid chromatography. *Anal. Biochem.*, **250**, 82–101.

Lamblin, G., Lhermitte, M., Boersma, A., Roussel, Ph., Rheinhold, V. (1980) oligosaccharides of human bronchial glycoproteins. Neutral di- and trisaccharides isolated from patient suffering from chronic bronchitis. *J. Biol. Chem.*, **255**, 4595–4598.

Lamblin, G., Rahmoune, H., Wieruszeski, J.M., Lhermitte, M., Strecker, G., Roussel, Ph. (1991) Structure of two sulfated oligosaccharides from respiratory mucins of a patient suffering from cystic fibrosis. *Biochem. J.*, **275**, 199–206.

Levery, S.B., Hakomori, S.I. (1987) Microscale methylation analysis of glycolipids using capillary gas chromatography chemical ionization mass fragmentography selected ion monitoring. *Methods Enzymol.*, **138**, 13–25.

Lis, H., Sharon, N. (1993) Protein glycosylation. Structural and function aspects. *Eur. J. Biochem.*, **218**, 1–27.

Lo Guidice, J.M., Herz, H., Lamblin, G., Plancke, Y., Roussel, Ph., Lhermitte, M. (1997) Structures of sulfated oligosaccharides isolated from the respiratory mucins of a non-secretor (O, Lea^{+}b^{-}) patient suffering from chronic bronchitis. *Glycoconjugate J.*, **14**, 113–125.

Lo Guidice, J.M., Wieruszeski, J.M., Lemoine, J., Verbert, A., Roussel, Ph., Lamblin, G. (1994) Sialylation and sulfatation of the carbohydrate chains in respiratory mucins from a patient with cystic fibrosis. *J. Biol. Chem.*, **269**, 18794–18813.

Lotan, R., Skutelsky, E., Damon, D., Sharon, N. (1975) The purification, composition and specificity of the anti-T lectin from peanut *(Arachis hypogaea). J. Biol. Chem.,* **250**, 8515–8523.

Maes, F., Florea, D., Delplace, F., Lemoine, J., Plancke, Y., Strecker, G. (1997) Structural analysis of oligosaccharide-alditols released by reductive β-elimination from oviducal mucins of *Rana temporaria. Glycoconj. J.,* **14**, 127–146.

Manzi, A.E., Varki, A. (1994) Compositional analysis of glycoproteins. In Fukuda, M., Kobata, A. (eds) Glycobiology a practical approach. IRL Press, pp. 27–77.

Mawhinney, T.P., Adelstein, E., Morris, D.A., Mawhinney, A.H., Barbero, G.J.—1987) Structure determination of five sulfated oligosaccharides derived from tracheobronchial mucus glycoproteins. *J. Biol. Chem.,* **262**, 2994–3001.

Mawhinney, T.P., Landrum, D.C., Gayer, D.A., Barbero, G.J. (1992) Sulfated sialyloligosaccharides derived from tracheobronchial mucus glycoproteins of a patient suffering from cystic fibrosis. *Carbohydr. Res.,* **235**, 179–197.

Michalski, J.C., Alonso, C. (1998) HPLC of oligosaccharides and glycopeptides. Oliver, R.W.A. (ed) *HPLC of macromolecules.* Second edition. IRL Press, pp. 171–202.

Michalski, J.C., Peter-Katalinic, J., Paz-Parente, J., Montreuil, J., Strecker, G. (1984) Behavioue of 2-acetamido-2-deoxy-beta-D-glucopyranosyl residue during sequential hydrazinolysis, N-reacaetylation, reduction and methylation of glycoasparagines. *Carbohydr. Res.,* **134**, 177–189.

Moloney, D.J., Lin, A.L., Haltiwanger, R.S. (1987) The O-linked fucose glycosylation pathway, *J. Biol. Chem.,* **272**, 19046–19050.

Montreuil, J. (1982) Glycoproteins (Neuberger, A., Van Deenen, L.L.M. eds) In Comprehensive Biochemistry, Vol 19B/II, Elsevier, Amsterdam, pp. 1–188.

Montreuil, J. (1995) Glycoprotein structure and conformation. Verbert, A., (ed) *Methods on Glycoconjugates, a laboratory manual.* Harwood Acad. Pub. CH, pp. 1–26.

Montreuil, J., Bouquelet, S., Debray, H., Lemoine, J., Michalski, J.C., Spik, G., Strecker, G. (1994) In *Glycoproteins,* Chaplin, M.F., Kennedy, J.F. (eds) Carbohydrate analysis, A practical approach, Second edition, IRL Press, pp. 181–294.

Oates, M.D.G., Robsbottom, A.C., Schrager, J. (1974) Further investigation into the structure of human gastric mucin: the structural confirmation of the oligosaccharide chains. *Carbohydr. Res.,* **34**, 115–137.

Osawa, T., Tsuji, T. (1987) Fractionation and structural assessment of oligosaccharides and glycopeptides by use of immobilized lectins. *Ann. Rev. Biochem.,* **56**, 21–42.

Patel, T.J., Bruce, J., Merry, A., Bigge, C., Wormald, M., Jaques, A., Parekh (1993) Use of hydrazinolysis to release in intact and unreduced form both N- and O-linked oligosaccharides from glycoproteins. *Biochemistry,* **32**, 679–693.

Rademacher, T.W., Parekh, R.A., Dwek, R.A. (1988) Glycobiology. *Ann. Rev. Biochem.,* **57**, 785–838.

Reid, P.E., Culling, C.F.A., Dunn, W.L., Ramey, C.W., Clay, M.G. (1984) Chemical and histochemical studies of normal and diseased human gastrointestinal tract. A comparison between histologically normal colon, colonic tumors, alterative colitis and diverticular disease of the colon. *Histochem. J.,* **16**, 235–251.

Roussel, Ph., Lamblin, G. (1996) Human mucosal mucins in diseases. In *Glycoproteins and Disease* (Montreuil, J., Vliegenthart, J.F.G., Schachter, H. eds). New comprehensive Biochemistry. Vol 30. pp. 351–383. Elsevier NL.

Roussel, Ph., Lamblin, G., Lhermitte, M., Houdret, N., Lafitte, J.J., Perini, J.M., Klein, A., Scharfman, A. (1988) The complexity of mucins. *Biochimie,* **70**, 1471–1482.

Rudd, P.M., Dwek, R.A. (1997a) Rapid, sensitive sequencing of oligosaccharides from glycoproteins. *Curr. Opin. Biotechnol.,* **8**, 488–497.

Rudd, P.M., Dwek, R.A. (1997b) Glycosylation: heterogeneity and the 3D structure of proteins. *Crit. Rev. Biochem. Mol. Biol.,* **32**, 1–100.

Sangadala, S., Bhat, U.R., Mendicino, J. (1993) Structures of sulfated oligosaccharides in human trachea mucin glycoproteins. *Mol. Cell. Biochem.,* **126**, 37–47.

Sasaki, H., Ochi, N., Dell, A., Fukuda, M. (1988) Site specific glycosylation of human recombinant erythropoietin: analysis of glycopeptides or peptides at each glycosylation site by FAB-MS. *Biochemistry,* **27**, 8618–8626.

Schachter, H. & Brockhausen, I. (1992) *In Glycoconjugates, Composition, Structure and Function.* H. Allen & E. Kisailus Eds, Marcel Dekker, N.Y., pp. 263–332.

Schauer, R. (1991) Biosynthesis and function of N- and O-substituted sialic acid. *Glycobiology,* **1**, 449–452.

Shakin-Eshleman, S.H., Spitalnik, S.L., Kasturi, L. (1996) The amino-acid at the Xposition of an Asn-X-Ser sequensis an important determinant of N-linked are glycosylation efficiency. *J. Biol. Chem.,* **271**, 6363–6366.

Smythe, C., Candwell, F.B., Ferguson, M., Cohen, P. (1988) Isolation and structural analysis of a peptide containing the novel tyrosyl-glucose linkage in glycogenin. *EMBO J.,* **7**, 2681–2686.

Strecker, G., Wieruszeski, J.M., Martel, C., Montreuil, J. (1987) Structure of sulfated tetra- and pentasaccharides obtained by alkaline borohydride degradation of hen ovomucin. A fast atom bombardment–mass spectrometry and ^{1}H NMR analysis—*Glycoconjugate J.*, **4**, 329–337.

Strecker, G., Wieruszeski, J.M., Michalski, J.C., Alonso, C., Leroy, Y., Boilly, B., Montreuil, J. (1992) Primary structure of neutral and acidic oligosaccharide-alditols derived from the jelly coat of the *Mexican axolotl. Eur. J. Biochem.*, **207**, 995–1002.

Tomiya, N.J., Awaya, M., Kurono, S., Endo, Y., Arata, N., Takahashi, A (1988) Analysis of N-linked oligosaccharides using two-dimensional mapping technique. *Anal. Biochem.*, 171, 73–90.

Townsend, R.R. (1995) Analysis of glycoconjugates using high-pH anion exchange chromatography. In El Rassi, Z., (ed), *Carbohydrate analysis: High-Peformance Liquid Chromatography and Capillary Electrophoresis*, Elsevier, N.Y., pp. 181–209.

Troy, F.A. (1992) Polysialylation from bacteria to brains. *Glycobiology*, **2**, 5–23.

Van den Steen, P., Rudot, P.M., Dwek, R.A., Opdenakker, G. (1998) Concepts and principles of O-linked glycosylation. *Crit. Rev. Biochem. Mol. Biol.*, **33**, 151–208.

Van Halbeek, H. (1994) ^{1}H Nuclear magnetic resonance spectroscopy of carbohydrate chains of glycoproteins. *Methods Enzymol.*, **230**, 132–168.

Van Rooijen, J.J.M., Kamerling, J.P., Vliegenthart, J.F.G. (1998) Sulfated di- tri- and tetraantennary N-glycans in human Tamm Horsfall glycoprotein. *Eur. J. Biochem.*, **256**, 471–487.

Varki, A. (1992) Diversity of the sialic acid. *Glycobiology*, **2**, 25–40.

Vliegenthart, J.F.G., Casset, F. (1998) Novel forms of protein glycosylation. *Curr. Op. Struct. Biol.*, **8**, 565–571.

Vliegenthart, J.F.G., Dorland, L., Van Halbeek, H. (1983) High resolution ^{1}H nuclear magnetic resonance spectroscopy as a tool in the structural analysis of carbohydrates related to glycoproteins. *Adv. Carbohydr. Chem. Biochem.*, **41**, 209–374.

Vošhol, H., Van Zuylen, C.W.E.M., Orberger, G., Vliegenthart, J.F.G., Schachter, H. (1996) Structure of the HNK-1 carbohydrate epitope on bovine peripheral myelin glycoprotein PO. *J. Biol. Chem.*, **271**, 22957–22960.

Wieruszeski, J.M., Michalski, J.C., Montreuil, J., Strcker, G., Peter-Katalinic, J., Egge, H., Van Halbeek, H., Mutsaers, J.H.G.M., Vliegenthart, J.F.G. (1987) Structure of the monosialyl oligosaccharides derived from salivary gland mucin glycoproteins of the chinese swiftler (*Genus collocalia*). Characterization of novel types of extended core structure. Gal (β1-3)[GlcNAc(β1-6)]GalNAc(α1-3)GalNAc-ol and to chain termination [(α1-4)]$_{O-1}$[(α1-4)$_2$GlcNAc(β1-). *J. Biol. Chem.*, **262**, 6650–6657.

Yamashita, K., Ueda, I., Kobata, A. (1983) Sulfated-asparagine linked sugars chains of hen egg albumin. *J. Biol. Chem.*, **258**, 14144–14147.

A3. General Overview of the Structure, Synthesis and Degradation of Glycosaminoglycans and Glycolipids

Jean-Pierre Zanetta

1. GLYCOSAMINOGLYCANS

This introductory chapter will summarise the most striking properties of glycosaminoglycans, and additional information can be obtained from exhaustive reviews (Silbert *et al.*, 1997; McKusik, 1988; Neufeld, 1991; von Figura and Klein, 1981; Fluharty, 1981). The biological function of these compounds largely depends on the nature of the polypeptide bearing the GAGs (cell adhesion) or of the GAG chain itself (binding/trapping of growth factors).

1.1. Structures of Glycosaminoglycans

Glycosaminoglycans are acidic glycans formed of repetitive disaccharide structures. Most of them are the essential constituents of proteoglycans. Only one of them is never attached to protein: hyaluronic acid, a polymer made of repetitive (up to several tens of thousands of the disaccharide units of GlcAβ1-3GlcNAcβ1-4. In contrast with the other GAGs this compound is never sulphated. The other compounds comprise the family of chondroitin sulphate A, B and C, heparan sulphates, heparins and keratan sulphates. Except keratan sulphate which can be attached also to N-glycans, these GAGs are attached to a protein backbone through a O-glycosidic bond to a Ser/Thr residue of the polypeptide chain. The GAGs contain a common sequence in the attachment point involving one Xyl residue, two Gal residues and a GlcA residue (sequence Ser/Thr-O-Xyl-Gal-Gal-GlcA; Figure 1). In contrast, keratan sulphate constitute a terminal part of complex type N-glycans, similar to poly-N-acetyl-lactosaminic sequences found in glycoproteins.

Chondroitin sulphate are made of repetitive disaccharide units GlcAβ1-3GalNAcβ1-4, where GalNAc is sulphated on the carbon 4 for chondroitin sulphate A (Figure 1) and on the carbon 6 for chondroitin sulphate C. Chondroitin sulphates with mixed composition exist as well as chondroitin sulphate having some GalNAc residues sulphated both on position 4 and 6. These compounds could also present different degrees of sulphatation on the carbon 2 of GlcA. The length of these polymers may vary from hundreds to thousand disaccharide units.

Chondroitin sulphate B (or dermatan sulphate), in contrast, is made of the disaccharide unit containing iduronic acid (IdoA, the epimer in position 6 of GlcA) IdoAβ1-3GalNAcβ1-4, the GalNAc residue being generally sulphated in position 4 (Figure 1). It

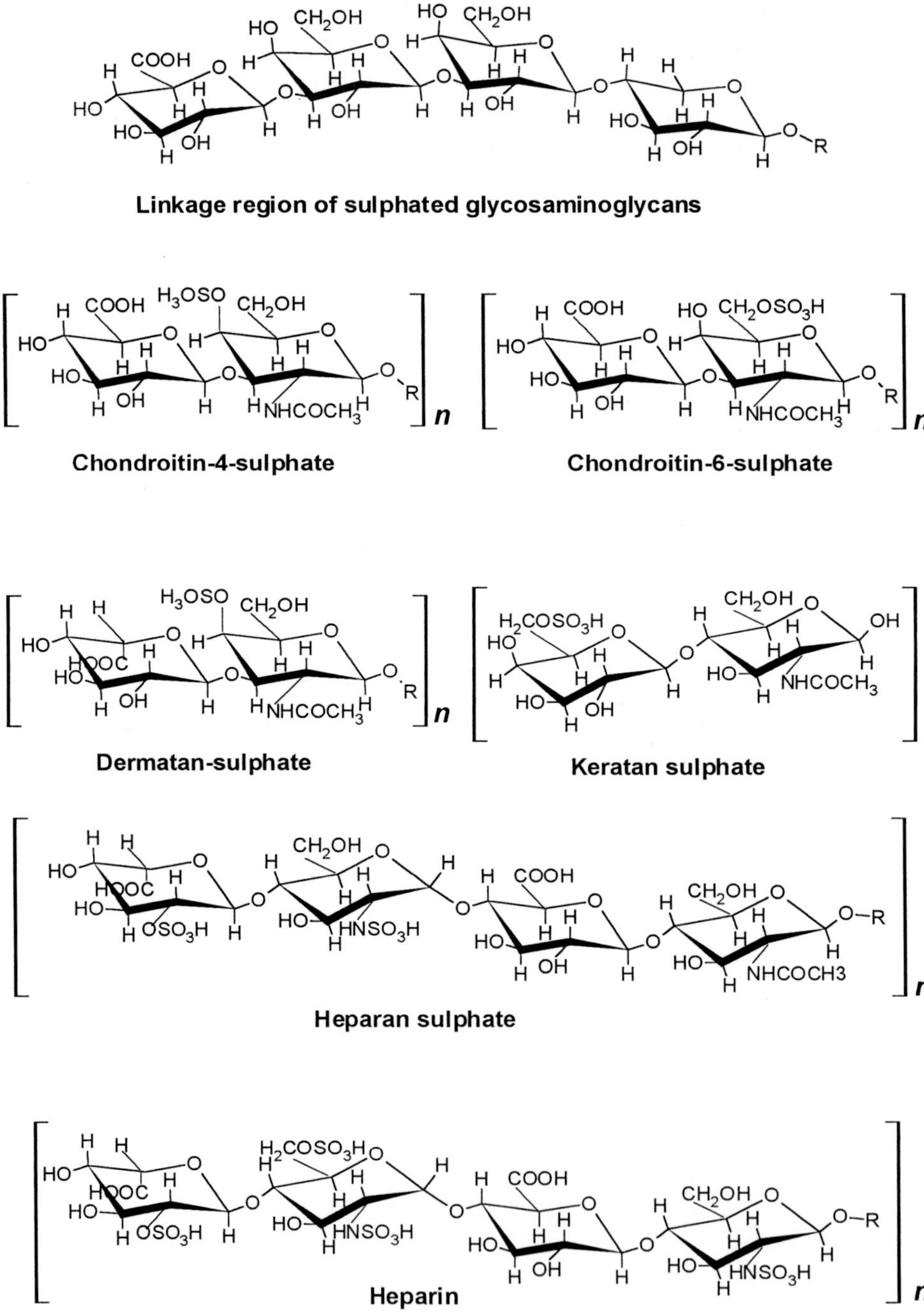

Linkage region of sulphated glycosaminoglycans

Chondroitin-4-sulphate

Chondroitin-6-sulphate

Dermatan-sulphate

Keratan sulphate

Heparan sulphate

Heparin

Figure 1: Example of structures of sulphated glycosaminoglycans and of their linkage region (R represents Ser or Thr residues).

should be stressed that IdoA is a very unstable free compound. It is not incorporated as such in the polymers but as GlcA, this addition being or not followed by the epimerisation on the C6. IdoA can be also sulphated in position 2.

Heparan sulphates constitute a heterogeneous family of compounds characterised by the fact that the hexosamine of the disaccharides is glucosamine. This glucosamine could be GlcN, GlcNAc, GlcN-N-SO₃H and one of these derivatives sulphated in other positions. The uronic acids could be either GlcA or IdoA, these uronic acids being themselves sulphated potentially in different positions. Consequently, these polymers present an extreme heterogeneity, the different compounds being formed from a succession of domains with different constitutions and different properties.

Heparin is one compound of this family, but in contrast with other proteoglycans, heparin is secreted as such. The bonds between the different constituents are generally β1-3 or β1-4, except when the uronic is IdoA in which the bond is α1-4. The degree of sulphatation may vary strongly from one component to the other. Heparin, in contrast with other heparan sulphates possesses some residues of 3-O-SO3H on N-sulphated GlcN.

Keratan sulphates are made of repetitive (up to 50) disaccharide units Galβ1-4GlcNAcβ1-3 in which Gal is sulphated in the 6 position. These chains are present on complex type N-glycans and are in fact similar to poly-N-acetyl-lactosaminic sequences, except that Gal is sulphated. These chains are generally much shorter that those of the other GAGs.

1.2. General Schemes of Biosynthesis of Glycosaminoglycans

The biosynthetic pathways of these compounds have been extensively studied and several enzymes involved in this synthesis have been identified and cloned. Briefly, the synthesis involves first the addition of the constituents of the linkage region to specific Ser or Thr residues of the polypeptide chain. The addition of these monosaccharides from UDP-precursors UDP-Xyl, UDP-Gal, UDP-GlcA (it should be stressed that UDP-Xyl results from the decarboxylation of UDP-GlcA) occurs progressively during the traffic of the protein in the ER and in the Golgi apparatus. The subsequent synthesis of the rest of the chain, involves multi-enzymatic complexes adding the different backbone monosaccharides then adding the sulphate substituents in the right position. The GlcNAc-transferase and GlcA-transferase are apparently coupled. The epimerisation of GlcA into IdoA occurs after the incorporation of GlcA into the chain.

Sulphatation occurs through the transfer of 3′-phosphoadenylyl 5′-phosphosulphate (PAPS) during and after polymerisation. Specific 4 and 6 sulphotransferases are present, acting or not on different segments of the chain. The epimerisation of GlcA into IdoA in dermatan sulphate is dependent on the 4-O-sulphatation of the GalNAc adjacent to the neo-formed IdoA.

For the synthesis of heparin, the initially incorporated GlcNAc has to be first de-acetylated, then N-sulphated, followed by the epimerisation of GlcA into IdoA coupled with 2-O-sulphation of IdoA and finally followed by 6-O-sulphatation and 3-O-sulphatation of GlcN-SO3H residues.

1.3. Degradation of Glycosaminoglycans

The degradation of these compounds occurs intracellularly, essentially in lysosomes. This asks the question of how these essentially extracellular compounds (extracellular matrix) are internalised into cells. This degradation, contrasts with glycoprotein catabolism endoglycosidases. However, these glycosidases cannot cleave heparan sulphate until the N-sulphate group of GlcN has been removed and re-acetylated. Therefore, the degradation of these compounds involves very complex mechanisms and involves several enzymes. Similarly, 2-O-SO3H IdoA cannot be cleaved by the iduronidase until the 2-O-SO3H group is not removed by a iduronate 2-sulphate sulphatase. The absence of one of these enzymes provokes the blockage of the degradation process with accumulation of undegraded material into lysosomes. Their absence is at the origin of several severe storage diseases known as mucopolysaccharidoses (von Figura and Klein, 1981; Fluharty, 1981).

2. GLYCOLIPIDS

Glycolipids constitute a very heterogeneous category of glycoconjugates since they are defined by the covalent coupling of different glycans to different lipid prosthetic groups. In fact, at least in higher organisms, three major types of glycolipids can be discerned: the glycoglycerolipids, the glycosphingolipids and the glycosyl-phosphoinositides. This classification favours the structure of the oligosaccharide chain relative to the lipid moiety because although the lipids in glycosphingolipids and in glycosyl-phosphatidy-linositol could be similar, their glycan structures and their functions are unrelated.

2.1. The Glycoglycerolipids

These compounds constitute a family of components built on an acylglycerol (two hydroxyl groups of glycerol are esterified by two fatty acids; diacylglycerol, DAG) or on an alkyl-acylglycerol (one hydroxyl group is esterified by a fatty acid, the other being etherified by a fatty alcohol, alkyl-acylglycerol, AAG) moiety (Ishizuka and Yamakawa, 1985). Most of these compounds particularly abundant in plants (but also abundant in myelin and testis of mammals) derive from the monogalactosyl-AAG or DAG in which the βGal is attached to the C3 of glycerol, the two other carbon being acylated by fatty acids. Higher members of this family are the digalactosyldiacylglycerol in which the second Gal residue is α1-6 linked to the former and the 6-O-sulfated monogalactosyl diacylglycerol (Figure 2). Sulphated compounds exist, the sulphate group being attached to carbon 3 or carbon 6 of Gal. HSO3-6-quinovose has been reported (Joyard *et al.*, 1986), but it is not clear if this compound correspond or not to a 6-sulfono-Glc (quinovose is the 6-deoxy-glucose). The modalities of the biosynthesis of these compounds appeared only recently in the literature (Dörman *et al.*, 1995)

2.2. The Glycosphingolipids

These compounds constitute a family of constituents built on a ceramide portion. The ceramides are constituted by a long-chain base (sphingoid base or LCBs) acylated on its amino group in position 2 by a fatty acid. Sphingoid bases constitute themselves a heterogeneous family. The most common member is sphingosine, or 2-amino-1,3-diol, 4-5ene octadecene (Figure 2), the configuration of carbon 2 being *erythro* and the double bond being in the *cis* configuration. Compounds with different number of carbons, ramification of the chain have being described as members of the sphingosine family. The

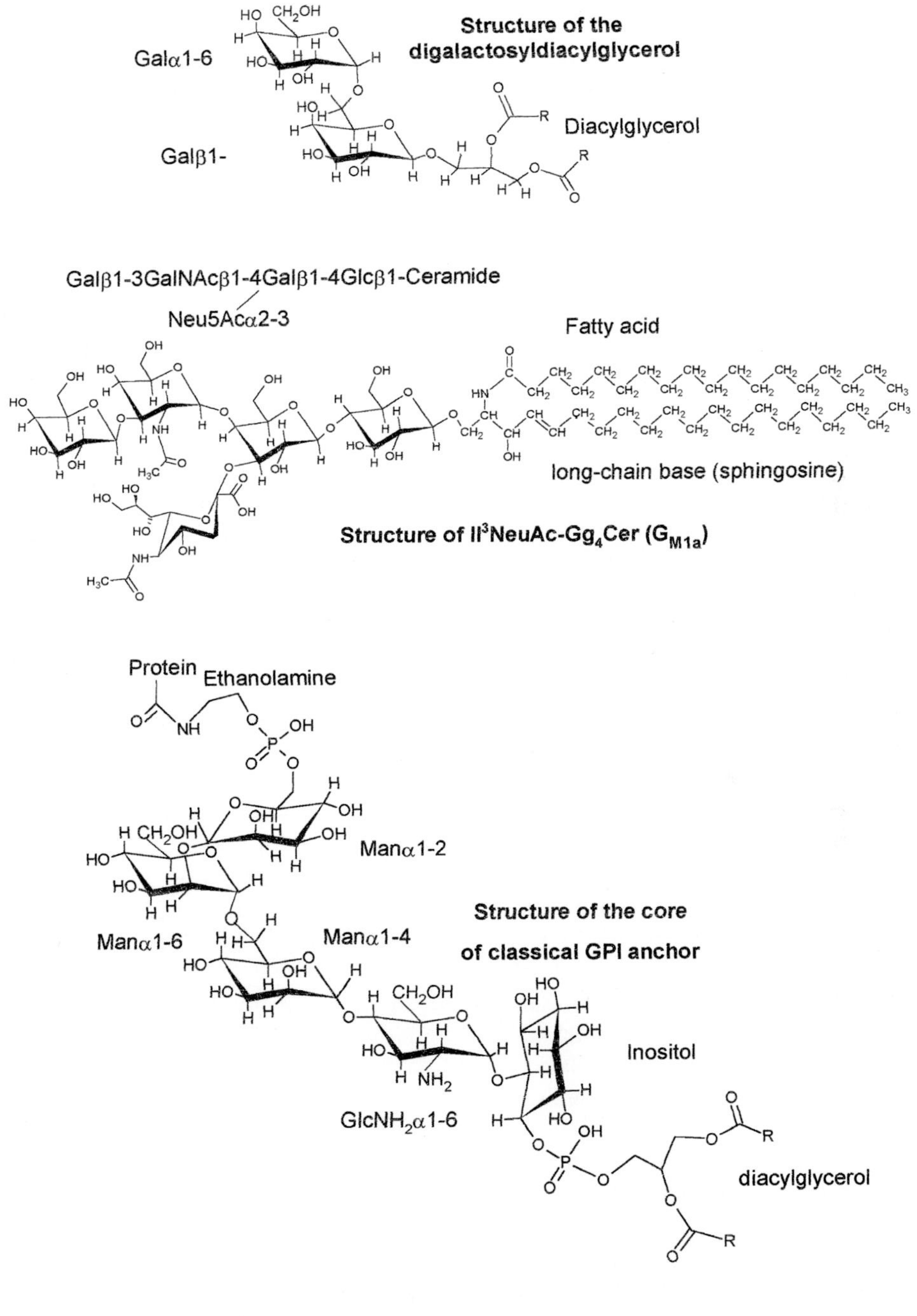

Figure 2: Schematic drawing of classical glycolipids.

saturated equivalents are termed sphinganines, whereas the sphinganines with an additional hydroxyl group on the carbon 4 are termed phytosphingosines. The latter compounds were abundant in plants (thus the terminology), but they are present in all organisms. Different studies indicated that these compounds might be much more heterogeneous than previously suggested, including compounds with double bonds in different position, additional hydroxyl groups, ethers in different position, *etc.* This heterogeneity of LCBs could have important biological functions since they are able to modulate the cell behaviour (protein kinase C inhibition, apoptosis). The fatty acid moiety could be also extremely heterogeneous with the same types of ramifications, hydroxylations, *etc.* which could change the properties of the molecules.

The carbohydrate moiety is attached through its reducing end to the carbon 1 of the sphingoid base. The first monosaccharide could be either Glc (glucosyl-ceramides) or Gal (galactosylceramides) giving the first members of two different families of compounds. When Gal is involved in the bound with the ceramide, the compounds belong to the family of galactosylceramides. This family is relatively restricted to a few compounds GalCer itself, its 3-O-sulphated derivative, the sulfatide (these two components are major lipid constituents of myelin) and a single sialylated minor compound known as GM4 or G7 in the literature, having apparently a unique localisation in some astrocytes.

When Glc is the first monosaccharide, subdivisions were concerned with the nature of the second monosaccharide and the following monosaccharides (Table 1). Different families of compounds are identified, the specific nomenclature allowing assigning the nature and the configuration from the two to the four first residues in the chain.

Based on their charge properties these glycosphingolipids can be devised into two families, the neutral compounds and the acidic ones, the latter being frequently gangliosides. The family of the neutral compounds is extremely heterogeneous and important compounds bear the blood group substance determinant (A, B, O, Lewis a, b, x, y; for review see Marita and Taniguchi, 1985).

The most abundant series is constituted by the ganglioside series (Figure 2; see for review, Wiegandt, 1985). These compounds are characterised by the core oligosaccharide structure GalNAcβ1-4Galβ1-4-Glcβ1-Cer. The compounds of this family could be sialylated on each Gal residue, giving a series of compounds having between 1 and 5 sialic acid residues in mammals (these compounds are essential constituents of the neuronal membranes). Besides these variations of the oligosaccharide chains, the compounds can present an extreme heterogeneity of the long-chain base and fatty acid composition. Furthermore, sialic acids can also present an extreme heterogeneity (Figure 3) since beside the existence of the N-acetyl (Neu5Ac) and the N-glycolyl (Neu5Gc) neuraminic acids, these compounds could be O-acylated (acetyl, lactyl, sulphate, phosphate) or alkylated (methyl, *etc.*) in the different positions, and that the compound having a OH group on the C5 carbon atom, KDN (3-deoxy-D-glycero-D-galacto-nonulosonic acid) presents also a similar type of heterogeneity (for review, see Schauer and Kamerling, 1997).

The biosynthesis of the ceramide portion involves first the synthesis of a long chain base having a keto group in position 3, the 3-keto-sphinganine from serine and palmitoyl-CoA. The keto-sphinganine is reduced into sphinganine (dihydrosphingosine) by the 3-keto-sphinganine reductase. The fatty acid is then transferred to the NH2 group of the C2 carbon atom from an acyl-CoA precursor by an acyl-CoA-transferase. The dihydrocer-amide could be further reduced by a dihydroceramide reductase to give the ceramide in which the LCB is sphingosine.

Table 1. Nomenclature proposed for defining the different families of glycosphingolipids. The sequences in bold characters are sufficient for assigning a compound to a family. The other sequences correspond to the core of the most abundant compounds of the family. Full names and the corresponding abbreviations are indicated as examples.

Family	Abbreviation	Structures	Name and abbreviation
Globo	**Gb**	**Galα1-4Galβ1-4Glc**	
		GalNAcβ1-3Galα1-4Galβ1-4Glc	
		Galβ1-3GalNAcβ1-3Galα1-4Galβ1-4Glc	Globopentaosylceramide **Gb$_5$Cer**
Isoglobo	**iGb**	**Galα1-3Galβ1-4Glc**	
		GalNAcβ1-3Galα1-3Galβ1-4Glc	
Lacto	**Lc**	**Galβ1-3GlcNAcβ1-3Galβ1-4Glc**	Lactotetraosylceramide **Lc$_4$Cer**
		(Galβ1-3GlcNAcβ1-3)nGalβ1-4Glc	
Neolacto	**nLc**	**Galβ1-4GlcNAcβ1-3Galβ1-4Glc**	Neolactotetraosylceramide **nLc$_4$Cer**
		(Galβ1-4GlcNAcβ1-3)nGalβ1-4Glc	
Ganglio	**Gg**	**GalNAcβ1-4Galβ1-4Glc**	
		Galβ1-3GalNAcβ1-4Galβ1-4Glc	Gangliotetraosylceramide **Gg$_4$Cer**
Gala	**Ga**	**Galα1-4Gal**	
Arthro	**Ar**	**GlcNAcβ1-3Manβ1-4Glc**	Arthrotriosylceramide **Ar$_3$Cer**
		GalNAcβ1-4GlcNAcβ1-3Manβ1-4Glc	
Mollu	**Ml**	**Manα1-3Manβ1-4Glc**	Mollutriosylceramide **Ml$_3$Cer**
		GlcNAcβ1-2Manα1-3Manβ1-4Glc	
		Fucα1-4GlcNAcβ1-2Manα1-3Manβ1-4Glc	

The additional monosaccharides were named according to their position and the nature of the bond. For example

Galβ1-3GalNAcβ1-4Galβ1-4Glc   is **IV^{3}NeuAc-,II^{3}NeuAc$_2$-Gg$_4$Cer**
3   3
| |
2αNeuAc   2αNeuAc8-2αNeuAc

The incorporation of the first sugar to the ceramide portion involves a specific transferase of the ceramide portion from UDP-monosaccharide precursors. The addition of more peripheral monosaccharides occurs from UDP-sugar precursors except for sialic acids, the precursor being CMP-Neu5Ac (or CMP-Neu5Gc or CMP-KDN).

The degradation of these constituents occurs first by internalisation into the endosomal compartment then in the lysosomal compartment. The degradation involves essentially

Figure 3: Diversity of sialic acids.

exoglycosidases, i.e., sialic acids when present are first terminal monosaccharide removed (except for the Neu5Acα2-3 linked characteristic of GM1, its elimination following the elimination of the neighbouring GalNAc residue). The ceramide portion is also destroyed in the lysosomal compartment. It should be stressed that products of the catabolism of glycosphingolipids (long-chain bases) are remodelled and that these products as well as LCBs are potent inhibitors of cell signalling.

2.3. Glycosyl-Phosphatidylinositides

Compounds of this family were only recently discovered (Ferguson *et al.*, 1988) as anchoring molecules for proteins in many organisms including man and as free compounds (essentially in parasites). Two different families of compounds could be distinguished depending on the structure of their lipid moieties: the true glycosyl-phosphatidylinositol (GPI) and the glycosyl phosphoinositide ceramides (GPI-Cer), the structure of the carbohydrate moieties responding to the same general organisation of the core (Figure 2).

In the true GPI, the lipid moiety is phosphatidylinositol, *i.e.* an Ino residue attached by its carbon 1 to the carbon 3 of a diacyl or alkyl-acyl glycerol through a phospho-di-ester bond. In the GPI-Cer, the phospho-inositol is bond to the carbon 1 of the sphingoid base of a ceramide.

The carbohydrate moieties of all these compounds have common features (Figure 2; Cole and Hart, 1997). The Ino is substituted in position 6 by a GlcN residue itself substituted by a Manα1-4 and Manα1-6 and Manα1-2. This last mannose residue is substituted by an ethanolamine residue through a phospho-diester bound. The NH2 group of ethanolamine can be involved in the formation of an amide bound with the C-terminal carboxyl group of a protein. Besides these core structures, additional monosaccharides, phospho-ethanolamine, phosphono-ethanolamine, fatty acids, phosphate groups could be present. Because of the amphiphilic properties, these compounds are extremely difficult to handle because of strong micelle formations and of difficulties to have a true solubilisation.

Anchoring of proteins and glycoproteins and proteoglycans can constitute an interest when two structures (cells) have to move the one relative to the other but still being attached together. The GPI anchoring is independent of the cytoskeleton elements. Besides anchoring of proteins, the GPI can be found as free lipids at the surface of parasites.

The first step in the biosynthesis of GPI is the addition of a GlcNAc residue through an UDP-GlcNAc precursor on a phosphatidyl-inositol moiety or a phosphoinositide-ceramide. This is followed by the rapid de-acetylation of GlcNAc, giving GlcN. Three successive Man residues are added, Manα1-4, Manα1-6 and Manα1-2 from a dolichol-P-Man precursor, forming the core oligosaccharide structures common to these compounds. Phospho-ethanolamine residues are subsequently added in the position 6 of either Manα1-6 or/and Manα1-2. The amino group of the later is involved in the formation of the amide bond with the C-terminal residue of the protein through the transamidase. Indeed, the attachment of the GPI moiety to proteins involves a transamidase reaction. Proteins to be anchored through GPI possess two different signal sequences one, N-terminal and the other C-terminal, which are removed after the GPI anchoring. These glycans can be further completed by successive addition of more peripheral sugars and/or by acylation with fatty acids.

The degradation of GPI anchor has not been studied extensively.

REFERENCES

Cole, R.N. and Hart, G.W. (1997) in "Glycoproteins II" (J. Montreuil, J.F.G. Vliegenhart and H. Schachter eds.) Elsevier, Amsterdam, pp. 69–88.

Dörman, P., Hoffman-Benning, S., Balbo I. and Benning, C. (1995) Plant Cell, 7, 1801–1807.

Ferguson, M.A.J., Homans, S.W., Dwek, R.A. and Rademacher, T.W. (1988) Science 239, 753–759.

Fluharty, A.L. (1981) in "Lysosomes and lysosomal storage disease" (J.W. Callahan and J.A. Lowden eds) Raven Press, New York, pp. 249–261.

Ishizuka, I. and Yamakawa T. (1985) in Glycolipids (H. Wiegandt ed.) New comprehensive Biochemistry vol. 10, Elsevier, Amsterdam, pp. 101–197.

Joyard J., Blée E., and Douce R. (1986) Biochim. Biophys. Acta 879, 78–87.

Marita, A. and Taniguchi, N. (1985) in Glycolipids (H. Wiegandt ed.) New comprehensive Biochemistry vol. 10, Elsevier, Amsterdam, pp. 1–99.

McKusik, V.A. (1988) Mendelian inheritance in man, 8th ed. John Hopkins Press, Baltimore, pp. 1072–1080.

Neufeld, E.F. (1991) Annu. Rev. Biochem. 60, 257–280.

Schauer, R. and Kamerling, P. (1997) in "Glycoproteins II" (J. Montreuil, J.F.G. Vliegenhart and H. Schachter eds.) Elsevier, Amsterdam, pp. 243–402.

Silbert, J.E., Bernfield, M. and Kokenyesi, R. (1997) in Glycoprotein II (J. Montreuil, J.F.G. Vlieganhart and H. Schachter eds.) Elsevier, Amsterdam, pp. 1–54.

von Figura, K. and Klein, U. (1981) in "Lysosomes and lysosomal storage diseases" (J.W. Callahan and J.A. Lowden eds) Raven Press, New York, pp. 229–248.

Wiegandt, H. (1985) in Glycolipids (H. Wiegandt ed.) New comprehensive Biochemistry vol. 10, Elsevier, Amsterdam, pp. 199–260

A4. Basement Membrane and Extracellular Matrix Organization

Monique Aumailley

The extracellular matrix provides cells with a structural and mechanical scaffold and critical information. Determination of the highly complex composition of the extracellular matrix began first with the purification of the constitutive molecules and was later accelerated by the development of molecular biology techniques allowing identification and characterization of new full-length polypeptides which have not been purified from tissues. The major extracellular matrix components are collagens, non-collagenous glycoproteins, and proteoglycans. Most of these molecules are chimaeras sharing one or more different structural domains and evolve from the combination of a relatively small repertoire of genes. Interactions studies with authentic or genetically engineered molecules and fragments have led to the realization that extracellular matrix components are organized into precisely ordered architectures. Gene targetting and identification of molecular defects in aquired or inherited human diseases have highlighted the pivotal and indispensable role of several components of the extracellular matrix in development and homeostasis.

1. INTRODUCTION

Multicellular organisms are maintained together by the presence in the extracellular space of a matrix constituted essentially by collagens, non collagenous proteins and proteoglycans. Due to specific interactions, these components are arranged according to precisely defined patterns which determine tissue-specific architecture and functions. Most of the extracellualar matrix components belong to large families of proteins and are constituted by multiple structural domains. This endows each member of a family with common and unique properties.

Collagens are characterized by the presence in their structure of the collagen (COL) motif which is a triple helix formed by three identical or different polypeptide chains, the α chains. The sequence involved in the formation of this motif is constituted by the repetition of a triplet, Gly-X-Y, allowing helical folding of three α chains together (Rich and Crick, 1961). Up to now 20 different collagen types (I-XX) are known, resulting from homo- or heterotrimeric combinations of 34 different α chains which are distinct gene products, some being alternatively spliced. The different collagen types vary by the number of Gly-X-Y triplets and by the presence of one or more non collagenous (NC) regions (for review see Shaw and Olsen, 1991; van der Rest and Bruckner, 1993; Brown and Timpl, 1996). On this basis two main classes of collagens can be distinguished, the fibril-forming collagens (I, II, III, V, and XI) with a single

Table 1. *Major protein families of the extracellular matrix*

Collagens (20 types known)
Fibrillins (Fibrillin 1 and 2)
Fibronectins (20 alternatively spliced variants in human)
Fibulins (Fibulin 1 and 2)
Laminins (more than 50 predicted members)
Matrilins (Matrilin 1, 2, 3 and 4)
Nidogens (Nidogen 1 and 2)
Tenascins
Thrombospondins (TSP1, 2, 3, 4 and 5)
Proteoglycans

COL domain of 300 nm, and the other collagens with one or several COL domains which size varies between collagen types. The latter comprises network-forming collagens (IV, VIII and X), microfibrillar collagen (VI), fibril-associated collagens with interrupted triple helix or FACIT (IX, XII, XIV, XVI, XIX), collagens with multiple triple-helix domains and interruptions (XV and XVIII) and transmembrane collagens (XIII and XVII).

The best characterized non-collagenous proteins of the interstitial matrix (Table 1) are fibronectins, the elastic fibril-associated molecules elastin and fibrillins, thrombospondins, matrilins, tenascins, vitronectin, and fibulins and in the basement membranes they are the laminins and nidogens. They contain specific and rather frequent motifs which are shared by other proteins of the matrix and of the intracellular compartment. Series of heptad repeats, like in laminins, thrombospondins, or matrilins serve for trimer formation by folding three polypeptide chains into an α-helical coiled-coil structure (Timpl and Brown, 1996). Other motifs like the FnIII, EGF-like, or vWA modules are separately folded. The FnIII motif was first described for fibronectin, in which three types of homologies, FnI, FnII, and FnIII, have been delineated in the amino acid sequence (Hynes, 1990). The FnIII module has an arrangement similar to that of the immunoglobulin fold with two layers of β sheets, one with three anti-parallel strands and the other with four anti-parallel strands, enclosing a hydrophobic core (Main *et al.*, 1992; Dickinson *et al.*, 1994). Interestingly, despite the complex folding into β sheets, the structure can undergo conformational changes by reversible unfolding a property which could regulate protein biological activity by masking or unmasking functional sites depending under the circumstances (Erickson, 1994; Plaxco *et al.*, 1996). The sequence of another frequent motif presents homology to that of the epidermal growth factor (EGF). It is a cysteine-rich motif which contain either 6 cysteine like in EGF (EGF-like) or in most of the motifs present in fibrillins or 8 cysteine like in laminin (LE motif). These motifs are arranged in arrays forming rod-like structures (Engel, 1991). The vWA module was first recognized as a 200-residue globular domain, the A domain, in the von Willebrand factor. This module is rather typical of extracellular proteins and it is presumably involved in many interactions (Colombatti *et al.*, 1993).

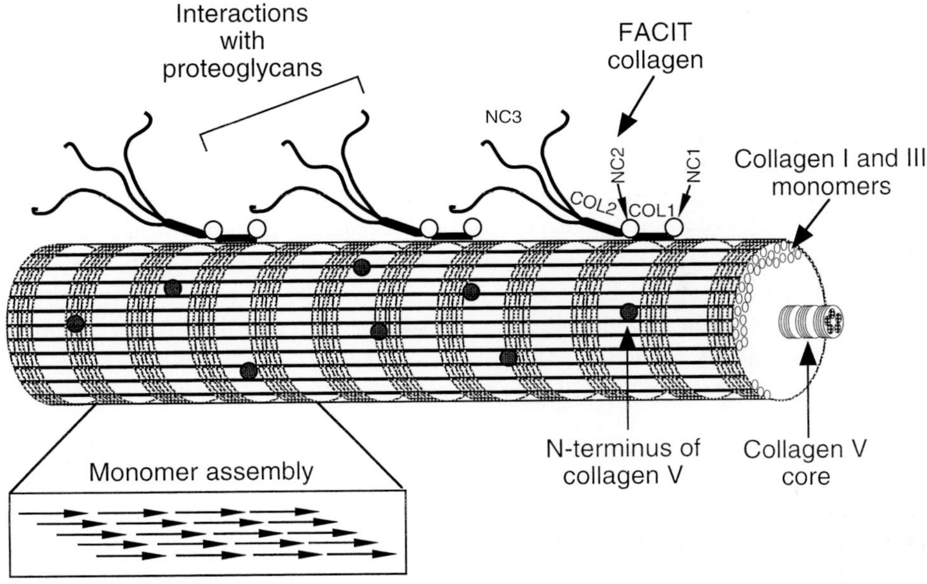

Figure 1: Model for collagen I-containing fibril formation and interactions.
In collagen II-containing fibrils, collagens I and III are replaced by collagen II, while collagen IX, instead of collagen XII or XIV, is associated with the fibril surface.

2. MOLECULES AND ASSEMBLIES IN THE INTERSTITIAL EXTRACELLULAR MATRIX

2.1. Collagen-Based Fibrils and Microfibrils

The fibril-forming collagens are the most abundant in the interstital connective tissue. After secretion and removal of large globular polypeptides from the precursors, the 300 nm-long helical rods assemble head-to-end longitudinally and aggregate laterally in a quarter-staggered manner to form the typical 67 nm banded fibrils (Figure 1). Mixing of different proportions of collagens V and/or XI with collagens II or with collagens I and III give rise to fibrils characterisitic of cartilage and vitreous or of the other connective tissues, respectively (Keene *et al.*, 1987; Mendler *et al.*, 1989). Collagens V and XI retain part of their non-collagenous terminal portions and can form heterotypic V/XI molecules (for review see Fichard *et al.*, 1994). They presumably form the fibril core and play a role in regulating fibril diameter (Linsenmayer *et al.*, 1993; Andrikopoulos *et al.*, 1995; Marchant *et al.*, 1996; van Steensel *et al.*, 1997). Over time the fibrils become stabilized by the formation of inter-molecular covalent cross-links.

Additional collagens, the FACITs, are associated to the surface of the fibrils. These collagens have a similarly short helical rod at the carboxy-terminus, the COL 1 domain, mediating the association of collagen IX or collagens XII and XIV with collagen II- or collagen I-containing fibrils, respectively (Wu *et al.*, 1992; Keene *et al.*, 1991). The

FACIT have one or more short helical rods interrupted by small non-collagenous domains and quite divergent amino-terminus. In collagen IX, the latter is relatively small and has a glycosaminoglycan chain (Yada *et al.*, 1990). For collagens XII and XIV the amino-terminal portions of the α chains are separately folded into extended arms of variable lenghts according to differently spliced variants. The corresponding sequences contain several single or successive motifs with homology to the type III repeats of fibronectin (FnIII) interspaced by vWF-A motifs (Gordon *et al.*, 1989, 1991; Yamagata *et al.*, 1991). A function of collagens XII and XIV may be to regulate the growth and the spatial organization of collagen fibrils via further interactions with heparin, proteoglycans, and the microfibrillar network of collagen VI (Font *et al.*, 1993; Brown *et al.*, 1993; Nishiyama *et al.*, 1994; Koch *et al.*, 1995). How two further members of the FACIT sub-family, collagens XVI and XIX (Pan *et al.*, 1992; Yamaguchi *et al.*, 1992; Yoshioka *et al.*, 1992; Myers *et al.*, 1993; Inoguchi *et al.*, 1995) are integrated into the extracellular matrix is not yet known.

Collagen VI forms a separate microfibrillar network of beaded filaments (Kielty *et al.*, 1992). Collagen VI is composed by the association 1:1:1 of the α1(VI), α2(VI) and α3(VI) chains folded in a dumbell-shape molecule, with an helical central domain of 105 nm constituting one-third of the protein, the two other thirds being formed by N- and C-terminal globular regions (von der Mark *et al.*, 1986). The latter consists of repetitive motifs with homology to the A domain of von Willebrand factor (vWF-A), with two motifs at the C-terminus of all three chains and one, one and nine motifs at the N-terminus of α1(VI), α2(VI) and α3(VI) chains, respectively, and one Fn III module in the α3(VI) chain (Chu *et al.*, 1989, 1990; Bonaldo *et al.*, 1990). The helical portion of two collagen VI monomers associate in an anti-parallel fashion and two dimers associate into tetramers which in turn associate end-to-end to form microfibrillar structures corresponding to the 100 nm periodicity microfibrils of the dermis and other interstitial connective tissues.

2.2. Non-Collagenous Microfibrillar Meshworks of the Interstitial Matrix

Several non-collagenous glycoproteins such as elastin and fibrillins are present in the interstitial matrix only, while others are also found in the plama, like vitronectin, thrombospondins and fibronectins. The latter are dimeric molecule which subunits (250–280 kDa) are disulfide-linked at the carboxy-terminus and form arms of 50 nm. There is only one fibronectin gene coding for about 20 different alternatively spliced variants in human. The sequence of the fibronectin subunit consists of twelve FnI, two FnII, and, depending on splicing variants, 15–18 FnIII motifs, arranged like pearls on a necklace (Hynes, 1990). Fibronectin forms a polymeric fibrillar network stabilized by disulfide bridges, and several domains of the molecule are involved in the assembly process, including the amino-terminal 29 kDa heparin-binding domain, the RGD-containing cell-binding domain and the first FnIII repeat (Chernousov *et al.*, 1991; Morla and Ruoslahti, 1992). Other fibronectin domains interact with heparin, collagen, and fibrin. In particular, next to the amino-terminal heparin binding region, there is a collagen binding site likely involved in the association of the fibronectin polymers with various collagenous structures and which could play a role in vivo in maintening tissue architecture. Thrombospondins too are localized to microfibrils (Arbeille *et al.*, 1991). They are large homotrimeric proteins resulting from the association of different gene products of 50, 140 and 180 kDa. The amino-terminal sequence are folded into globular domains and the adjacent stretch of residues is involved in trimer formation. The rest of the chains is characterized by the presence of FnIII and EGF-like motifs forming extended arms and by a globular carboxy-terminal domain (Adams and Lawler, 1993; Bornstein *et al.*, 1993). The molecules contain calcium binding sequences and interaction sites for heparin, proteoglycans such as decorin, collagens, and fibronectin (Frazier, 1991) likely involved in anchoring tissular thrombospondins to fibronectin microfibrillar networks or collagen fibers. Furthermore collagen fibrillogenesis could be controled by thrombospondins since mice that lack thrombospondin 2 display irregular collagen fibers (Kyriakides *et al.*, 1998).

Another dense microfibrillar meshwork, the so-called elastic fibers, is constituted by elastin, fibrillins, and others microfibril-associated glycoproteins (Cleary and Gibson, 1996). The elastin precursor polypeptide, tropoelastin, consists of alternating hydrophobic and cross-linking regions and due extensive cross-linking, elastin forms large insoluble polymers. Fibrillins belong to a family of single polypeptide chain proteins, which sequences are almost completely formed by EGF-like motifs, many of them bind calcium, a property presumably involved in maintaining protein structure (Corson *et al.*, 1993; Zhang *et al.*, 1994; Reinhardt *et al.*, 1997). The monomers appear like flexible and extended entities of 148 nm (Sakai *et al.*, 1991). How these monomers polymerize into the 10–12 nm diameter beaded microfibrils associated with elastin fibrils is still a matter of debate and other proteins may be involved in microfibril formation (Cleary and Gibson, 1996; Brown-Augsburger *et al.*, 1996). In particular, a calcium-binding homodimeric molecule, fibulin 2, rich in EGF-like motifs (Pan *et al.*, 1993), displays affinity for fibrillins or fibronectins and in vivo or in cell culture it co-localizes with fibronectin or fibrillin in dense microfibrillar meshworks (Sasaki *et al.*, 1996). Moreover, a 21 kDa microfibril-associated glycoprotein, MAPG-1, binds to tropoelastin (Brown-Augsburger *et al.*, 1996), fibrillin (Henderson *et al.*, 1996), and collagen VI (Finnis and Gibson, 1997) which suggests that different microfibrillar meshworks could be connected.

Additional non-collagenous glycoproteins are present in the extracellular matrix, but their architectural integration is not yet fully understood, although their structure appears well-suited for interactions with other matrix molecules. This is the case, for example, of tenascins and matrilins. In their carboxy-terminus, these glycoproteins contain short stretches of heptad repeats involved in folding three chains into an α-helical coiled-coil domain. In tenascins the assembly region is followed by successive EGF-like motifs and a serie of FnIII repeats in variable number according to alternative spliced variants, forming extended arms of 75 nm which could interact with neighbouring extracellular matrix components (Chiquet-Erhisman, 1990; Sage and Bornstein, 1991). The matrilins, which arms are constituted by EGF-like motifs and vWA modules, could be involved in bridging collagens and proteoglycans (Wagener *et al.*, 1998). Finally, several classes of proteoglycans with different sizes and protein cores bind to many of the structural proteins (Iozzo and Murdoch 1997). They probably regulate the formation and the stabilization of the various networks and add a further degree of diversity in extracellular matrix architecture.

3. BASEMENT MEMBRANE-ASSOCIATED MOLECULES AND NETWORKS

3.1. Collagens Associated with the Basement Membranes

The most abundant collagen in all basement membranes are the collagens IV. Six genetically different α chains, α1(IV), α2(IV), α3(IV), α4(IV), α5(IV), and α6(IV), are known (Leinonen *et al.*, 1994) that form several different heterotrimeric molecules, the most common being [α1(IV)]2α2(IV) and the others having a restricted and tissue-specific distribution (Brown and Timpl, 1995). All collagen IV molecules contain a 400 nm-long COL domain with several small interruptions giving flexibility to the helical rod. A large non collagenous domain, NC1, present at the carboxy-terminal end of the rod participates in the formation of collagen IV dimers, while at the amino-terminus a stretch of four anti-parallel helical rods, the 7S domain, overlap and mediate collagen IV tetramerization. In addition, parts of the helical rod from adjacent molecules form supercoiled structures. This results in the formation of a stable and large network likely to constitute the backbone of the basal lamina and an anchoring support for cells and other constituants (Brown and Timpl, 1995).

In basement membranes underlying squamous epithelia where additional collagen types are present, collagen IV interacts with collagen VII (Burgeson, 1993). The latter is an homotrimer with the longest triple helix, 450 nm, described among vertebrate collagens and it is the major constituant of the anchoring fibrils typical of these specialized basement membranes (Sakai *et al.*, 1986; Bruckner-Tuderman *et al.*, 1987). The COL sequence of the α1(VII) chain presents several imperfections and is bordered at the N- and C-terminus by non-collagenous sequences, the NC1 and NC2 domains, respectively. The former is constituted by two vWF-A motifs, ten FnIII repeats and a region homologous to cartilage matrix protein (Christiano *et al.*, 1994). The NC1 domains are separately folded into extended arms of 36 nm. The smaller NC2 domain present in the precursor form, is proteolytically removed in mature collagen VII allowing anti-parallel dimerization over 60 nm of the collagen rods (Bruckner-Tuderman *et al.*, 1995). The dimers are in turn laterally aggregated with the NC1 domains at both ends and form bundles of molecules constituting the anchoring fibrils. Another collagen, collagen XVII, is also specific of these specialized basement membranes. Collagen XVII, known as the bullous pemphigoid antigen of 180 kDa, is a transmembrane molecule with a type II orientation, i.e. the carboxy-terminus is extracellular. The amino-terminal region consists of eight heptad repeats forming a globular head and the ectodomain is composed by 13–15 distinct collagenous segments interspaced by small non collagenous sequences forming a 60 nm rod adjacent to the transmembrane region and a distal extended 100–130 nm flexible tail (Guidice *et al.*, 1992; Li *et al.*, 1993; Hirako *et al.*, 1996). Since collagens associate in homo or heterotypic oligomers or polymers through interactions between their triple helices, the helical rods of collagen XVII adjacent to the cell membrane could, therefore, laterally interact, but experimental evidence is lacking.

Other collagens, like collagen XVIII are present or associated with many but not all basement membranes. In particular, collagen XVIII is found in endothelial basement membranes, a crucial location for the antiangiogenic function of its proteolytically cleaved 20 kDa C-terminal fragment, endostatin (O'Reilly *et al.*, 1997).

3.2. Network-Forming Glycoproteins of the Basement Membranes

The most abundant non-collagenous glycoproteins of basement membranes are the laminins, a family of probably more than 50 members (Aumailley and Smyth, 1998). They are large molecules constituted by three polypeptides, the α, β, and γ chains, maintained together by folding of a large portion of their carboxy-teminal part into a triple stranded coil-coiled conformation stabilized by disulfide bridges (Beck *et al.*, 1990). The first described laminin molecule was isolated from the EHS tumour transplantable to mouse (Figure 2). It is constituted by the $\alpha 1$ (400 kDa), the $\beta 1$ and the γ 1 chains (each of ~200 kDa) associated to form a cruciform molecule with three short arms-one with a lenght of 43 nm and two of 48 nm-and a 77 nm long arm. Later many additional laminin chains were

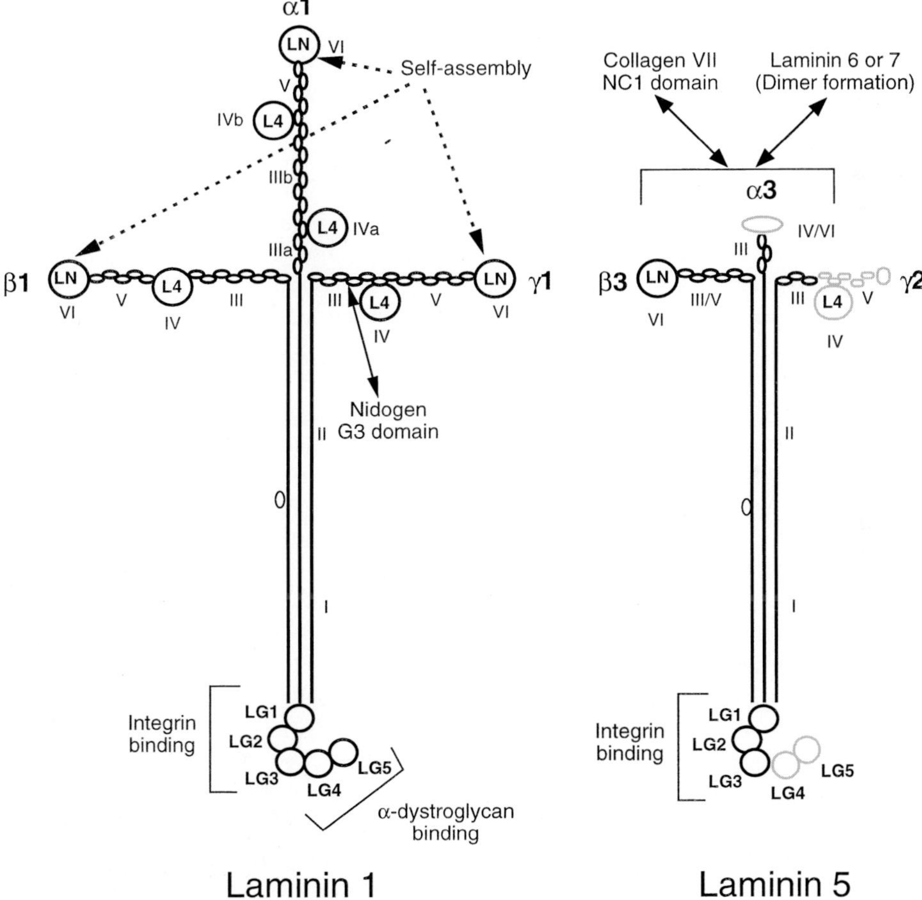

Figure 2: Schematic representation of two selected laminin isoforms.
Laminin 1 (left), was the first laminin isoform discovered, and laminin 5 (rigth) is the smallest isoform known. The characterictic laminin modules are indicated on the figure (L4, LN, and LG) and the roman numerals indicate the different structural domains of the proteins. For laminin 5, the portions of the molecule presumably cleaved off in the mature form are indicated by dotted lines. The most important interaction sites are shown.

discovered and up to now five α, four β and three γ genetically different chains are known, some being full-length (~380–450 kDa for the α chains, and ~200 kDa for the β and chains), and others being truncated (~180–200 kDa for the α chains and 140–150 kDa for the β and γ chains) due to either a unique shorter coding sequence, alternative splicing, or extracellular proteolytic processing. Full-length and truncated laminin chains differ by the length of their amino-terminal regions (Figure 2). The long forms are characterized by the presence of 11–21 LE motif rows interspaced by one, in the β and γ chains, or two, in the α chains, globular domains (L4) and by the presence of an amino-terminal domain (LN) involved in laminin polymerization. The short variants (Figure 2) have a reduced number of LE motifs interrupted by one L4 globular domain in the chain only and terminated by either a LN motif in the β chain or a rudimentary atypical globular domain in the α chain (see for review Aumailley and Smyth, 1998). The carboxy-terminal regions of the different α, β, and γ chains, which form the long arm of the molecules, present more structural homologies than the amino-terminal portions. They are constituted by a ~600-residue-stretch of heptad repeats allowing folding of three chains into a coiled-coil α helix (Beck *et al.*, 1990). The carboxy-terminus the α chain is longer than that of the β and γ chains and forms a pentalobular structure by folding five subdomains, the LG domains. Expression of laminin chain variants is tissue- and development-specific leading to the presence of tissue-specific isoforms. For example, laminins 2 and 4 are predominant in the basal lamina of muscle and motor neurons synapses while laminin 5 is associated with the anchoring filaments spanning the basement membrane below the hemidesmosomes of basal cells in squamous epithelia (Ekblom, 1996; Aumailley and Rousselle, 1998). In addition, some laminin chains, such as the $\alpha2$, $\alpha3$ and $\gamma2$ chains, are extracellularly processed to shorter polypeptides (Paulsson *et al.*, 1991; Marinkovich *et al.*, 1992), a process which could modify laminin function (Goldfinger *et al.*, 1998).

Laminins with three full-length chains self-associate into polymers by mean of interactions between their LN amino-terminal globular domains (Yurchenco *et al.*, 1992; Cheng *et al.*, 1997). The process is calcium dependent and reversible allowing extraction of laminins by neutral buffers containing chelating agents (Paulsson *et al.*, 1987) and it presumably plays a role in basement membrane remodelling. In addition, laminins containing the $\gamma1$ chain forms stable equimolecular complexes with nidogen/entactin, a smaller glycoprotein of 150 kDa (Paulsson *et al.*, 1987). Nidogen is formed by two globular amino-terminal domains, G1 and G2, linked by a short rod to the G3 carboxy-terminal globular domain (Fox *et al.*, 1991). The G3 domain interacts with high affinity with residues exposed at the surface of the 4th LE motif of the laminin $\gamma1$ chain (Baumgartner *et al.*, 1996) while the G2 domain binds to collagen IV or to the proteoglycan perlecan (Fox *et al.*, 1991; Aumailley *et al.*, 1993). In vivo, these nidogen interactions presumably result in the stabilization of basement membrane architecture by linking together the laminins and collagen IV polymeric networks (Figure 3). They are also crucial for the formation of new basement membranes and remodelling. This is supported by experiments with function-blocking antibodies interfering with nidogen binding to the laminin $\gamma1$ chain which dramatically perturb branching epithelial morphogenesis (Kadoya *et al.*, 1997).

An alternative assembly model has been proposed for laminin isoforms composed of truncated chains, as it is the case of laminin 5. The $\gamma2$ chain of the latter has not an affinity high enough for nidogen due to replacement of two residues in its 4th LE motif (Mayer *et al.*, 1995). Moreover, the truncated $\alpha3$ and $\gamma2$ chains lack domain LN and consequently

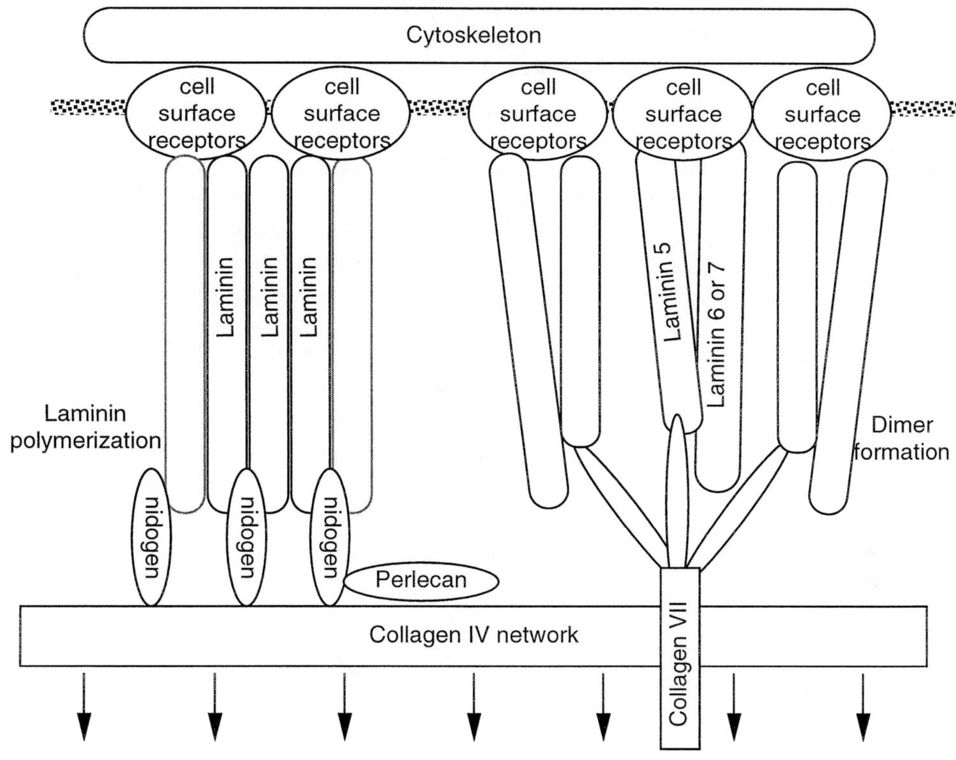

Figure 3: Models for supramolecular assemblies in basement membranes.
Laminins are central elements for anchoring cells to the underlying extracellular matrix. The model at the left applies to full-length chain laminins while the model at the right was developed for laminins with truncated chains.

the isoforms containing one of these chains cannot polymerize according to the three arm-based model (Cheng *et al.*, 1997). However, several interactions specific to these chains presumably allow integration in a different basement membrane architecture. Laminins 5, 6, and 7, which contain the α3 chain, form covalent dimers (Champliaud *et al.*, 1996) and laminin 5 or laminin 5/6 dimers have affinity for the non-collagenous NC1 domain of collagen VII (Chen *et al.*, 1997; Rousselle *et al.*, 1997). Since the latter binds to collagen IV (Burgeson, 1993), it suggests that there is a physical continuity between epithelial laminins and collagens IV and VII (Figure 3). Furthermore, collagen XVII co-localizes with laminin 5 in the anchoring filaments (Masunaga *et al.*, 1997), where they could interact. This assembly model is strongly supported by the phenotypes developed by patients with inborn defects in the genes coding for laminin 5, collagen VII, or collagen XVII chains or with auto-immune disorders associated with auto-antibodies against either one of these proteins and by the phenotypes of mice harboring site-directed mutations of the corresponding genes (for review see Aumailley and Smyth, 1998). Lack or alteration of laminin 5, collagen VII or XVII is associated with major alterations of the basement membranes where these proteins are normally present and with a split within the basal or sub-basal lamina causing the formation of blisters and a loss of cohesion between the basal epithelial cells and the underlying interstital matrix.

In conclusion, by mixing a few different structural domains, a very large diversity is found in extracellular matrix proteins. Some of these domains are involved in highly specific interactions leading to a well ordered organization of the components into unique architectures. This contributes to bring each individual component in a defined conformation and to affect the spatial orientation of their different domains rendering them accessible or not for biological interactions. After being considered for a long time as a mechanical scaffold for tissue and cells, the extracellular matrix is now well recognized to control cellular functions via specific interactions with cell surface receptors. Besides transfering signals to the cell interior, binding of these receptors are very important also in maintaining stability of the architecture and any defect in the interactions can result in major mechanical dysfunctions and perturb the dynamic connections between cells and the extracellular matrix as well as gene expression. The rapidly expanding knowledge on extracellular matrix structures and organization opens perspectives for new potential therapeutic approaches including gene therapy to replace defective molecules and the use of antibodies or of synthetic or recombinant peptides to compete with or to mimic specific extracellular matrix functions.

REFERENCES

Adams, J.C., and Lawler, J. (1993) The thrombospondin family. *Curr. Biol.*, **3**, 188–190.

Andrikopoulos, K., Liu, X., Keene, D.R., Jaenisch, R., and Ramirez, F. (1995) Targeted mutation in the col5a2 gene reveals a regulatory role for type V collagen during matrix assembly. *Nat. Genet.*, **9**, 31–36.

Arbeille, B., Fauvel-Lafève, F., Lemesle, M., Tenza, D., and Legrand, Y.J. (1991) Thrombospondin: a component of microfibrils in various tissues. *J. Histochem. Cytochem.*, **39**, 1367–1375.

Aumailley, M., Battaglia, C., Mayer, U., Reinhardt, D., Nischt, R., Timpl, R., and Fox, J. (1993) Nidogen mediates the formation of ternary complexes of basement membrane components. *Kidney Intern.*, **43**, 7–12.

Aumailley, M., and Rousselle, P. (1998) Laminins of the dermo-epidermal junction. *Matrix Biol.*, in press.

Aumailley, M., and Smyth, N. (1998) The role of laminins in basement membrane functions. *J. Anat.*, **193**, 1–21.

Baumgartner, R., Czisch, M., Mayer, U., Pöschl, E., Huber, R., Timpl, R., and Holak, T.A. (1996) Structure of the nidogen binding LE module of the laminin $\gamma1$ chain in solution. *J. Mol. Biol.*, **257**, 658–668.

Beck, K., Hunter, I., and Engel, J. (1990) Structure and function of laminin: anatomy of a multidomain glycoprotein. *FASEB J.*, **4**, 148–160.

Bonaldo, P., Russo, V., Bucciotti, F., Doliana, R., and Colombatti, A. (1990) Structural and functional features of the $\alpha3$ chain indicate a bridging role for chicken collagen VI in connective tissues. *Biochemistry*, **29**, 1245–1254.

Bornstein, P., Devarayalu, S., Edelhoff, S., and Disteche, C.M (1993) Isolation and characterization of the mouse thrombospondin 3 (Thbs3) gene. *Genomics*, **15**, 607–613.

Brown, J.C., Mann, K., Wiedeman, H., and Timpl, R. (1993)Structure and binding properties of collagen type XIV isolated from human placenta. *J. Cell Biol.*, **120**, 557–567.

Brown, J.C., and Timpl, R. (1995) The collagen superfamily. *Int. Arch. Allergy Immunol.*, **107**, 484–490.

Brown-Augsburger, P., Broekelmann, T., Rosenbloom, J., and Mecham, R.P. (1996) Functional domain on elastin and microfibril-associated glycoprotein involved in elastic fibre assembly. *Biochem. J.*, **318**, 149–155.

Bruckner-Tuderman, L., Schnyder, U.W., Winterhalter, K.H., and Bruckner, P. (1987) Tissue form of type VII collagen from human skin and dermal fibroblasts in culture. *Eur. J. Biochem.*, **165**, 607–611.

Bruckner-Tuderman, L., Nilssen, O., Zimmermann, D.R., Dours-Zimmermann, M.T., Kalinke, D.U., Gedde-Dahl, T. Jr, and Winberg, J.O. (1995) Immunohistochemical and mutation analyses demonstrate that procollagen VII is processed to collagen VII through removal of the NC-2 domain. *J. Cell Biol.*, **131**, 551–559.

Burgeson, R.E. (1993) Type VII collagen, anchoring fibrils, and epidermolysis bullosa. *J. Invest. Dermat.*, **101**, 252–255.

Champliaud, M.F., Lunstrum, G.P., Rousselle, P., Nishiyama, T., Keene, D.R., and Burgeson, R.E. (1996) Human amnion contains a novel laminin variant, laminin 7, which like laminin 6, covalently associates with laminin 5 to promote stable epithelial-stromal attachment. *J. Cell Biol.*, **132**, 1189–1198.

Chen, M., Marinkovich, M.P., Veis, A., Cai, X., Rao, C.N., O'Toole, E.A., and Woodley, D.T. (1997) Interactions of the amino-terminal noncollagenous (NC1) domain of type VII collagen with extracellular matrix components. A potential role in epidermal-dermal adherence in human skin. *J. Biol. Chem.*, **272**, 14516–14522.

Cheng, Y.S., Champliaud, M.F., Burgeson, R.E., Marinkovich, M.P., and Yurchenco, P.D. (1997) Self-assembly of laminin isoforms. *J. Biol. Chem.*, **272**, 31525–31532.

Chernousov, M.A., Fogerty, F.J., Koteliansky, V.E., and Mosher, D.F. (1991) Role of the I–9 and III-1 modules of fibronectin in formation of an extracellular fibronectin matrix. *J. Cell Biol.*, **266**, 10851–10858.

Chiquet-Erhisman, R. (1990) What distinguishes tenascin from fibronectin. *FASEB J.*, **4**, 2598–2604.

Christiano, A.M., Rosenbaum, L.M., Chung-Honet, L.C., Parente, M.G., Woodley, D.T., Pan, T.C., Zhang, R.Z., Chu, M.L., Burgeson, R.E., and Uitto, J. (1992) The large non-collagenous domain (NC-1) of type VII collagen is amino-terminal and chimeric. Homology to cartilage matrix protein, the type III domains of fibronectin and the A domains of von Willebrand factor. *Hum. Mol. Genet.*, **7**, 475–480.

Chu, M.L., Pan, T.C., Conway, D., Kuo, H.J., Glanville, R.W., Timpl, R., Mann, K., and Deutzmann, R. (1989) Sequence analysis of α1(VI) and α2(VI) chains of human type VI collagen reveals internal triplication of globular domains similar to the A domains of von Willebrand factor and two α2(VI) chain variants that differ in the carboxy terminus. *EMBO J.*, **7**, 1939–1946.

Chu, M.L., Zhang, R.Z., Pan, T.C., Stokes, D., Conway, D., Kuo, H.J., Glanville, R., Mayer, U., Mann, K., Deutzmann, R., and Timpl, R. (1990) Mosaic structure of globular domains in the human type VI collagen α3 chain: similarity to von Willebrand factor, fibronectin, actin, salivary proteins and aprotinin type protease inhibitors. *EMBO J.*, **9**, 385–393

Cleary, E.G., and Gibson, M.A. (1996) Elastic tissue, elastin and elastin associated microfibrils. In: Comper WD (ed) *Extracellular Matrix*, Volume 2, Molecular Components and Interactions. Harwood Academic Publishers, Amsterdam, pp. 95–140.

Colombatti, A., Bonaldo, P., and Doliana, R. (1993) Type A modules: interacting domains found in several non-fibrillar collagens and in other extracellular matrix proteins. *Matrix*, **13**, 297–306.

Corson, G.M., Chalberg, S.C., Dietz, H.C., Charbonneau, N.L., and Sakai, L.Y. (1993) Fibrillin binds calcium and is coded by cDNAs that reveal a multidomain structure and alternatively spliced exons at the 5' end. *Genomics*, **17**, 476–484.

Dickinson, C.D., Veerapandia, B., Dai, X.P., Hamlin, R.C., Xuong, N.H., Ruoslahti, E., and Ely, K.R. (1994) Crystal structure of the tenth type III cell adhesion module of human fibronectin. *J. Mol. Biol.*, **236**, 1079–1092.

Ekblom, P. (1996) Receptors for laminins during epithelial morphogenesis. *Curr. Op. Cell Biol.*, **8**, 700–706.

Engel, J. (1991) Common structural motifs in proteins of the extracellular matrix. *Curr. Op. Cell Biol.*, **3**, 779–785.

Erickson, H.P. (1994) Reversible unfolding of fibronectin type III and immunoglobulin domains provides the structural basis for stretch and elasticity of titin and fibronectin. *Proc. Natl. Acad. Sci. U.S.A.*, **91**, 10114–10118.

Fichard, A., Kleman, J.P., and Ruggiero, F. (1995) Another look at collagen V and XI molecules. *Matrix Biol.*, **14**, 515–531.

Finnis, M.L., and Gibson, M.A. (1997) Microfibril-associated glycoprotein-1 (MAPG-1) binds to the pepsin resistant domain of the α3(VI) chain of type VI collagen. *J. Biol. Chem.*, **272**, 22817–22823.

Font, B., Aubert-Foucher, E., Goldschmidt, D., Eichenberger, D., and van der Rest, M. (1993) Binding of collagen XIV with the dermatan sulfate side chain of decorin. *J. Biol. Chem.*, **268**, 25015–25018.

Fox, J., Maye,r U., Nischt, R., Aumailley, M., Reinhardt, D., Wiedemann, H., Mann, K., Timpl, R., Krieg, T., Engel, J., and Chu, M.-L. (1991) Recombinant nidogen consists of three globular domains and mediates binding of laminin to collagen IV. *EMBO J.*, **10**, 3137–3146.

Frazier, W.A. (1991) Thrombospondins. *Curr. Op. Cell Biol.*, **3**, 792–799.

Goldfinger, L.E., Stack, M.S., and Jones, J.C.R. (1998) Processing of laminin-5 and its functional consequences: role of plasmin and tissue-type plasminogen activator. *J. Cell Biol.*, **141**, 255–265.

Gordon, M.K., Gerecke, D.R., Dublet, B., van der Rest, M., and Olsen, B.R. (1989) Type XII collagen. A large multidomain molecule with partial homology to type IX collagen. *J. Biol. Chem.*, **264**, 19772–19778.

Gordon, M.K., Castagnola, P., Dublet, B., Linsenmayer, T.F., van der Rest, M., Mayne, R., and Olsen, B.R. (1991) Cloning of a cDNA for a new member of the class of fibril-associated collagens with interrupted triple helices. *Eur. J. Biochem.*, **201**, 333–338.

Guidice, G.J., Emery, D.J., and Diaz, L.A. (1992) Cloning and primary structural analysis of the bullous pemphigoid autoantigen BP 180. *J. Invest. Dermatol.*, **99**, 243–250.

Henderson, M., Polewski, R., Fanning, J.C., and Gibson, M.A. (1996) Microfibril-associated glycoprotein-1 is specifically located on the beads of the beaded-filament structure of fibrillin-containing microfibrils as visualized by the rotary shadowing technique. *J. Histochem. Cytochem.*, **44**, 1389–1397.

Hirako, Y., Usukura, J., Nishizawa, Y., and Owaribe, K. (1996) Demonstration of the molecular shape of BP180, a 180-kDa bullous pemphigoid antigen and its potential trimer formation. *J. Biol. Chem.*, **271**, 13739–13745.

Hynes, R.O. (1990) *Fibronectins*. Springer Verlag, New York, 546 pages

Inoguchi, K., Yoshioka, H., Khaleduzzaman, M., and Ninomiya, Y. (1995) The mRNA for $\alpha1(XIX)$ collagen chain, a new member of FACITs, contains a long unusual 3' untranslated region and displays many unique splicing variants. *J. Biochem.*, **117**, 137–146.

Iozzo, R.V., and Murdoch, A.D. (1996) Proteoglycans of the extracellular environment: clues from the gene and protein side offer novel perspectives in molecular diversity and fumction. *FASEB J.*, **10**, 598–614.

Kadoya, Y., Salmivirta, K., Talts, J.F., Kadoya, K., Mayer, U., Timpl, R., and Ekblom, P. (1997) Importance of nidogen binding to laminin $\gamma1$ for branching epithelial morphogenesis of the submandibular gland. *Developmemt*, **124**, 683–691.

Keene, D.R., Lunstrum, G.P., Morris, N.P., Stoddard, D.W., and Burgeson, R.E. (1991) Two type XII-like collagens localize to the surface of banded collagen fibrils. *J. Cell Biol.*, **113**, 971–978.

Keene, D.R., Sakai, L.Y., Bachinger, H.P., and Burgeson, R.E. (1987) Type III collagen can be present on banded collagen fibrils regardless of fibril diameter. *J. Cell Biol.*, **105**, 2393–2402.

Kielty, C.M., Whittaker, S.P., Grant, M.E., and Shuttlewoort,h C.A. (1992) Type VI collagen microfibrils: evidence for a structural association with hyaluronan. *J. Cell Biol.*, **118**, 979–990.

Koch, M., Bohrmann, B., Matthison, M., Hagios, C., Trueb, B., and Chiquet, M. (1995) Large and small splice variants of collagen XII: differential expression and ligand binding. *J. Cell Biol.*, **130**, 1005–1014.

Kyriakides, T.R., Zhu, Y.H., Smith, L.T., Bain, S.D., Yang, Z., Lin, M.T., Danielson, K.G., Iozzo, R.V., LaMarca, M., McKinney, C.E., Ginns, E.I., and Bornstein, P. (1998) Mice that lack thrombospondin 2 display connective tissue abnormalities that are associated with disordered collagen fibrillogenesis, an increased vascular density, and a bleeding diathesis. *J. Cell Biol.*, **140**, 419–430.

Leinonen, A., Mariyama, M., Mochizuki, T., Tryggvason, K., and Reeders, S.T. (1994) Complete primary structure of the human type IV collagen $\alpha4(IV)$ chain. Comparison with structure and expression of the other $\alpha(IV)$ chains. *J. Biol. Chem.*, **269**, 26172–26177.

Li, K., Tamai, K., Tan, E.M., and Uitto, J. (1993) Cloning of type XVII collagen. Complementary and genomic DNA sequences of mouse 180-kilodalton bullous pemphigoid antigen (BPAG2) predict an interrupted collagenous domain, a transmembrane segment, and unusual features in the 5'-end of the gene and the 3'-untranslated region of the mRNA. *J. Biol. Chem.*, **268**, 8825–8834.

Linsenmayer, T.F., Gibney, E., Igoe, F., Gordon, M.K., Fitch, J.M., Fessler, L.I.,. and Birk, D.E. (1993) Type V collagen: molecular structure and fibrillar organization of the chicken $\alpha1(V)$ NH2-terminal domain, a putative regulator of corneal fibrillogenesis. *J. Cell Biol.*, **121**, 1181–1189.

Main, A.L., Harvey, T.S., Baron, M., Boyd, J., and Campbell, I.D. (1992) The three-dimensional structure of the tenth type III module of fibronectin: an insight into RGD-mediated interactions. *Cell*, **71**, 671–678.

Marchant, J.K., Hahn, R.A., Linsenmayer, T.F., and Birk, D.E. (1996) Reduction of type V collagen using a dominant-negative strategy alters the regulation of fibrillogenesis and results in the loss of corneal-specific fibril morphology. *J. Cell Biol.*, **135**, 1415–1426.

Marinkovich, M.P., Lunstrum, G.P., and Burgeson, R.E. (1992). The anchoring filament protein kalinin is synthesized and secreted as a high molecular weight precursor. *J. Biol. Chem.*, **267**, 17900–17906.

Mayer, U., Pöschl, E., Gerecke, D.R., Wagman, D.W., Burgeson, R.E., and Timpl, R. (1995) Low nidogen affinity of laminin-5 can be attributed to two serine residues in EGF-like motif γ 2III4. *FEBS Lett.*, **365**, 129–132.

Mendler, M., Eich-Bender, S.G., Vaughan, L., Winterhalter, K.H., and Bruckner, P. (1989) Cartilage contain mixed fibrils of collagen type II, IX and XI. *J. Cell Biol.*, **108**, 191–197.

Morla, A., and Ruoslahti, E. (1992) A fibronectin self assembly site involved in fibronectin matrix assembly: reconstruction a synthetic peptide. *J. Cell Biol.*, **118**, 421–429.

Myers, J.C., Sun, M.J., D'Ippolito, J.A., Jabs, E.W., Neilson, E.G., and Dion, A.S. (1993) Human cDNA clones transcribed from an unusually high-molecular-weight RNA encode a new collagen chain. *Gene*, **123**, 211–217.

Nishiyama, T., McDonough, A.M., Bruns, R.R., and Burgeson, R.E. (1994) Type XII and XIV collagens mediate interactions between banded collagen fibers in vitro and may modulate extracellular matrix deformability. *J. Biol. Chem.*, **269**, 28193–28199.

O'Reilly, M.S., Boehm, T., Shing, Y., Fukai, N., Vasios, G., Lane, W.S., Flynn, E., Birkhead, J.R., Olsen, B.R., and Folkman, J. (1997) Endostatin: an endogenous inhibitor of angiogenesis and tumor growth. *Cell*, **88**, 277–285.

Pan, T.C., Zhang, R.Z., Mattei, M.G., Timpl, R., and Chu, M.L. (1992) Cloning and chromosomal location of human $\alpha1(XVI)$ collagen. *Proc. Natl. Acad. Sci. U.S.A.*, **89**, 6565–6569.

Pan, T.C., Sasaki, T., Zhang, R.Z., Fässler, R., Timpl, R., and Chu, M.L. (1993) Structure and expression of fibulin-2, a novel extracellular matrix protein with multiple EGF-like repeats and consensus motifs for calcium binding. *J. Cell Biol.*, **123**, 1269–1277.

Paulsson, M., Aumailley, M., Deutzmann, R., Timpl, R., Beck, K., and Engel, J. (1987) Laminin-nidogen complex. Extraction with chelating agents and structural characterization. *Eur. J. Biochem.*, **166**, 11–19.

Paulsson, M., Saladin, K., and Engvall, E. (1991) Structure of laminin variants. The 300-kDa chains of murine and bovine heart laminin are related to the human placenta merosin heavy chain and replace the α chain in some laminin variants. *J. Biol. Chem.*, **266**, 17545–17551.

Plaxco, K.W., Spitzfaden, C., Campbell, I.D., and Dobson, C.M. (1996) Rapid refolding of a proline-rich all-β-sheet fibronectin type III module. *Proc. Natl. Acad. Sci. U.S.A.*, **93**, 10703–10706.

Reinhardt, D.P., Ono, R.N., and Sakai, L.Y. (1997) Calcium stabilizes fibrillin-1 against proteolytic degradation. *J. Biol. Chem.*, **272**, 1231–1236.

Rich, A., and Crick, F.H.C. (1961) The molecular structure of collagen. *J. Mol. Biol.*, **3**, 483–506.

Rousselle, P., Keene, D.R., Ruggiero, F., Champliaud, M.F., van der Rest, M., and Burgeson, R.E. (1997) Laminin 5 binds the NC-1 domain of type VII collagen. *J. Cell Biol.*, **138**, 719–728.

Sage, E. H., and Bornstein, P. (1991) Extracellular proteins that modulate cell-matrix interactions. SPARC, tenascin, and thrombospondin. *J. Biol. Chem.*, **266**, 14831–14834.

Sakai, L.Y., Keene, D.R., and Engvall, E. (1986) Fibrillin, a new 350-kD glycoprotein, is a component of extracellular microfibrils. *J. Cell Biol.*, **103**, 2499–2509.

Sakai, L.Y., Keene, D.R., Glanville, R.W., and Bächinger, H.P. (1991) Purification and partial characterization of fibrillin, a cysteine-rich structural component of connective tissue microfibrils. *J. Biol. Chem.*, **266**, 14763–14770.

Sasaki, T., Wiedemann, H., Matzner, M., Chu, M.L., and Timpl, R. (1996) Expression of fibulin-2 by fibroblasts and deposition with fibronectin into a fibrillar matrix. *J. Cell Sci.*, **109**, 2895–2904.

Shaw, L.M., and Olsen, B.R. (1991) FACIT collagens: diverse molecular bridges in extracellular matrix. *TIBS*, **16**, 191–194.

Timpl, R., and Brown, J.C. (1996) Supramolecular assembly of basement membranes. *BioEssays*, **18**, 123–132.

van Steensel, M.A., Buma, P., de Waal Malefijt, M.C., van den Hoogen, F.H., and Brunner, H.G. (1997) Oto-spondylo-megaepiphyseal dysplasia (OSMED): clinical description of three patients homozygous for a missense mutation in the COL11A2 gene. *Am. J. Med. Genet.*, **70**, 315–323.

von der Mark, H., Aumailley, M., Wick, G., Fleischmajer, R., and Timpl, R. (1984) Immunochemistry, genuine size and tissue localization of collagen VI. *Eur. J. Biochem.*, **142**, 493–502.

Wagener, R., Kobbe, B., and Paulsson, M. (1998) Matrilin-4, a new member of the matrilin family of extracellular matrix proteins. *FEBS Lett.*, **436**, 123–127.

Wu, J.J., Woods, P.E., and Eyre, D.R. (1992) Identification of cross-linking sites in bovine cartilage type IX collagen reveals an antiparallel type II-type IX molecular relationship and type IX to type IX bonding. *J. Biol. Chem.*, **267**, 23007–23014.

Yada, T., Suzuki, S., Kobayashi, M., Hoshino, T., Horie, K., and Kimata, K. (1990) Occurence in chick embryo vitreous humor of a type IX collagen proteoglycan with an extraordinarily large chondroitin sulfate chain and short α1 polypeptide. *J. Biol. Chem.*, **265**, 6992–6999.

Yamagata, M., Yamada, K.M., Yamada, S.S., Shinomura, T., Tanaka, H., Nishida, Y., Obara, M., and Kimata, K. (1991) The complete primary structure of type XII collagen shows a chimeric molecule with reiterated fibronectin type III motifs, von Willebrand factor A motifs, a domain homologous to a noncollagenous region of type IX collagen, and short collagenous domains with an Arg-Gly-Asp site. *J. Cell Biol.*, **115**, 209–221.

Yamaguchi, N., Kimura, S., McBride, O.W., Hori, H., Yamada, Y., Kanamori, T., Yamakoshi, H., and Nagai, Y. (1992) Molecular cloning and partial characterization of a novel collagen chain, α1(XVI), consisting of repetitive collagenous domains and cysteine-containing non-collagenous segments. *J. Biochem.*, **112**, 856–863.

Yoshioka, H., Zhang, H., Ramirez, F., Mattei, M.G., Moradi-Ameli, M., van der Rest, M., and Gordon, M.K. (1992) Synteny between the loci for a novel FACIT-like collagen locus (D6S228E) and α1(IX) collagen (COL9A1) on 6q12-q14 in humans. *Genomics*, **13**, 884–886.

Yurchenco, P.D., Cheng, Y.-S., and Colognato, H. (1992) Laminin forms an independent network in basement membranes. *J. Cell Biol.*, **117**, 1119–1133.

Zhang, H., Apfelroth, S.D., Wu, W., Davis, E.C., Sanguineti, C., Bonadio, J., Mecham, R.P., and Ramirez, F. (1994) Structure and expression of fibrillin-2, a novel microfibrillar component preferentially located in elastic matrices. *J. Cell Biol.*, **124**, 855–863.

A5. Membrane Lectins as Adhesion Receptors

Michèle Aubery

1. INTRODUCTION

Several biological events including pathological ones, such as regulation of gene expression, cell proliferation, cell differentiation, cell migration, apoptosis, and formation of metastases, are dependent upon cell adhesion, and different types of adhesion receptors have been described in the literature, including cadherins, integrins, mucins, selectins (Engel, 1992; Hynes, 1992; Paulsson, 1992; Mercurio, 1995; Crocker and Feizi, 1996; Laflamme and Auer, 1996; Ruoslahti and Öbrink, 1996; Sanchez-Mateos *et al.*, 1996; Penberthy *et al.*, 1997; Aumailley and Gayraud, 1998; Pignatelli, 1998). The cell–extracellular matrix interaction is a multistep phenomenon including cell attachment to extracellular matrix components, such as laminin, collagen and fibronectin, *via* specific cell surface-receptors, followed by rearrangements of plasma membrane and cytoskeletal elements which trigger cell spreading. Most surface receptors are glycoproteins, and it is now clear that the carbohydrates carried by these receptors play a central role in cell-recognition processes (Hughes, 1992a; Olden, 1993). Carbohydrates can interact with other carbohydrates, as well as with certain proteins called lectins (Hakomori, 1992; Fukuda, 1995). The affinity of such interactions is low and may serve as regulatory mechanisms in cell-adhesion phenomena giving rise to transient, specific and reversible interactions, such as that observed during the rolling of leukocytes on endothelial cells or during the formation of cancer cell metastases. However, carbohydrates spaced closely together on a polypeptide chain might become packed tightly together when the molecule is folded, thereby forming a clustered carbohydrate patch and increasing the binding strength (Spillmann, 1994; Yi *et al.*, 1998). Thus, cell adhesion is a dynamic phenomenon which implicates not only one receptor, but a variety of different receptors each of which is characterized by its affinity binding (more or less high) for a given cell type leading to the consolidation or not of the cell interactions.

Some adhesion receptors, like integrins, VCAM-1, ICAM-1 and CD44, and certain cell-surface glycoproteins involved in cell adhesion perform lectin-like function(s) in cellular recognition (Chammas *et al.*, 1991; Bennet *et al.*, 1996); they have multiple ligand-binding sites, one that recognizes a specific peptide sequence in protein, such as the Arg-Gly-Asp recognition sequence common to many integrins, and a second site that shows marked homology with a lectin-like protein that recognizes specific carbohydrates.

Cell adhesion is modified during physiological and pathological processes, as are the expression patterns of adhesion receptors, glycosylation and endogenous animal lectins (Albelda, 1993; Hébert and Monsigny, 1993; Sobel, 1993).

Lectins are proteins that bind to specific carbohydrate structures. Animal lectins constitute a large family with members having distinct cellular localizations (nucleus, cytoplasm, membrane) (Sève *et al.*, 1994; Drickamer, 1995; Hadj-Sahraoui *et al.*, 1996; Perillo *et al.*, 1998). They have primarily been classified according to the nature of their saccharide ligand, and to their dependence upon divalent cations. However, determination of the primary structure of numerous lectins has enabled their classification based on the sequence homology of the carbohydrate-recognition domain responsible for direct interaction with the carbohydrate ligand.

The two major groups of animal lectins are the C-type lectins, which bind to carbohydrates of different specificities in a Ca^{2+}-dependent manner (for example: selectins, collectins, calreticulins), and the S-type lectins or galectins which specifically recognize β-galactosides in a Ca^{2+}-independent manner and make up various categories according to their overall organization. Other groups have been described, such as the immunoglobulin-like lectins, and, recently, additional groups of novel animal lectins have been identified. Animal lectins are involved in various biological functions; they regulate cell growth and trigger or inhibit apoptosis (Hughes, 1992b; Barondes *et al.*, 1994; Chammas *et al.*, 1994; Hughes, 1994; Zanetta *et al.*, 1994; Drickamer, 1995; Perillo *et al.*, 1995; Poirier and Kimber, 1997; Kieda, 1998; Perillo *et al.*, 1998). In addition, lectin expression is modified during development, and is also altered at sites of inflammation and in various carcinomas (Hébert and Monsigny, 1993; Lotan *et al.*, 1994; Welphy *et al.*, 1994; Perillo *et al.*, 1995; Kannagi, 1997; Leffler, 1997; Perillo *et al.*, 1998; Weiss *et al.*, 1998).

Altered lectin expression in tumor cells has been associated with their metastatic potential, thereby suggesting a role of animal lectins in cell-adhesion processes. Indeed, the localization of certain lectins at sites of adhesion supports their potential function in cell–extracellular matrix and cell–cell interactions (Hughes, 1992b; Chiu *et al.*, 1994; Huflejt *et al.*, 1997). Taking into consideration the specific binding of animal lectins to various cell-surface glycoconjugates (Hébert and Monsigny, 1993; Powell and Varki, 1994; Zanetta *et al.*, 1994; Bresalier *et al.*, 1996; Poirier and Kimber, 1997; Liu *et al.*, 1998; Perillo *et al.*, 1998), their roles in cell-recognition processes have been widely suggested; however, it appears that among the animal lectins, only a few of them have been directly implicated in cell-adhesion mechanisms as supported by experimental adhesion assays.

This review will focuses on our current understanding of the demonstrated role of animal lectins, and the potential function(s) of the lectin domain of lectin-like receptors in adhesion of normal and pathological cells.

2. ROLE OF ANIMAL LECTINS IN CELL ADHESION

The best documented example in which lectin activity has been clearly defined in cell adhesion is that of the **selectins** (Welphy *et al.*, 1994; McEver *et al.*, 1995; Whealan, 1996; McEver, 1997; Kieda, 1998). Selectins are calcium-dependent membrane-anchored, C-type lectins that initiate adhesion and rolling of circulating leukocytes on endothelial cells, platelets or other leukocytes during the inflammation process (Hogg and Landis, 1993; Kannagi, 1997; McEver, 1997; Kieda, 1998). Each selectin is composed of an amino-terminal carbohydrate-recognition domain characteristic of C-type lectins, followed by an epidermal growth factor (EGF)-like domain, short consensus repeats, a transmembrane domain, and a short cytoplasmic tail. L-Selectin is expressed on most leukocytes and mediates their adhesion to and rolling on the endothelium; E-selectin is expressed on activated endothelial cells and recognizes leukocytes; P-selectin is expressed on activated platelets and endothelial cells, and plays a role in the interaction between activated platelets or endothelial cells and leukocytes. The minimal carbohydrate structure involved in recognition is sialylated, fucosylated carbohydrate chains, such as sialyl Lewis[x] or sialyl Lewis[a] (Kannagi, 1997). Moreover, sialyl Lewis[x] and sialyl Lewis[a] are frequently expressed on human cancer cells, are involved in their adhesion to vascular endothelial cells and thus contribute to the metastasis of cancer (Kannagi, 1997).

2.1. The Calreticulins

The **calreticulins,** first isolated from the endoplasmic reticulum where they play a role of molecular chaperone for immature N-linked glycoproteins including integrins (Lenter and Vestweber, 1994), exhibit extensive molecular identity with calnexin. A cell-surface form of calreticulin has been also detected in various cell types, such as human leukocytes and endothelial cells (Burns *et al.*, 1994; Dedhar, 1994). In addition, White *et al.* (1995) used murine melanoma B16 cells and demonstrated that calreticulin is responsible for their spreading on laminin. Specific oligomannosides, especially the Man6-9 present on laminin, are recognized by B16 cell-surface calreticulin. In agreement with the previous reports by Chandrasekaran *et al.* (1994a, b), this interaction must occur either when the integrin $\beta1$ is engaged or afterwards, suggesting that the cross-talk between cell-surface calreticulin and the ectodomain of the integrin is necessary to mediate the cell signaling events which trigger cell spreading. Moreover, Coppolino *et al.* (1995) noted that the integrin—calreticulin interaction depends up the activation state of the integrin. In contrast, the cell attachment mediated by $\beta1$ integrins is independent of the lectin. Leung-Hagesteijn *et al.* (1994) observed the colocalization of calreticulin and $\beta1$ integrins at focal adhesion sites in prostatic carcinoma PC-3 cells. Moreover, these authors demonstrated that calreticulin specifically bound to the N-terminal domain of integrins *via* the consensus sequence KXGFFKR of the cytoplasmic domain of their α subunit. Although these data are not in agreement with those of White *et al.* (1995) concerning the integrin domain interacting with calreticulin, it is clear that the calreticulin interacts directly with the integrin to initiate the latter's cell signaling (Sastry and Horwitz, 1993; Schwartz and Ingber, 1994; Schwartz, 1995; Yamada and Miyamoto, 1995) and thus triggers cell spreading on the extracellular matrix.

2.2. Galectin-1

Galectin-1 (L-14, galaptin) is an homodimer present in various tissues, and it was shown to either promote or inhibit cell adhesion to the extracellular matrix. Galectin-1 binds to membrane receptors and extracellular matrix components, such as laminin and fibronectin. Addition of exogenous recombinant galectin-1 to myoblasts inhibits their adhesion to laminin (Cooper *et al.*, 1991). Gu *et al.* (1994) reported that galectin-1 inhibits rat myoblast adhesion to laminin, thereby allowing the myoblast fusion and subsequent muscle differentiation. This effect results from galectin-1 binding to laminin poly-*N*-acetyllactosamine chains which interferes with $\alpha7\beta1$–laminin interaction; $\alpha7\beta1$ is the major laminin receptor on myoblasts. The galectin-1 effect is selective, since its binding to fibronectin does not interfere with fibronectin–$\alpha7\beta1$ interaction. In contrast, galectin-1 promotes the adhesion of various cells, e.g., Chinese hamster ovary CHO cells, teratocarcinoma F9 cells, olfactory neurons, melanoma cells, to laminin by cross-linking cell-surface receptors and laminin carbohydrates (Zhou and Cummings, 1993; van den Brûle *et al.*, 1995). Ozeki *et al.* (1995), studied rhabdosarcoma cells and showed that cell adhesion to fibronectin occurs even in the presence of the specific GRGDS peptide if galectin-1 is also present; this galectin-1 effect is inhibited by lactose. The galectin-1 concentrations at the adhesion site, relative to those of integrin and laminin were implicated in the lectin effects. In addition to its contribution to cell–matrix adhesion, galectin-1 also appears to be involved in cell–cell interactions since Lotan *et al.* (1994) used anti-galectin-1 antibody to show that galectin-1 is expressed at the liver microvessel endothelial cell surface and participates in the tumor cell–endothelial cell adhesion.

2.3. Galectin-3

Galectin-3 (Mac-2, CBP-35, CBP-30), a monomer expressed on various cell types, exhibited, like galectin-1, either an adhesive or an anti-adhesive function. Sato and Hughes (1992) reported that baby hamster kidney (BHK) cell adhesion to laminin or its E8 fragment was not altered by anti-galectin-3 antibodies. However, a high concentration of galectin-3 (45 μg/ml) was shown to reduce the adhesion and spreading of BHK on laminin (Sato and Hughes 1992). More recently, Ochieng *et al.* (1998) analyzed the role of galectin-3 in the adhesion of various cell types (human breast carcinoma cell line, human fibrosarcoma, mouse spindle-cell-carcinoma cell line, human prostate-carcinoma-cell line PC-3)) to laminin, collagen IV or fibronectin and demonstrated that the incubation of these cells with galectin-3 abolished their adhesion to the matrix molecules. It was also observed that galectin-3 bound to the $\alpha1\beta1$ integrin in a lactose-dependent manner, thereby preventing their interactions with extracellular matrix. The galectin-3 effect appears to be specific because it is abolished in the presence of lactose, whereas the lectin remains effective in the presence of sucrose. In contrast, addition of purified galectin-3 has been shown to promote the adhesion of human neutrophils to laminin (Kuwabara and Liu, 1996), to mediate rapid cell adhesion of a human breast carcinoma cell line to laminin, and possibly to be implicated in their invasive capacity through matrigel (Warfield *et al.*, 1997).

2.4. Galectin-4

Galectin-4 is localized mainly at adhesion sites (Huflejt *et al.*, 1997) and these authors demonstrated that human colon adenocarcinoma T84 cells adhered strongly (19% *versus* 1% to bovine serum albumin) to recombinant galectin-4-coated substrate in a lactose-dependent manner. Cell adhesion to laminin was higher than that observed on galectin-4, but was only partly inhibited (10%) in the presence of lactose, and was not affected by cellobiose.

2.5. The Cerebellar Soluble Lectin (CSL)

The **cerebellar soluble lectin** (CSL) is a mannose-binding protein that is externalized at certain stages of nervous tissue development, and it has been shown to be involved in cell adhesion when it was used as the substrate for oligodendrocytes (Zanetta *et al.*, 1994). In addition, CSL clustering on the cell surface could play a role in signal transduction.

2.6. The Mouse Macrophage Galactose/N-Acetylgalactosamine C-Type Lectin (MMGL)

The **mouse macrophage galactose/*N*-acetylgalactosamine C-type lectin** (MMGL) is involved in malignant cell–activated macrophage adhesion (Yamamoto *et al.*, 1994). These authors also showed that carbohydrate–protein interaction is an initial step in tumor cell attachment and migration.

2.7. The Versican C-Type Lectin

The **versican C-type lectin** binds to insolubilized fucose and *N*-acetylglucosamine, and also binds to tenascin-R (Aspberg *et al.*, 1995), suggesting this lectin's role in the modulation of cell adhesion because tenascin is an anti-adhesive molecule (Chiquer-Ehrismann, 1995). Table 1 summarizes animal lectins involved in cellular adhesion.

3. ADHESION RECEPTORS WITH A LECTIN-LIKE FUNCTION

3.1. The 67-kDa Receptor

The functional characterization of the **67-kDa elastin/laminin receptor** demonstrated the efficiency of both protein- and carbohydrate-binding domains. The 67-kDa receptor binds to laminin and to elastin with similar affinity. The receptor affinity is greatly influenced by its lectin domain, since the receptor binds to laminin and/or elastin with higher affinity in the absence of galactosides, whereas in the presence of these carbohydrates the affinity is dramatically reduced and the receptor–ligand interaction is abolished (Mecham *et al.*, 1991; Hinek, 1996). In addition, the 67-kDa receptor could be involved in the regulation of laminin–$\alpha 6\beta 4$ integrin interactions, since the expressions of both the 67-kDa receptor and the $\alpha 6$ subunit were coregulated (Ardini *et al.*, 1997), and consequently influenced the signaling pathway mediated by the $\alpha 6\beta 4$ integrin.

3.2. Integrins

Integrins are a large family of heterodimeric membrane receptors (Hynes, 1992). They are able to interact with cytoskeletal proteins and to transmit signals to the nucleus. The potential role(s) of the lectin-like function(s) of **integrin(s)** has been evidenced by Chammas *et al.* (1991, 1994), who demonstrated the influence of carbohydrate moieties on the $\alpha 6\beta 1$ integrin—laminin interaction-mediated murine melanoma cell adhesion using enzymatic deglycosylation experiments and glycosylation inhibition assays. Their results suggested that the $\alpha 6\beta 1$ integrin and laminin act like lectin-like molecules, with the carbohydrate moieties of each ligand interacting with amino-acid residues from the other,

Table 1. Animal lectins involved in cellular adhesion

Lectins	Specificity	Role in cell interaction	Cell types
Selectins (L-selectin, E-selectin, P-selectin)	fucosylated, sialylated polylactosaminoglycans (sialyl Lewis[x], sialyl Lewis[a])	cell–cell adhesion	leukocytes, endothelial cells, platelets
Calreticulins	oligomannoside chains	cell–ECM adhesion	leukocytes, endothelial cells, melanoma cells
Galectin-1	β-galactosides	cell–ECM cell–cell, adhesion, de-adhesion	myoblasts, CHO, teratocarcinoma F9, olfactory neurons, melanoma, rhabdosarcoma cells. tumor cells, endothelial cells
Galectin-3	β-galactosides	cell–ECM cell–cell, adhesion, de-adhesion	BHK, neutrophils, breast carcinoma, fibrosarcoma cells
Galectin-4	β-galactosides	cell–ECM adhesion	colon adenocarcinoma cells
CSL	N-glycans	cell–ECM adhesion	oligodendrocytes
MMGL	galactose/N-acetylglucosamine	cell–cell adhesion	tumor cells, macrophages

BHK: baby hamster kidney cells; ECM: extracellular matrix; CHO: Chinese hamster ovary cells; CSL: cerebellar soluble lectin; MMBL: mouse macrophage galactose/N-acetylgalactosamine C-type lectin.

thereby initiating a lectin–lectin interaction. Our group reported that glycosylation of the $\beta1$ subunit of the $\beta1$ integrin family is implicated in the modulation of keratinocyte–lymphocyte adhesion in pathological situations (Boukhelifa *et al.*, 1998; Braut-Boucher *et al.*, 1998). Moreover, these data indicated that glycosylation could modulate the lectin-like binding and function of $\beta1$ integrins. Zheng *et al.* (1994) reported that N-glycosylation of both the α and β subunits of the $\alpha5\beta1$ integrin, a fibronectin receptor, is necessary for the integrin to bind to fibronectin and could interfere with cytoplasmic domain functions of the integrins and consequently in subsequent cell signaling.

3.3. Cbg72

The membrane concanavalin-A binding glycoprotein **cbg72** is a laminin receptor that mediates chick-embryo fibroblast spreading on laminin (Moutsita *et al.*, 1991), and a lectin function specific to N-acetylglucosamine was identified recently (Botti *et al.*, unpublished results). Table 2 summarizes adhesion receptors with a lectin-like function.

Table 2. Adhesion receptors with a lectin function

Adhesion receptors	*Specificity*	*Role in cell interaction*	*Cell types*
67 kD	lactose	cell–ECM adhesion	fibroblasts
$\alpha6\beta1$	*N*-glycans	cell–ECM adhesion	murine melanoma cells
cbg72	*N*-acetylglucosamine	cell–ECM adhesion, spreading	fibroblasts

cbg: Concanavalin A-binding glycoprotein; ECM: extracellular matrix.

4. CONCLUSION

Lectin–carbohydrate interactions are important in cell adhesion and could result in the transduction of various signals depending up the pathophysiological state of the cells. The apparent disparity between the results reported for different cellular models highlights the complexity surrounding the lectin functions. Indeed, lectins seem to play contradictory roles in cellular adhesion; they can promote or inhibit the adhesion of cells to the extracellular matrix. Lectins can block adhesion by binding to molecules in the extracellular matrix, thereby masking the domains of these molecules that interact with the integrins present at the cell surface. Lectins can favor adhesion of cells to the extracellular matrix by forming bridges between the latter two. These various properties were demonstrated by using different experimental conditions: anti-lectin antibodies to block the effects of lectins, lectins as adhesion substrates, lectins to interact directly with either the cell-surface receptors or extracellular matrix molecules. In addition, the properties observed, attachment or spreading of the cells on extracellular matrix or cell–cell adhesion, can also be differently modulated by the same lectin. It is important to keep in mind that glycosylation modifications during the course of physiological and/or pathological processes engender changes in the ligands of these lectins, with, as a consequence, altered lectin–carbohydrate interactions. Finally, most of the experimental studies were conducted on malignant or transfomed cell lines and, therefore, it is now time to investigate normal cells under conditions approximating as closely as possible the *in vivo* situation and then to approach the study of the role(s) of these lectins *in situ*, because the anti-adhesive and adhesive properties of lectins should improve our understanding of the metastatic process and might provide a therapeutic means to their prevention (Heavner, 1996).

ACKNOWLEDGMENTS

I would like to express my gratitude to my colleagues particularly Drs. F. Braut-Boucher, C. Derappe, J. Font and J. Pichon for their valuable comments and discussions during the preparation of this manuscript. This work was supported by the Institut National de la Santé et de la Recherche Médicale, by the Fondation pour la Recherche Médicale and by the Université René-Descartes, Paris V, France.

REFERENCES

Albelda, S.M. (1993) Role of integrins and other cell adhesion molecules in tumor progression and metastasis. *Lab. Invest.*, **68**, 4–17.

Ardini, E., Tagliabue, E., Magnifico, A., Buto, S., Castronovo, V., Colnaghi, M.I., Menard, S. (1997) Co-regulation and physical association of the 67 kDa monomeric laminin receptor and the $\alpha 6 \beta 4$ integrin. *J. Biol. Chem.*, **272**, 2342–2345.

Aspberg, A., Binkert, C., Ruoslahti, E. (1995) The versican C-type lectin domain recognizes the adhesion protein tenascin-R. *Proc. Natl. Acad. Sci. USA*, **92**, 10590–10594.

Aumailley, M., Gayraud, B. (1998) Structure and biological activity of the extracellular matrix. *J. Mol. Med.*, **76**, 253–265.

Barondes, S.H., Cooper, D.N.W., Gitt, M.A., Leffler, H. (1994) Galectins: structure and function of a large family of animal lectins. *J. Biol. Chem.*, **269**, 20807–20810.

Bennet, K.L., Modrell, B., Greenfield, B., Bartolazzi, A., Stamenkovic, L., Peach, R., Jackson, D.G., Spring, F., Araffo, A. (1996) Regulation of CD44 binding by glycosylation of variably sliced exon. *J. Cell Biol.*, **131**, 1623–1633.

Boukhelifa, M., Paulin, Y., Font, J., Pichon, J., Giner, M., Wantyghem, J., Aubery, M., Braut-Boucher, F. (1998) Integrins of the $\beta 1$ family influence keratinocyte—lymphocyte interaction. *J. Invest. Dermatol.*, **111**, 650–655.

Braut-Boucher, F., Font, J., Pichon, J., Paulin, Y., Boukhelifa, M., Aubery, M., Derappe, C. (1998) T lymphocytes from Sézary syndrome patients express $\beta 1$ integrins whose $\beta(1–6)$-branched *N*-linked oligosaccharides reflect their adhesive capacity. *Leuk. Res.*, **22**, 947–952.

Bresalier, R.S., Byrd, J.C., Raz, A. (1996) Colon cancer mucins a new ligand for the beta-galactoside-binding protein galectin-3. *Cancer Res.*, **56**, 4354–4367.

Burns, K., Atkinson, E.A., Bleackley, R.C., Michalak, M. (1994) Calreticulin: from Ca^{2+} binding to control of gene expression. *Trends Cell Biol.*, **4**, 152–154.

Chammas, R., Veiga, S.S., Line, S., Potocnjak, P., Brentoni, R.R. (1991) Asn-linked oligosaccharide-dependent interaction between laminin and gp 120/140. An $\alpha 6 \beta 1$ integrin. *J. Biol. Chem.*, **266**, 3349–3355.

Chammas, R., Jasiulionis M.G., Jin, F., Villa-Verde, D.M.S., Reinhold, V.N. (1994) Carbohydrate-binding proteins in cell-matrix interactions. *Braz. J. Med. Biol. Res.*, **27**, 2169–2179.

Chandrasekaran, S., Tanzer, M.L., Giniger, M.S. (1994a) Oligomannosides intiate cell spreading of laminin-adherent murine melanoma cells. *J. Biol. Chem.*, **269**, 3356–3366.

Chandrasekaran, S., Tanzer, M.L., Giniger, M.S. (1994b) Characterization of oligomannoside binding to the surface of murine melanoma cells. Potential relationship to oligomannoside-initiated cell spreading. *J. Biol. Chem.*, **269**, 3367–3373.

Chiquet-Ehrismann, R. (1995) Inhibition of cell adhesion by anti-adhesive molecules. *Curr. Opin. Cell Biol.* 7, 715–719.

Chiu M.L., Parry D.A.D., Feldman S.R., Klapper D.G., O'Keefe E.J. (1994) An adherens junction protein is a member of the family of lactose-binding lectins. *J. Biol. Chem.*, **269**, 31770–31776.

Cooper, D.N.W., Massa, S.M., Barondes, S.H. (1991) Endogenous muscle lectin inhibits myoblast adhesion to laminin. *J. Cell. Biol.* **115**, 1437–1448.

Coppolino, M., Leung-Hagesteijn, C., Dedhar, S., Wilkins, J. (1995) Inducible interaction of integrin $\alpha 2 \beta 1$ with calreticulin. Dependence on the activation state of the integrin. *J.Biol.Chem.*, **270**, 23132–23138.

Crocker, P.R., Feizi, T. (1996) Carbohydrate recognition systems: functional triads in cell—cell interaction. *Curr. Opin. Struct. Biol.*, **6**, 879–891.

Dedhar, S. (1994) Novel functions for calreticulin: interaction with integrins and modulation of gene expression. *Trends Biochem. Sci.*, **19**, 269–271.

Drickamer, K. (1995) Increasing diversity of animal lectin structures. *Curr. Opin. Struct. Biol.*, **5**, 612–616.

Engel, J. (1992) Laminins and other strange proteins. *Biochemistry*, **44**, 10643–10651.

Fukuda, M. (1995) Carbohydrate-dependent cell adhesion. *Bioorg. Med. Chem.*, **3**, 207–215.

Gu, M., Wang, W., Song, W.K., Cooper, D.N.W., Kaufman, S.J. (1994) Selective modulation of the interaction of $\alpha 7 \beta 1$ integrin with fibronectin and laminin by L-14 lectin during skeletal muscle differentiation. *J. Cell. Sci.*, **107**, 175–181.

Hadj-Sahraoui, Y., Sève, A.P., Doyennette-Moyne, M.-A., Saffar, L., Felin, M., Aubery, M., Gattegno, L., Hubert, J. (1996) Nuclear and cytoplasmic expression of the carbohydrate-binding protein CBP70 in tumoral and healthy cells of the macrophage lineage. *J. Cell. Biochem.*, **62**, 529–542.

Hakomori, S.-I. (1992). Carbohydrates as adhesion molecules. In C.G. Gahnberg, T. Mondrup-Poulsen, L. Wogensen Bach; B. Hokfelt (eds.) *Leukocyte Adhesion, Basic and Clinical Aspects*. Elsevier Science Publishers, Amsterdam.

Heavner, G.A. (1996) Active sequences in cell adhesion molecules: targets for therapeutic intervention. *Drug Discovery Today*, **1**, 295–304.

Hébert, E., Monsigny, M. (1993) Oncogenes and expression of endogenous lectins and glycoconjugates. *Biol. Cell.*, **79**, 97–109.

Hinek, A. (1996) Biological roles of the non-integrin elastin/laminin receptor. *Biol. Chem.*, **377**, 471–480.

Hogg, N., Landis, R.C. (1993) Adhesion molecules in cell interactions. *Curr. Opin. Immunol.*, **5**, 383–390.

Huflejt M.E., Jordan E.T., Gitt M.A., Barondes S.H., Leffler H. (1997) Strikingly different localization of galectin-3 and galectin-4 in human colon adenocarcinoma T84 cells. *J. Biol. Chem.*, **272**, 14294–14303.

Hughes, R.C. (1992a) Role of glycosylation in cell interaction with extracellular matrix. *Biochem. Soc. Trans.*, **20**, 279–284.

Hughes, R.C. (1992b) Lectins as cell adhesion molecules. *Curr. Opin. Struct. Biol.*, **2**, 682–692.

Hughes, R.C. (1994) Mac-2: a versatile galactose-binding protein of mammalian tissues. *Glycobiology*, **4**, 5–12.

Hynes, R.O. (1992) Integrins: versatility, modulation and signaling in cell adhesion. *Cell*, **69**, 11–25.

Kannagi, R. (1997) Carbohydrate-mediated cell adhesion involved in hematogenous metastasis of cancer. *Glycoconjugate J.* **14**, 577–584.

Kieda, C. (1998) Role of lectin—glycoconjugate recognition in cell-cell interactions leading to tissue invasion. In J. Adford (ed.) *Advances in Experimental Medicine and Biology* **435**, *Glycoimmunology 2*, Plenum Press, New York and London. pp. 75–82.

Kuwabara, I., Liu, F.T. (1996) Galectin-3 promotes adhesion of human neutrophils to laminin. *J. Immunol.*, **156**, 3939–3944.

Laflamme, S.E., Auer, K.L. (1996) Integrin signaling. *Semin. Cancer Biol.*, **7**, 111–118.

Leffler, H. (1997) Introduction to galectins. *Trends Glycosci. Glycotechnol.*, **9**, 1–11.

Lenter, M., Vestweber, D. (1994) The integrin chains $\beta1$ and $\alpha6$ associate with the chaperone calnexin prior to integrin assembly. *J. Biol. Chem.*, **269**, 12263–12268.

Leung-Hagesteijn, C.Y. Milankov, K., Michalak, M., Wilkins, J., Dedhar, S. (1994) Cell attachment to extracellular matrix substrates is inhibited upon down-regulation of expression of expression of calreticulin, an intracellular integrin α-subunit-binding protein. *J. Cell. Sci.*, **107**, 589–600.

Liu, S.S., Parker, W., Everett, M.L., Platt, J.L. (1998) Differential recognition by proteins of α-galactosyl residues on endothelial cell surfaces. *Glycobiology*, **8**, 433–443.

Lotan, R., Belloni, P.N., Tressler, R.J., Lotan, D., Xu, X.-C., Nicolson, G.L. (1994) Expression of galectins on microvessel endothelial cells and their involvement in tumour cell adhesion. *Glycoconjugate J.*, **11**, 462–468.

McEver, R.P. (1997) Selectin-carbohydrate interactions during inflammation and metastasis. *Glycoconjugate J.*, **14**, 585–591.

McEver, R.P., Moore, K.L., Cummings, R.D. (1995) Leukocyte trafficking mediated by selectin-carbohydrate interaction. *J. Biol. Chem.*, **270**, 11025–11028.

Mecham, R.P., Whitehouse, L., Hay, M., Hinek, A., Sheer, M.S. (1991) Ligand affinity of the 67-kD elastin/laminin binding protein is modulated by the protein's lectin domain: visualization of elastin/laminin receptor complexes with gold-tagged ligand. *J. Cell. Physiol.* **113**, 187–194.

Mercurio, A.M. (1995) Laminin receptors: achieving specificity through cooperation. *Trends Cell. Biol.*, **5**, 419–423.

Moutsita, R., Aubery, M., Codogno, P. (1991) A Mr 72 k cell surface concanavalin A binding glycoprotein is specifically involved in the spreading of chick embryo fibroblasts onto laminin substrate. *Exp. Cell. Res.*, **192**, 236–242.

Ochieng, J., Leitebrowning, M.L., Warfield, P. (1998) Regulation of cellular adhesion to extracellular matrix proteins by galectin-3. *Biochem. Biophys. Res. Commun.*, **246**, 788–791.

Olden, K. (1993) Adhesion molecules and inhibition of glycosylation in cancer. *Semin. Cancer Biol.*, **4**, 269–276.

Ozeki, Y., Matsui, T., Yamamoto, Y., Funahashi, M., Hamako, J., Titani, K. (1995) Tissue fibronectin is an endogenous ligand for galectin-1. *Glycobiology*, **5**, 255–261.

Paulsson, M. (1992) Basement membrane proteins: structure, assembly and cellular interactions. *Crit. Rev. Biochem. Mol. Biol.*, **27**, 93–127.

Penberthy, T.W., Jiang, Y.L., Graves, D.T. (1997) Leukocyte adhesion molecules. *Crit. Rev. Oral Biol. Med.*, **8**, 380–388.

Perillo, N.L., Pace, K.E., Seilhamer, J.J., Baum, L.G. (1995) Apoptosis of T cells mediated by galectin-1. *Nature (London.)*, **378**, 736–769.

Perillo, N.L., Marcus, M.E., Baum, L.G. (1998) Galectins: versatile modulators of cell adhesion, cell proliferation, and cell death. *J. Mol. Med*, **76**, 402–412.

Pignatelli, M. (1998) Integrins, cadherins, and catenins-molecular cross-talk in cancer cells. *J. Pathol.* **186**, 1–2.

Poirier, F., Kimber, S. (1997) Cell surface carbohydrates and lectins in early development. *Mol. Hum. Reprod.*, **3**, 907–918.

Powell, L.D., Varki, A. (1994) The oligosaccharide binding specificities of CD22β, a sialic acid-specific lectin of B cells. *J. Biol. Chem.*, **269**, 10628–10636.

Ruoslahti, E., Öbrink, B. (1996) Common principles in cell adhesion. *Exp. Cell Res.*, **227**, 1–11.

Sanchez-Mateos, P., Cabanas, C., Sanchez-Madrid, F. (1996) Regulation of integrin function. *Semin.Cancer Biol.*, **7**, 99–109.

Sastry, S.K., Horwitz, A.F. (1993) Integrin cytoplasmic domains mediators of cytoskeletal linkages and extra- and intracellular initiated transmembrane signaling. *Curr. Opin. Cell Biol.*, **5**, 919–931.

Sato, S., Hughes, R.C. (1992) Binding specificity of a baby hamster kidney lectin for H type I and II chain poly lactosamine glycans, and appropriately glycosylated forms of mlaminin and fibronectin. *J. Biol. Chem.* **267**, 6983–6990.

Schwartz, M.A. (1995) Signaling by integrins: implications for tumorigenesis. *Cancer Res.*, **53**, 1503–1506.

Schwartz, M.A., Ingber, D.E. (1994) Integrating with integrins. *Mol. Biol. Cell*, **5**, 389–393.

Sève, A.P., Hadj-Sahraoui, Y., Felin, M., Doyennette-Moyne, M.-A., Aubery, M., Hubert, J. (1994) Evidence that lactose binding to CBP35 disrupts its interaction with CBP70 in isolated HL60 cell nuclei. *Exp. Cell Res.*, **213**, 191–197.

Sobel, M.E. (1993) Differential expression of the 67 kDa laminin receptor in cancer. *Semin. Cancer Biol.*, **4**, 311–317.

Spillmann, D. (1994) Carbohydrates in cellular recognition: from leucine-zipper to sugar-zipper? *Glycoconjugate J.*, **11**, 169–171.

van den Brûle, F.A., Buicu, C., Baldet, M.C., Sobel, M.E., Cooper D.N.W., Marschal, P., Castronovo, V. (1995) Galectin-1 modulates human melanoma cells adhesion to laminin. *Biochem. Biophys. Res. Commun.*, **209**, 760–766.

Welphy, J., Keene, J.L., Schmuke, J.J., Howard, S.C. (1994) Selectins as potential targets of therapeutic intervention in inflammatory diseases. *Biochim. Biophys. Acta*, **1197**, 215–226.

Warfield, P.R., Makker, P.N., Raz, A., Ochieng, J. (1997) Adhesion of human breast carcinoma cells to extracellular matrix proteins is modulated by galectin-3. *Invasion Metastasis*, **17**, 101–112.

Weiss, W.I., Taylor, M.E., Drickamer, K. (1998) The C-type lectin superfamily in the immune system. *Immunol. Rev.*, **163**, 19–34.

Whealan, J. (1996) Selectin synthesis and inflammation. *Trends Biochem. Sci.*, **21**, 65–69.

White, T.K., Zhu, Q., Tanzer, M.L. (1995) Cell surface calreticulin is a putative mannoside lectin which triggers mouse melanoma cell spreading. *J. Biol. Chem.*, **270**, 15926–15929.

Yamada, K.M., Miyamoto, S. (1995) Integrin transmembrane signaling and cytoskeletal control. *Curr. Opin. Cell. Biol.*, **7**, 681–689.

Yamamoto, K., Ishida, C., Shinohara, Y., Hasegawa, Y., Konami, Y., Osawa, T., Irimura, T. (1994) Interaction of immobilized recombinant mouse C-type macrophage lectin with glycopeptides and oligosaccharides. *Biochemistry*, **33**, 8159–8166.

Yi, D., Lee, R.T., Longo, P., Beger, E.T., Lee, Y.C., Petri, W.A.Jr., Schnaar, R.L. (1998) Structural specificity and polyvalent carbohydrate recognition by the *Entamoeba histolytica* and rat hepatic *N*-acetylgalactosamine/galactose lectins. *Glycobiology*, **8**, 1037–1043.

Zanetta, J.P., Badache, A., Maschke, S., Marschal, P., Kuchler, S. (1994) Carbohydrates and soluble lectins in the regulation of cell adhesion and proliferation. *Histol. Histopathol.*, **9**, 385–412.

Zheng, M., Fany, H., Hakomori, S.-I. (1994) Functional role of *N*-glycosylation in $\alpha5\beta1$ integrin receptor. De-*N*-glycosylation induces dissociation or altered association of $\alpha5$ and $\beta1$ subunits and concomitant loss of fibronectin binding activity. *J. Biol. Chem.*, **269**, 12325–12331.

Zhou, Q., Cummings, R.D. (1993) L-14 lectin recognition of laminin and its promotion of *in vitro* cell adhesion. *Arch. Biochem. Biophys.*, **300**, 6–17.

Part B

Relevance to Pathology and Therapeutic Aspects

B1. The Involvement of Bisecting N-Acetylglucosamine in Cancer

Naoyuki Taniguchi, Cong-Xiao Gao, Yoshita Ihara, Eiji Miyoshi, Marafumi Yoshimura, Yin Sheng, Ahmed S. Sultan and Yoshitaka Ikeda

1. INTRODUCTION

The sugar chains of complex carbohydrates are generally thought to play a role in a variety of biological and pathophysiological processes such as adhesion, sorting, cell growth, differentiation and carcinogenesis. Cell surface membrane glycoproteins undergo cancer-associated changes with respect to their carbohydrate moieties and the structural changes may occur as a result of enhancement of glycosyltransferases or glycosidases which are associated with the biosynthetic or degradation processes, respectively. In most of cases, however, the structural changes in glycoproteins in cancer tissues are due to the activation of glycosyltransferases.

The biosynthesis of sugar chains in N-glycans is catalyzed by several glycosyltransferases which catalyze the addition of fucose, N-acetylglucosamine, galactose and sialic acid residues to carbohydrate chains. The bisecting N-acetlyglucosamine(bisecting GlcNAc) is a product of N-acetylglucosaminyltransferase III (GnT-III) which results in its attachment via β1-4 linkage to the core mannose structure of the biantennary sugar chains (Figure 1).

The bisecting GlcNAc has been found in various complex and hybrid types of oligosaccharides of various glycoproteins such as IgG and γ-glutamyltranspeptidases

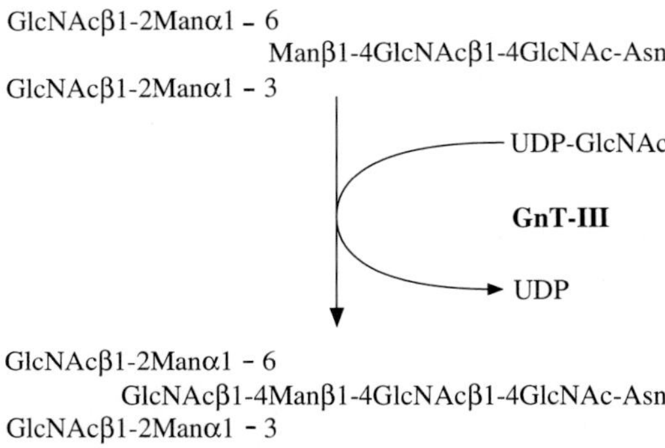

Figure 1: Enzymatic catalysis of N-acetylglucosaminyltranferase III.

(Kornfeld and Kornfeld, 1976; Baenziger and Kornfeld, 1974; Irimura *et al.*, 1981; Carver *et al.*, 1981; Yamashita *et al.*, 1983; Yamashita *et al.*, 1983; Fujii *et al.*, 1990). Yamashita *et al.* first showed that the bisecting GlcNAc is abundant in the γ-glutamyltranspeptidase purified from rat primary hepatoma but was nearly lacking in the same enzyme isolated from normal rat liver (Yamashita *et al.*, 1983). γ-Glutamyltranspeptidase is a typical membrane glycoprotein (Tate and Meister, 1985; Taniguchi and Ikeda, 1998) and is known to be activated during the process of hepatocarcinogenesis in rat hepatoma tissues during chemical carcinogenesis (Fiala *et al.*, 1976; Taniguchi *et al.*, 1974). Subsequent studies have shown that GnT-III is also highly activated in hepatoma tissues in terms of enzymatic activity (Nishikawa *et al.*, 1988) as well as mRNA levels (Miyoshi *et al.*, 1993), suggesting that γ-glutamyltranspeptidase may be one of the endogenous acceptor molecules for GnT-III. Our unpublished data also indicate that parenchymatous cells of rat liver contains very little γ-glutamyltranspeptidase and GnT-III whereas non-parenchymal cells contain both enzymes (Ohno unpublished data).

2. BIOSYNTHESIS AND ROLES OF BISECTING N-ACETYLGLUCOSAMINE

2.1. Enzymatic Basis for the Formation of Bisecting GlcNAc Residues in Glycoproteins.

The bisecting GlcNAc is attached to the β1-4 mannose in the core region of N-glycans and is formed by the enzymatic reaction of UDP-N-acetylglucosamine:β-D-mannoside β1-4 N-acetylglucosaminyltransferase III (GnT-III). The enzyme was first described in hen oviduct (Narashimhan, 1982) and a high level of activity of the enzyme has been also reported in hepatic nodules of rat liver during hepatocarcinogenesis (Narashimhan and Schachter, 1988), Novikoff ascites tumor cells (Koenderman *et al.*, 1989), and human serum (Ishibashi *et al.*, 1989). At least six N-acetylglucosaminyltransferases (Gleeson and Schachter, 1983) are known, and shown in Figure 2.

The enzymatic properties and substrate specificity of GnT-I, GnT-II, GnT-III, GnT-V (Taniguchi and Ihara, 1995) and recently GnT-IV (Oguri *et al.*, 1997) have been characterized and all these cDNAs have been cloned. Quite recently GnT-VI, and its enzyamtic properties has been also characterized (Taguchi *et al.*, 1998 and unpublished data).

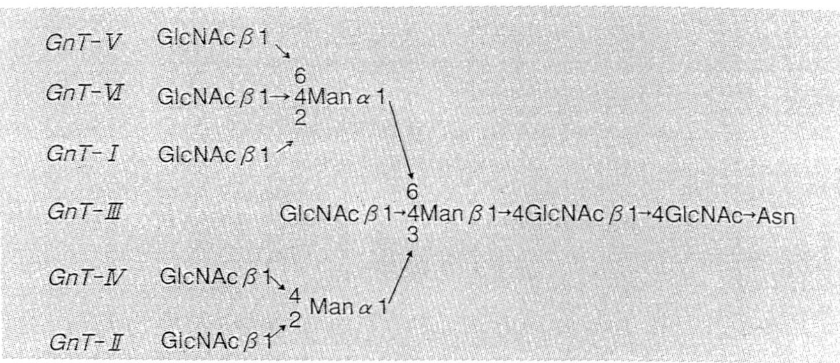

Figure 2: Various N-acetylglucosaminyltransferases.

Each N-acetylglucosaminyltransfearse is different in terms of protein structure as well as in enzymatic properties including their substrate specificity. No sequence homology between these N-acetylglucosaminyltransferase is evident, suggesting that the divergent points of 6 N-acetylglucosaminyltransferase are quite old in the phylogenetic tree and the substrate specificities of these enzymes also support these observations.

GnT-III was purified from rat kidney by affinity chromatography using the substrate, biantennary sugar chain as a ligand (Nishikawa *et al.*, 1992) and a partial amino acid sequence has been obtained. Oligonucleotide primers designed from our knowledge of the amino acid sequence allowed for the cloning of its cDNA. The rat cDNA encodes 536 amino acids and the molecule contains a EGF-like motif which is in structure similar to those found in human β4 integrin. There is no sequence homology to the other cloned glycosyltransferases including GnT-I, II, IV, V and VI. The enzyme is a typical type II transmembrane protein with a cytoplasmic domain, a transmembrane anchor domain, a stem region and a catalytic domain.

It is well known that the bisecting GlcNAc structure affects the conformation of sugar chains and once GnT-III acts on the biantennary sugar chains, other glycosyltransferase such as GnT-II, GnT-IV, GnT-V, α-1-6Fucosyltransferase and β 1-4 Galactosyltransferase are inactive (Schachter, 1986; Gu *et al.*, 1993). Therefore, GnT-III is a key enzyme in the biosynthesis of N-glycans. GnT-III is abundant in normal brain and kidney whereas no message is found in the adult rat liver.

2.2. Molecular Model for Substrate Specificity

Molecular modeling was carried out on INDIGO work stations (Silicon Graphics) using the insight II/discover (Biosym. Technologies) software. Each sugar unit was assumed to adopt the standard conformation and the torsion angles between two sugar units were obtained from nuclear magnetic resonance data (Fujii *et al.*, 1990). Based on nuclear magnetic resonance data, the biantennary structure of a core mannose is twisted in the presence of bisecting GlcNAc. Therefore, for example, GnT-III acts on the biantennary structure to form a bisecting GlcNAc, and as a result, other enzymes such as GnT-1 dependent α3-6 mannosidase, GnT-II, IV, V and α1-6 Fucosyltransferase which share the common substrate are not able to act on the same substrate.

2.3. Bisecting GlcNAc May Act as a Sorting Signal by Binding to an Endogenous Lectin Molecule?

In order to determine whether or not bisecting GlcNAc residues are involved in the sorting of N-glycans, a hepatoma cell line, mRLN32 was chosen and used in conjunction with forskolin, a unique reagent which is widely used as an adenylyl cyclase activator to induce different types of proteins in many cell lines via the accumulation of cAMP. Forskolin was found to induce GnT-III in normal hepatocytes as well as in hepatoma cells and resulted in an increase in bisecting GlcNAc in various glycoproteins in the tissues. Furthermore, to determine the correlation between the oligosaccharide structure of a specific glycoprotein and its distribution, especially after the addition of the bisecting GlcNAc residue, we examined a number of glycoproteins that have different roles and expression sites, such as Lamp-1 and β-glucuronidase (lysosomal proteins), γ-glutamyltranspeptidase (plasma membrane protein), ceruloplasmin and α-fetoprotein (secretary proteins). Lamp-1 is expressed on the surface of many tumor cells, although the majority of the molecule inside the lysosome. Our finding showed that forskolin resulted in a significant enhancement in GnT-III at the transcriptional level and an increase in bisecting GlcNAc residues in various glycoproteins as judged by lectin binding to erythroagglutinin phytoahemagglutinin (E-PHA). However, even though such an incerase is observed, secretory glycoproeins such as ceruloplasmin and α-fetoprotein are secreted normally whereas the extent of sorting of γ-glutamyltranspeptidase, Lamp-1 and β-glucuronidase which are localized on the cell surface are decreased (Sultan *et al.*, 1997). There are several possible explanations for these events. One possibility is that the specific structure of the bisecting GlcNAc provides a negative sorting signal for certain proteins. The other possibility is that the existence of some endogenous binding proteins, such as lectin-like molecule, which specifically recognize the bisecting structure of the glycoproteins and which may interrupt the sorting of the glycoproteins to the cell surface or to lysosomal membranes. Quite recently Gao in this laboratory purified a protein which binds to bisecting GlcNAc from bovine spleen using a specific ligand column and found that this protein has the characteristics of a lectin-like protein and found that the protein is annexin V. Probably annexin V may recognize the bisecting GlcNAc residues of glycoproteins and thus regulate the sorting mechanism of glycoproteins.

2.4. Bisecting GlcNAc Inhibits the Elongation of the β1-6 GlcNAc Structure and Suppresses the Lung Metastasis in Mice

The predominant surface glycopeptide of baby hamster kidney cells transformed by Rous sarcoma virus has been shown to be a triantennary, completely sialylated, complex glycopeptide containing a core region of Man, GlcNAc, and Fuc. Synthesis of these elongated branches is initiated by β-1,6N-acetylglucosaminyltransferase (GnT-V) which catalyzes the formation of the β1-6 branch (Van den Eijenden *et al.*, 1988). GnT-V activity is correlated with the metastatic potential of ras-transformed Rat2 fibroblasts, SPA mammary cancer cells and MDAY-D2 lymphoma cell line. (Yousefi *et al.*, 1991; Dennis *et al.*, 1987), as well as human colon cancer cells (Saitoh *et al.*, 1992). Rat2 fibroblasts transfected with H-ras or v-fps exhibited metastatic potential and had elevated GnT-V activity and increased β1-6 branching, whereas a mutant with decreased GnT-V activity from a highly metastatic tumor cell line had a decreased potential for metastatis in mice (Dennis *et al.*, 1989). Very recently Dennis's group found that knock out mice of GnT-V did not develop the cancer (Granovsky *et al.*, 2000). These observations suggest that a positive correlation exists between β1-6 branching and metastic capacity. We have focused on the substrate specificity of both GnT-V and GnT-III because these two enzymes compete for the same biantennary structure of N-linked oligosaccharides as their substrate, and once a bisecting GlcNAc residue is added to the core mannose by GnT-III, GnT-V is no longer able to form any further triantennary structure (Schachter *et al.*, 1986; Gu *et al.*, 1993). We established a highly metastatic subclone, B16-hm, from low metastatic B16-F1 murine melanoma cells. The gene which encodes GnT-III was introduced into the B16-hm cells, and three clones that stably expressed high levels of GnT-III activity were obtained (Yoshimura *et al.*, 1995).

Lectin blotting was performed to analyze the alterations in carbohydrate structures on the cell surface of parental and transfected cells. The two lectins used were L-PHA and E-PHA. L-PHA binds preferentially to GlcNAc residues on β1-6 branches of tri- or tetraantennary sugar chains, and the binding is interrupted in the presence of a bisected biantennary glycopeptide. E-PHA has high affinity for bisected oligosacchrides but does not interact strongly with β1-6 structures. Whole cell lysates from B16-hm cells and negative controls were highly reactive to L-PHA. Both a 95 kDa protein and a 80kDa protein showed a particularly strong L-PHA binding. The proteins from positive transfectants showed reduced reactivity to L-PHA. In contrast, staining with E-PHA yielded only a few faint signals for proteins from B16-hm cells and negative transfectants, whereas strong signals were observed for proteins with the 95 kDa and 80 kDa proteins from the positive transfectants. These data showed that the expression of GnT-III led to increased synthesis of bisecting GlcNAc, which, in turn, suppressed the formation of β1-6 tri- and tetraantennary N-linked oligosaccharides. In addition, structural analysis of the transfectants indicated that the level of bisecting GlcNAc was markedly increased and that tri- and tetraantennary structures, including β1-6 structure, were decreased as the result of competition for substrate between intrinsic GnT-V and ectopically expressed GnT-III (Taniguchi *et al.*, 1996).

3. USE OF GnT-III TRANSFECTANTS

3.1. Lung Colonization in Experimental Metastasis is Decreased in GnT-III Transfectants

When positive transfectants were injected into syngeneic C57BL/mice, significantly fewer metastatic nodules were observed, compared to the parent cell and negative transfectants. The number of metastatic colonies observed in case of the negative transfectant was similar to that in the parental cells. These results demonstrate that GnT-III expression of the transfectants decreased the metastatic potential of B16-hm melanoma *in vivo* (Yoshimura *et al.*, 1995). To determine whether this was due to increased susceptibility of the GnT-III transfected cells to capture by the immune system, the B16-hm cells and GnT-III tranfectants were also intravenously injected into athymic BALB/c nude mice. The positive transfectants produced fewer metastatic nodules than the parent cells and produced negative transfectants to the same extent as was observed in syngeneic C57BL/6 mice, suggesting that the immune system had no effect on the metastatic process. Subcutaneous injection of B16-hm cells and GnT-III transfectants into C57BL/6 mice yielded fewer than 20 lung colonies even after 98 weeks, too few to evaluate the metastatic potential in the "spontaneous metastasis assay" which is in agreement with previous findings.

3.2. *In Vitro* Invasiveness and Cell Attachment are Decreased in GnT-III Transfectants Without Affecting Cell Growth.

In this type of experimental metastasis, the number of metastasized lung colonies is dependent on tumor invasiveness, adhesion to the endothelial cells, attachment to the extracellular matrix, and cell growth. Invasive potential was tested in vitro using a Boyden chamber coated with Matrigel, an extract prepared from mouse Engelbreth-Holm-Swarm sarcoma. The ability of GnT-III positive transfectants to reach the bottom of the well through the Matrigel was suppressed significantly compared to that of the B16-hm cells, and the negative transfectants for the first 30 and 60 min. At the end of 4-h of incubation, these cells attached equally to collagen and laminin. Cell attachment to fibronectin did not differ significantly among the B16-hm cells and the GnT-III transfectants. No difference in cell morphology was observed among parental B16hm cells and GnT-III transfectants when these cells were plated and spread on collagen, laminin, or fibronectin. The doubling times, as judged by cell growth kinetics, were nearly the same, indicating that ectopic expression of the GnT-III gene did not affect the proliferation of the cells. It is difficult to evaluate the adhesion of the GnT-III positive transfectants to the murine lung capillary endothelium in vitro, beacuse the capillary endothelial cells isolated from the murine lung are not available at present. The possibility, therefore, remains that the expression of GnT-III may also affect the adhesion to the endothelial cells, which is mediated, in part, by sialyl-Lewis X structures and selectin receptors.

3.3. Mechanism by which Metastatic Potential is Decreased by Transfection of GnT-III Gene to Melanoma B16 Hm Cells: Accumulation of E-Cadherin at the Cell-Cell Border in GnT-III Transfectants

Two clones with selected GnT-III activity, designated as B16hm-III-1 and -2 and one clone with no detectable GnT-III activity, designated as B16-hm-neo-1 were used as positive transfectants and a control transfectant, respectively. Morphologically, the B16-hm cells and the control transfectant appeared fibroid with loose cell-cell contacts, whereas the positive transfectants were epithelioid and proliferated in a compact organization (Yoshimura *et al.*, 1996). In order to examine whether the reduced metastatic potential of B16-hm melanoma cells, expressing ectopic GnT-III was due to the altered biological function of E-cadherin that mediates homotypic cell-cell adhesion, since the E-cadherin expression correlates inversely to metastatic phenotype in many cancer cells (Nagafuchi *et al.*, 1987). E-cadherin expression was examined by indirect immunofluorescence. In B16-hm cells and the control transfectant, E-cadherin was weakly expressed at the cell-cell contacts. Positive transfectants, however, showed intense fluorescence with condensation at the cell-cell contacts, indicating an elevated expression of E-cadherin at the cell-cell contacts of positive transfectants. This indicates that E-cadherin is accumulated at the cell-cell boarder in GnT-III transfectants.

3.4. Aberrant Glycosylation of E-Cadherin in GnT-III Tranfectants

E-PHA lectin was used to analyze the alterations of carbohydrate structures of E-cadherin. Western blotting studies showed that the expression of E-cadherin was increased more in positive transfectants than in B16-hm cells and control transfectants, which is consistent with the results of immunofluorescence microscopy. The signal for E-PHA binding to the positive cells was detected as bands at around 125 kDa, corresponding to immunoprecipitated E-cadherin, while these bands were barely detectable in the B16-hm cells and the control transfectant. The relative E-PHA binding ratio (%) normalized by E-cadherin signal density was markedly higher in the transfectants as compared to mock transfectants. This indicates that the E-cadherin in the positive transfectants was glycosylated by GnT-III and few, if any, bisected oligosaccharides were attached to E-cadherin in the parental cells and the control transfectant. Compared with the B16-hm cells and the control transfectant, E-cadherin transcripts were not increased in positive transfectants, suggesting that the increased E-cadherin expression in the GnT-III transfectants was not increased at the transcriptional level.

3.5. Turnover of the E-Cadherin is Prolonged because of a Decreased Release from the GnT-III Transfectants

The turnover rate of E-cadherin was examined by chase studies. In B16-hm cells, the immunoprecipitated surface E-cadherin showed the most intense signal at 4 h and then declined to an undetectable level at 12 h. The immunoprecipitated surface E-cadherin from the B16-hm-III-1 cells showed the most intensified signal at 8h, and the signal corresponding to the immunoprecipitated E-cadherin was maintained at the detectable level during the entire chase time periods. Released E-cadherin was detected as a band 104 kDa in the supernatants from both B1-hm cells and positive transfectants, and was less than that of the surface E-cadherin. In the B16-hm cell, supernatants, the released E-cadherin was first detected at 2h, gradually increased to a maximum at 8 h, and thereafter declined to a detectable level during the remainder of the incubation. In B16-hm-III-1 cells, the released E-cadherin was undetectable for the initial 4 h of the incubation. The signals of released E-cadherin at 4 and 8 h were faint and much weaker than that detected in the supernatants of the B16-hm cells and declined to an undetectable level at 12 and 24 hrs. Turnover and release of the E-cadherin of a control transfectant were similar to that of B16-hm cells, showing that the transfection procedures had no effect on the turnover of E-cadherin. Collectively, these results indicate that the expression of GnT-III prolonged the turnover of E-cadherin and inhibited the release of E-cadherin from cell surface, thus resulting in an increased level and accumulation of E-cadherin molecules at the cell-cell contacts.

3.6. Increased Cell Aggregation in GnT-III Transfectants

To determine whether the increased E-cadherin expression was involved in the homotypic cell adhesion of B16-hm cells, cell aggregation was assayed using an antibody to E-cadherin which specifically blocks E-cadherin-mediated adhesion. In B16-hm cells and a control transfectant, more than 90% of cell aggregation was inhibited by ECCD-2, a blocking antibody to E-cadherin. The inhibitory effect by ECCD-2 was, however, considerably lower in the positive transfectants, showing that increased E-cadherin expression was associated with increased cell aggregation and that the raised E-cadherin was biologically functional.

4. INVOLVEMENT OF GnT-III IN PATHOLOGY

4.1. Suppression of Hepatitis B Virus Related Antigens in GnT-III Transfected Human Hepatoma Cell Line, HB611

A HB611 cell line was established by transfecting three copies of the complete hepatitis B virus (HBV) genome arranged in tandem into a human hepatoblastoma cell line Huh6. The cells produce a large amount of hepatitis B surface antigen (HBsAG) and hepatitis B envelope antigen (HBeAg) and HB virion into the medium.

GnT-III gene was transfected into the cells and the positive transfectants were cloned by hygromycin resistant selection. Three clones have high activities of GnT-III and secreted lower levels of HBV-related proteins into the medium in comparison with other clones. these clones showed a significant suppression of Hepatitis B virus (HBV) -related mRNAs, i.e., hepatitis B e antigen and hepatitis B surface antigen and an increased binding with E-PHA as judged by lectin blot. Expression of β-actin, α-fetoprotein, albumin and prealbumin as not affected. Treatment of these cells with tunicamycin or swainsonin, oligosaccharide processing inhibitors resulted in an enhanced expression HBV-related mRNA (Miyoshi *et al.*, 1995). These data indicate that some glycoproteins whose sugar structures are changed by over expression of GnT-III suppress HBV gene expression. The mechanism by which HBV gene expression is suppressed by the sugar structure is not yet clear: a bisecting GlcNAc structure produced by ectopic expression of GnT-III may lead to changes of oligosaccharides in glycoproteins such as receptor and adhesion molecules. the second possibility is the some unidentified lectins in HB611 cells, which may directly recognize a bisecting GlcNAc may control the intracellular transport of certain proteins.

4.2. Bisecting GlcNAc on K562 Cells Suppresses Natural Killer Cytotoxicity and Promotes Spleen Colonization

Natural killer (NK) cells which comprise 10–15% of the lymphocytes in human peripheral blood, are morphologically large granular lymphocytes with CD3–, CD16+, CD56+ and are able to lyze target cells without prior sensitization or MHC restriction. Numerous molecules have been reported to be involved in this spontaneous non-MHC-restricted cytotoxicity, including the interaction of function-associated antigen or CD2 on the NK cell surface with other target cell ligands in the intercellular adhesion molecule 1 (CD 54), or leukocyte function associated antigen-3 (CD 58), respectively. These adhesion molecules are necessary for conjugate formation between NK cells and target cells in order for NK cells to recognize their targets. However, nature of specific NK receptors, or receptor ligands on the target cells remains obscure, since NK cells do not recognize and kill all tumor target cells. We and other investigators have reported that the potential target structures for the interaction of NK cells and target cells are not only proteins but also carbohydrate determinants including N-glycans. When target cells react with effector cells, NK activity is usually evaluated based on measurements of 51_{Cr} release. Because of the sensitivity to NK cytotoxicity and undetectable expression of MHC molecules, the K562 cell line has been commonly used as target cells for standard NK activity assays. We used an alternate approach including gene introduction to examine the effect of bisecting GlcNAc on NK cytotoxicity using the transfection of GnT-III. Our findings showed that GnT-III positive transfectants were more resistant to NK cytolysis, less tightly bound by NK cells, thus resulting in the ability to colonize the spleen in nude mice after subcutanous inoculation (Yohimura *et al.*, 1996).

4.3. Altered N-Glycans of Surface Glycoproteins on GnT-III Positive Tranfectants Decrease Susceptibility of NK Cytotoxicity and Decreased Effector Binding to K562 Target Expressing GnT-III

The nature of the alteration of glycoproteins on the cell surface was analyzed by flow cytometry using FITC-conjugated lectins. E-PHA binds preferentially to bisecting GlcNAc in N-glycans and the binding of E-PHA to the cell surface was increased in the positive transfectants. Because K562 cells have been used as the NK target, we analyzed the influence of the bisecting GlcNAc on the susceptibility to NK cytotoxicity using standard 51_{Cr} releasing assays. Six target lines including the positive transfectants of GnT-III, were exposed to the effector from the same individuals simultaneously, and % cytolysis at the various E:T ratios was the mean of five separate assay values with effector cells from five individuals. The NK effector cells were able to cytolyze control transfectants as well as original K562 cells, and the cytotoxicity correlated with the E:T ratio. Lysis by the effector cells was, however, completely blocked when the positive transfectants were exposed to the NK effector cells. The cytotoxicity to positive transfectants tended to be elevated at a higher E:T ratio. The binding of human effector cells to the original K562 cells and the GnT-III transfectants was also assayed, since NK cytolysis is initiated by the interaction of the effector cells with the targets. The binding to positive transfectants was significantly decreased, as compared to original K562 cells and control transfectants, suggesting that lowered binding of effector cells to positive transfectants was responsible for the decreased NK cytolysis of the positive transfectants.

4.4. Tumor Formation by GnT-III Positive K562 Cells in the Spleen of the Nude Mice

In athymic nude mice, T lymphocyte-mediated cytotoxicity is severely impaired because of a deficiency in T lymphocytes, whereas NK cells function normally. To evaluate the susceptibility to NK cytotoxicity *in vivo*, original K562 cells and GnT-III transfectants were subcutanously inoculated into nude mice. At 6 weeks after inoculation, the mice were sacrificed, and the sites and the number of tumor formations were examined in order to evaluate tumorgenicity and metastatic potential. No tumorigenicity was observed in lymphnodes, bone marrow, liver, and subcutaneous heel pad injections after the injection of the original K562 cells and positive and control transfectants. As listed in Table 2, all nude mice which were injected with positive transfectants developed tumor colonies on the surface of their enlarged spleens. In addition, the original K562 cells and the control transfectants displayed no splenic lesions. Microscopic analysis revealed that in all nude mice that had been injected with positive transfectants, the spleen contained numerous large neoplastic cells and that the spleen architecture, such as cortex and follicles in the medulla was destroyed. No neoplastic cells were detected in random sections of the spleens taken from nude mice which had been injected with original K562 cells or control transfectants. Neoplastic cells were not observed microscopically in the lymph nodes, liver, bone marrow and the subcutanous injection site at the heel pad in nude mice which had been injected with original K562 cells nor in random sections taken from positive and control transfectants. Spleen colonizations were also assayed by subcutanous inoculation of cells into nude mice that were depleted of NK cells by i.p administration of an antibody to asialo GM1. At 3 weeks after the inoculation of the cells, the mice were sacrificed, and the splenic lesions were evaluated. The number of splenic colonies in the original K562 cells, control transfectants and positive transfectants were the same as in the NK-depleted nude mice. *In vitro* growth rate determined by the MTT assay was also identical. These data indicate that the specific colonization of the spleen by positive transfectants was not due to cell proliferation.

5. GnT-III AND CELL SURFACE MOLECULES

5.1. CD44 Enhances Cell Adhesion to Hyaluronate, Tumor Growth and Metastasis in B16 Melanoma Cells Expressing GnT-III

CD44 is a cell-surface glycoprotein expressed on many different cell types and functions as an adhesion molecule for hyaluronate (Arruffo *et al.*, 1990). As deduced from cDNA sequences, the CD44 molecule contains several putative N-glycosylation sites and a few sites for chondroitin-sulfate attachment. The biological role of CD44 molecules in tumor metastasis and lymphocyte function has been analyzed. Positive correlation of CD44 with metastasis of several tumor cell lines has been reported in melanoma cells and lymphoma cells. In order to evaluate the effect of bisecting GlcNAc on the biological function of CD44 as a receptor for HA, we examined the adhesion of positive transfectants to immobilized HA (Shen *et al.*, 1997). The time course for adhesion to HA was significantly short, and increased positively, as compared with both parental cells and control. At 4 hr after incubation, approximately 57% of the positive transfectants had attached to HA, while only 23% of the native B16 hm cells were attached. An inhibitory antibody to CD44, KM 201 was found to block the adhesion of native B16 hm cells to HA and positive transfectants to an equal extent. In native B16 hm cells and in positive transfectants, adhesion to HA was completely blocked after treatment of immobilized HA with hyaluronidase. The results of the adhesion assay and the blocking study involving an inhibitory antibody suggest that CD44-mediated adhesion and that the attachment is enhanced in positive transfectants.

5.2. Binding of Hyarulonate (HA) on the Cell Surface was Increased in GnT-III Transfectants

As judged by flow cytometry using FITC-labeled HA, HA bound more strongly to positive transfectants than to native cells. These results suggest that the CD44-mediated attachment of HA was enhanced in positive transfectants. The surface CD 44 level, however, was not significantly different between parental cells and positive transfectants. Since the expression levels of sialyl-Lewis X and sialyl Lewis A have been reported to be positively correlated to metastasis and tumor progression, we also examined the surface amounts of sialyl Lewis X and sialyl Lewis A by flow cytometry. Compared with the controls, surface sialyl Lewis X expression was decreased in the positive transfectants. This indicates that an increase in bisecting GlcNAc residues leads to a decrease in the amount of sialyl Lewis X. The elongation of N-glycans is suppressed in the presence of bisecting GlcNAc.

5.3 Bisecting GlcNAc Residues are Responsible for Increase in Cell Adhesion to HA in Positive GnT-III Tranfectants.

Cell adhesion was evaluated after treatment of the cells with various glycosidases to determine the sugar residues required to regulate cell adhesion to HA in B16-hm cells and positive transfectants. In native B16-hm cells, treatment with β-galactosidase resulted in a significant decrease in cell adhesion, as compared with non treatment and the extent of this decrease was found to be of the same extent as was found for treatment with β-N-acetylhexosaminidase of N-glycosidase. In native B16 hm cells, galactose residues on N-glycans are critical for CD 44 mediated adhesion to HA. Treatment with neuraminidase resulted in increases in cell adhesion in native cells, showing that sialic acid residues inhibit the CD44-mediated adhesion to HA, as has been reported. The results found in control transfectants were similar to the result in native cells. In addition, positve transfectants showed a significant decrease in cell adhesion after the treatment with β-N-acetlyhexosaminidase. This indicates that the GlcNAc residues contribute to the increase in cell adhesion to HA in the GnT-III transfectants. The treatment of positive transfectants with an O-glycosidase resulted in a decrease in cell adhesion to the same extent as treatment of parental cells and control transfectants. Cell adhesion by native cells, control cells and positive transfectants was completely blocked on treatment with a combination of O-glycosidase and N-glycosidase, indicating that both N- and O-glycans participate in CD44 mediated adhesion to HA in these cell types. These data show that the GlcNAc residues were responsible for the increase in CD44-mediated adhesion to HA in positive transfectants.

5.4. Increased Local Tumor Growth and Metastatic Potential in Positive GnT-III Transfectants

Since CD44 has been reported to be involved in tumor metastasis (Sherman *et al.*, 1994), the enhanced cell attachment of HA in positive transfectants may also affect metastatic potential. We therefore examined local tumor growth and metastatic tumor developemnt after s.c. incoculation of native cells and positive transfectans into the backs of sygeneic mice. The tumor weights of native cells, control transfectants, and positive transfectants are determined. The local tumor growth of postive transfectants was promoted to a greater extent more than that of native cells and contral cells. In addition the repeated administration of KM 201, a blocking antibody to CD 44, suppressed local tumor growth of positvie transfectants to the same level as that of B16hm cells and control transfectants. These findings indicate that CD44-dependent local tumor growth was enhanced in positive transfectants.

5.5. Bisecting GlcNAc Binding Protein

Gao (Gao *et al.*, 2000) in our group quite recently identified the bisecting GlcNAc binding protein from porcine spleen microsome. The protein was extracted from porcine spleen microsome with Triton X-100 and then purified using a DEAE cellulose column and affinity chromatograph on a bisected biantennary sugar chain as a ligand. The purified protein binds to the surface of the GnT-III cDNA transfected K562 cells, which expressed a significant amount of bisecting GlcNAc, in a dose dependent manner. The protein also inhibited the binding of E-PHA to the transfected cells. Microsequencing analysis of the peptide fragments, obtained by digestion of the purified protein with Lysyl endopeptidases and staphylococcus aureus V8 protease revealed that the protein was Annexin V. This protein may function as a lectin which is capable of recognizing the bisecting GlcNAc containing oligosaccharides. Tanemura *et al.* (1997) reported that the GnT-III transfected porcine endothelial cells resulted in the down regulation of xenotransplantion antigen designated as α1-3 Gal epitope responsible for the superacute rejection from porcine to human xenotransplantation. The glycoproteins less than 80Kda may be a target binding protein for bisecting GlcNAc.

5.6. Ectopic Expression of GnT-III in Transgenic Hepatocyts Disrupts Apolipoprotein B Secretion and Induces Fatty Liver

Over the past few years, genes encoding glycosyltransferases (glycogenes) which are involved in the biosynthesis of complex carbohydrates have been cloned and characterized. However, our knowledge of the functional properties of these genes are quite limited except for the fact that the function is directly related to the onset of the disease. One strategy for analyzing the functional properties of the gene is to knock out the gene or overexpress it *in vitro* or *in vivo*. In these strategies, however, which are different from the genes encoding proteins, the resulting phenotypes produced as a result of knocking out or overexpression of the glycosyltranferase genes are sometimes due to a secondary effect of the modification of sugar chains. Therefore when the gene(s) are knocked out or overexpressed, it must always be borne in mind that the phenotypic changes are directly due to the genes of interest. In such cases, the identification of the target glycoproteins due to the gene knocked out or overexpressed is a prerequisite. We have taken the strategy for overexpression of the GnT-III gene in the mice or in several other types of cells to better understand the role of bisecting GlcNAc residue in glycoproteins. The bisecting GlcNAc is a unique sugar residue because, surprisingly, many different types of phenotypic changes result from it *in vivo* and *in vitro*. Quite recently we established a strain of transgenic mice that specifically express GnT-III in hepatocytes, since normal mice hepatocytes contain no GnT-III and it would be interesting to better understand the biological significance of the bisecting GlcNAc in hepatocytes, provided one can ectopically overexpress the gene (Ihara *et al.*, 1998). We found that the transgenic hepatocytes had a swollen oval-like morphology and contained numerous lipid droplets. In order to confirm the increase of bisecting GlcNAc in transgenic mice serum, N-glycans were extracted from sera through N-glycanase digestion and labeled with 2-aminopyridine. After glycosidase treatment the labeled glycans were separated by reverse-phase HPLC under the conditions for the separation of complex N-glycans. The peak area for biantennary and biantennary sugars with bisecting GlcNAc were calculated and the ratio of

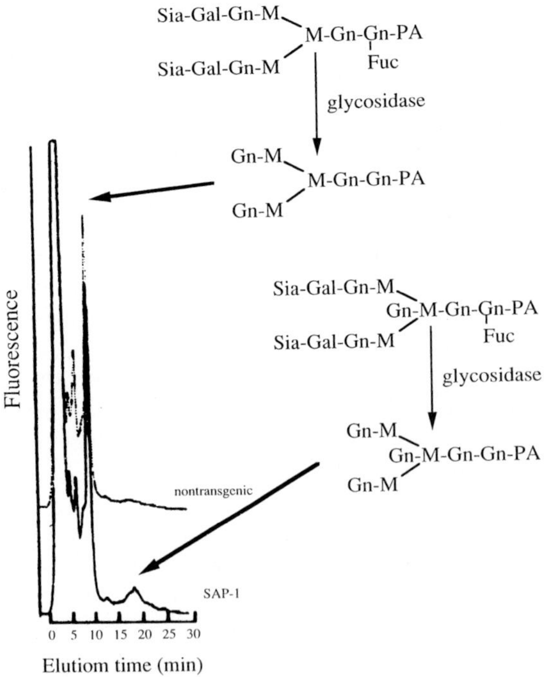

Figure 3: Detection of complex N-glycans bearing bisecting GlcNAc among the serum glycoproteins in GnT-III transgenic hepatocytes by HPLC.

bisecting GlcNAc sugars were determined. Serum glycoproteins with complex N-glycans bearing bisecting GlcNAc was significantly increased in the transgenic mice serum as shown in Figure 3. A possible target protein was determined to be apolipoprotein B100 because the bisecting GlcNAc has been accumulated and in the transgenic mice, serum triglyceride, β- and pre-β-lipoprotein formation as well as apolipoprotein B100 were significantly decreased compared with levels in nontransgenic mice. Apo-B100 is a glycoprotein and plays an important role in the formation of lipoprotein complexes. In the liver, apo-B100 is required for the assembly and secretion of very low density lipoproteins and low density lipoproteins (Yao and McLeod, 1994). Aberrant glycosylation of apo-B100 may cause a decrease in the release of lipoprotein and, hence, an accumulation of apo-B in the liver. Some fatty livers develop hepatoma and at present it remains unclear whether or not a fatty liver may develop a primary hepatoma in the case of trangenic mice.

5.7. Inhibiton of Growth Factor Signaling by GnT-III Gene Transfection

N-glycans appear to be important for the function of the epidermal growth factor (EGF) receptor (Soderquist and Carpenter, 1984) Rebaa *et al.* found that bisecting GlcNAc play a potential role in the EGF receptor function by transfection of GnT-III gene in to a human glioma cell line, U373 MG. Stable GnT-III transfectants were generated in a glioma cell line, U373 MG, to determine the consequences of overexpression of the bisecting GlcNAC on EGF receptor binding and function because E-PHA had such a profound effect on binding and function. In the transfected cells, a significant decrease in EGF binding and EGF receptor autophosphorylation were observed as compared to the control cells. In contrast, proliferation of the GnT-III transfected cells was stimulated by EGF. These data show that changes in EGF receptor glycosylation by GnT-III reduces the number of the active receptors in U373 MG cells and that this change lead to change in the cellular response to EGF (Rebaa *et al.*, 1997).

Ihara *et al.* (1997) also reported that GnT-III transfected rat phenochromocytoma PC 12 did not repond to nerve growth factor (NGF) (Ihara *et al.*, 1997). The PC12 cells normally shows neurite outgrowth for diffrentiation after treatment with nerve growth factor because NGF binds to its receptor, Trk, to form a complex and mediates the signal by autophosphorylation at tyrosine phosphorylation. These data indiate that overexpression of GnT-III regulates the intracellular signaling pathway of tyrosine phosphorylation modified by NGF or EGF.

6. PERSPECTIVES

This review focused on the role of bisecting GlcNAc in cancer. The bisecting GlcNAc appears to be a double edged sword in terms of metastatic potentials. In fact overexpression of GnT-III, namely an increase in bisecting GlcNAc residues in glycoproteins on the cell surface or on specific proteins may play a pivotal role in cancer metastasis. This is especially so for the case of melanoma metastasis to lung, where the bisecting GlcNAc functions in a beneficial role to the host because the addition of bisecting GlcNAc to E-cadherin alters its sensitivity to proteolysis and is retained on the cell-cell border to enhance cell-cell binding to suppress metastasis. However, in the case of melanoma metastasis to spleen, the overexpression of bisecting GlcNAc may bind to a lectin-like molecule in the spleen and

Table 1. Possible functions of bisecting GlcNAc residues and target proteins

Biological functions glycoproteins	Cells	Target
Suppression of metastasis	melanoma B16-hm	E-cadherin
Enhancement of metastasis	melanoma B16-hm	CD44
Resistance to NK cells	K562 cells	NK receptors?
Lipoprotein sorting	hepatocytes	apolipoprotein B 100
Sorting	hepatoma M31	cell surface proteins lysosomeal proteins
Signaling	glioma U373MG PC 12	EGF receptor NGF receptor
Regulation of gene expression	HB611	HBV related nuclear factor?
Down regulation of xenotransplantation antigen	porcine endotherial cells	glycoproteins

also modify the CD 44 protein, thus enhancing its adhesion to hyalulonate. This may result in an enhancement of metastatic potential in the spleen. In K562 cells, the bisecting GlcNAc on K562 cells suppresses natural killer cytotoxicity and spleen colonization. The discrepancy of metastatic potential between lung metastasis and spleen metastasis by melanoma cells or K562 cells can be explained by the fact that the spleen cells contain a lectin-like protein which specifically binds to bisecting GlcNAc. Therefore the overexpression of GnT-III, namely overexpression of bisecting GlcNAc in some glycoproteins may facilitate the binding of tumor cells to the spleen cell surface but not enhance metastasis in the other cells. Therefore the bisecting GlcNAc binding protein may regulate metastatic potential only in the spleen. We have quite recently identified the bisecting GlcNAc binding protein from bovine spleen and this may explain the discrepancy found between the lung metastasis and spleen metastasis.

Recent studies on the effect of GnT-III gene transfection on growth factor signaling indicated that growth factor receptors such as EGF and nerve growth factor receptor undergo aberrant glycosylation which resulted in the impairment of signaling pathway (Tsuda *et al.*, 2000). The role of bisecting GlcNAc on the function of growth factor signaling remains still unclear but this type of approach may open new insight into the role of glycosyltransferase gene in cancer cells.

In conclusion, bisecting GlcNAc residues in glycoproteins may play a crucial role in regulating various biological events as shown in Table 1.

REFERENCES

Arruffo, A., Stamenkovic, I., Melnick, N., Underhill, C.B., and Seed, B. (1990) CD44 is the principal cell surface receptor for hyaluronate. *Cell*, **61**, 1303–1313

Baenziger, J. and Kornfeld, S. (1974) Structure of the carbohydrate units of IgA 1 Immunoglobulin II. Structure of the O-glycosidically linked oligosaccharide units. *J. Biol. Chem.* **249**, 7260–7269

Carver, J.P., Grey, A.A., Winnik, F.M., Hakimi, J., Ceccarini, C., and Atkinson, P.H. (1981) Determination of the structure of four glycopeptides from hen ovalbumin using 360-MHz proton magnetic resonance spectroscopy. *Biochemistry* **20**, 6600–6606

Dennis, J.W., Laferte, S., Waghorne, C., Breitman, M.L. and Kerbel, R.S. (1987) β1-6 branching of Asn-linked oligosaccharides is directly associated with metastasis. *Science* **236**, 582–585

Dennis, J.W., Kosh, K., Bryce, D.M., and Breitmann, M.L. (1989) Oncogenes conferring metastatic potential induce increased branching of Asn-linked oligosaccharides in rat2 fibroblasts. *Oncogene* **4**, 853–860

Fiala, S., Mohindru, A., Kettering, W.G., Fiala, A.E., and Morris, H.P. (1976) Glutathione and γ-glutamyltranspeptidase in rat liver during chemical carcinogenesis. *J. Natl. Cancer. Inst.* **57**, 591–598

Fujii, S., Nishiura, T., Nishikawa, A., Miura, R., and Taniguchi, N. (1990) Structure heterogeneity of sugar chains in immunoglobulin G. Conformation of immunoglobulin G molecule and substrate specificities of glycosyltransferases. *J. Biol. Chem.* **265**, 6009–6018

Gao, C.X., Vozumi, N., Miyoshi, E., Nagai, K., Ikeda, Y., Teshima, T., Noda, K., Shiba, T., Honke, K., and Taniguchi, N. (2000) A novel carbohydrate binding activity of annexin V toward a bisecting N-acetylglucosamine. Glycobiology, in press

Gu J., Nishikawa, A., Tsuruoka, N., Ohno, M., Yamaguchi, N., Kangawa, K., and Taniguchi, N. (1993) Purification and characterization of UDP-N-acetylglucosamine: α-6-mannose β 1-6 N-acetylglucosaminyltransferase (N-acetylglucosaminyltransferase V) from a human lung cancer. *J. Biochem* **113**, 614–619

Gleeson and Schachter (1983) Control of glycoprotein synthesis. *J. Biol. Chem.* **258**, 6162–6173

Granovsky, M., Fata, J., Dawling, J., Muller, W.J., Khokha, R., and Dennis, J.W. (2000) Suppression of tumor growth and metastasis in Mgat5-deficient mice. *Nat Med.*, **6**, 306–12

Ihara Y., Sakamoto Y, Mihara, M., Shimizu, K. and Taniguchi, N. Overexpression of N-acetylglucosaminlytransferase III disrupts the tyrosine phosphorylation of Trk with resultant signaling dysfunction in PC 12 cells treated with nerve growth factor. *J. Biol. Chem.* **272**, 9629–9634, 1997

Ihara, Y., Yoshimura, M., Miyoshi, E., Nishikawa, A., Sultan S.A., Toyosawa s., Ohnishi,A., Suzuki, M., Yamamura K., Ijuhin N. and Taniguchi, N. (1998) Ectopic expression of N-acetylglucosaminlytranferase III in transgenic hepatocytes disrupts apolipoprotein B secretion and induces aberrant cellular morphology with lipid storage, *Proc. Natl. Acad. Sci. USA* **95**, 2526-2530

Irimura, T., Tsuji, T., Tagami, S., Yamamoto, K., and Osawa, T. (1981) Structure of a complex-type sugar chain of human glycophorin A. *Biochemistry* **20**, 560–566

Ishibashi, K., Nishikawa, A., Hayashi, N., Kasahara, A., Sato, N., Fujii, S., Kamada, T., and Taniguchi, N. (1989) N-Acetylglucosaminyl-transferase III in human serum, liver and hepatoma tissues: incresed activity in liver cirrhosis and hepatoma patients. *Clin. Chim. Acta.* **185**, 325–332

Koenderman, A.H.L., Koppen, P.L., Koeleman, C.A.M., and van den Eijnden, D.H. (1989) N-acetylglucosaminyltransferase III, IV and V activities in Novikoff ascites tumor cells, mouse lymphoma cells and hen oviduct. *Eur. J. Biochem.* **181**, 651–655

Kornfeld, R., and Kornfeld, S. (1976) Comparative aspects of glycoprotein structure. Annu. Rev. *Biochem.* **45**, 217–237

Miyoshi, E., Nishikawa, A., Ihara, Y., Gu, J., Sugiyama, T., Hayashi, H., Fusamoto, H., Kamamoto, T., and Taniguchi, N. (1993) N-Acetyl-glucosaminyltransferase III and V messenger RNA levels in LEC rats during hepatocarcinogenesis. *Cancer Res.* **53**, 3899–3902

Miyoshi, E., Ihara, Y., Hayashi, N., Fusamoto, H., Kamada, T., nad Taniguchi, N.(1995) tranfection of N-acetlyglucosaminyltranferase III gene suppresses expression of hepatitis B virus in a human hepatoma cell line, HB 611. *J. Biol. Chem.*, **270**, 28311–28315

Nagafuchi, A., Shirayoshi, Y., Okazaki, K., Yasuda, K., and Takeichi, M. (1987) Transformation of cell adhesion properties by exogenously introduced E-cadherin cDNA. *Nature*, **329**, 341–343

Narashima, S. (1982) Control of glycoprotein synthesis. UDP-GlcNAc: glycopeptide β4-N-acetylglucosaminyl-transferase III, an enzyme in hen oviduct which adds GlcNAc in β1-4 linkage to the β-linked mannose of the trimannosyl core of N-glycosyl oligosaccharide. *J. Biol. Chem.* **257**, 10235–10242

Narasimhan, S., Schachter, H. (1988) Expression of N-Acetyl-glucoaminyltransferase III in hepatic nodules during rat liver carcinogenesis promoted by orotic acid. *J. Biol. Chem.* **263**, 1273–81

Nishikawa, A., Fujii, S., Sugiyama, T., Hayashi, N., and Taniguchi, N. (1988) High expression of an N-acetylglucosaminyltr ansferase III in 3'-Methyl DAB-induced hepatoma and ascites hepatoma. *Biochem. Biophys. Res. Commun.* **152**, 107–112

Nishikawa, A., Ihara, Y., Hatakeyama, M., Kangawa, K., and Tanguchi, N. (1992) Purification, cDNA cloning, and expression of UDP-N-acetylglucosamine: β-D-mannoside β-1,4N-acetylglucosaminyl-transferase III from rat kidney. *J. Biol. Chem.* **267**, 18199–18204

Oguri, S., Minowa, T. Ihara, T., Taniguchi, N., Ikenaga, H., and Takeuchi, M. (1997) Purificaiton and characterization of UDP-N-acetlyglucosamine: α3-Dmannoside β1, 4-N-acetylglucosaminyltranferase (N-acetlyglucosaminyltranferase IV) from bovine small intestine. *J. Biol. Chem.* **272**, 22721–22727, 1997

Rebaa, A., Yamamoto H., Saito T., Meuillet, E., Kim P., Kersey D.S., Bremer, E.G., Taniguchi, N. and Moskal J.R. (1997) Gene transfection-mediated overexpression of β1,4N-acetylglucosamine bisecting oligosaccharides in glioma cell line U373 MG inhibits Epidermal growth factor receptor function. *J. Biol. Chem.* **272**, 9272–9279

Saitoh, O., Wang, W.-C., Lotan, R., and Fukuda, M. (1992) Differential glycosylation and cell surface expression of lysosomal membrane glucoproteins in sublines of a human colon cancer exhibiting destinct metastatic potentials. *J. Biol. Chem.* **267**, 5700–5711

Schachter, H. (1986) Biosynthetic controls that determine the branching and microheterogeneity of protein-bound oligosaccharides. *Biochem. Cell Biol.* **64**, 163–181

Sheng, Y., Yoshimura, M., Inoue, S., Oritani, K., Nishiura, T., Yoshida, H., Ogawa, M., Okajima, Y., Matsuzawa, Y., and Taniguchi, N. (1997) Remodeling of glcoconjugates on CD44 enhances cell adhesion to hyaluronate, tumor growth and metastasis in B 16 melanoma cells expressing β1,4-N-acetylglucosaminyltransferase III. *In. J. Cancer.* **73**, 850–858

Sherman, L., Sleeman, J., Herrlich, P., and Ponta, H. (1994) Hyaluronate receptors: key players in growth, differentiation, migration and tumor progression. *Curr. Opin. Cell Biol.* **6**, 726–733

Sultan, A.S., Miyoshi, E., Ihara, Y., Nishikawa, A., Tsukada, Y., and Taniguchi, N. (1997) Bisecting GlcNAc structures act as negative sorting signals for cell surface glycoproteins in forskolin-treated rat hepatoma cells. *J. Biol. Chem.* **272**, 2866–2872

Taguchi, T., Ogawa, T., Kitajima, K., Inoue, S., Inoue, Y., Ihara, Y., Sakamoto, Y., Nagai, K., and Taniguchi, N. (1998) A method for determination of UDP-GlcNAc: GlcNAc β 1-6(GlcNAc β 1-2) Man α 1-R [GlcNAc to Man] β 1-4 N-acetylglucosaminyltransferase VI activity using a pyridylaminated tetraantennary oligosaccharide as an acceptor substrate. *Analytical Biochem.* **255**, 155–157

Tanemura, M., Miyagawa, S., Ihara, Y., Matsuda, H., Shirakura, R. and Taniguchi, N. (1997) Signiicant down regulation of the major swine xenoantigen by N-acetyglucosaminylransferase III gene transfection. *Biochem. Biophys. Res. Commun.* **2354**, 359–362

Taniguchi, N. and Ikeda, Y. (1997) γ-Glutamyltranspeptidase: Mechanism and Gene Expression. *Advances in Enzymology* **72**, 239–278

Taniguchi, N., Tsukada, Y., Mukuo, K., and Hirai, H. (1974) Effect of Hepatocarcinogenic azo dyes on glutathione and related enzymes in rat liver. *GANN.* **65**, 381–387

Taniguchi, N., and Ihara, Y. (1995) Recent progress in the molecular biology of the cloned N-acetylglucosaminyltransferases. *Glycoconjugate J.* **12**, 733–738

Taniguchi, N., Yoshimura, M., Miyoshi, E., Ihara, Y., Nishikawa, A., and Fujii, S. (1996) Remodeling of cell surface glycoproteins by N-acetylglucosaminyltransferase III gene transfection: modulation of metastatic potentials and down regulation of hepatitis B virus replication. *Glycobiology* **6**, 691–694

Tate, S.S., and Meister, A. (1985) γ-glutamyltranspeptidase from kidney. *Methods Enzymol.* **113**, 400–419

Tsuda, T., Ikeda, Y., and Taniguchi, N., (2000) The Asn-420-linked sugar chain in human epidermal growth factor receptor suppresses ligand-independent spontaneous oligomerization: Possible role of a specific sugar chain in controllable receptor activation. *J. Biol. Chem.* in press.

van den Eijnden, D.H., Koenderman, A.H.L., and Schiphost, W.E.C.M. (1988) Biosynthesis of blood group i-active polyactosaminoglycans. Partial purification and properties of an UDP-GlcNAc: N-acetyllactosaminidase β1-3-N-acetylglucosaminyl-transferase from Novikoff tumor cell ascites fluid. *J. Biol. Chem.* **263**, 12461–12471

Yamashita, K., Hitoi, A., Matsuda, Y., Tsuji, A., Katunuma, N., and Kobata, A. (1983) Structure studies of the carbohydrate moieties of rat kidney γ-glutamyltranspeptidase. An extremely heterogeneous pattern enriched with non-reducing terminal N-acetyl-glucosamine residues. *J. Biol. Chem.* **258**, 1098–1107

Yamashita, K., Hitoi, A., Taniguchi, N., Yokosawa, N., Tsukada, Y., and Kobota, A. (1983) Comparative study of the sugar chains of γ-glutamyltranspeptidase purified from rat liver and rat AH-66 hepatoma cells. *Cancer Res.* **43**, 5059–5063

Yao, A., and McLeod, R.S. (1994) Synthesis and secretion of hepatic apolipoprotein B-containing lipoproteins. *Biochim. Biophys. Acta.* **1212**, 152–166

Yoshimura, M., Ihara, Y., Ohnishi, A., Ijuhin, N., Nishiura, T., Kanakura, Y., Matsuzawa, Y. and Taniguchi, N. (1996) Bisecting N-acetlyglucosamine on K552 cells suppresses Natural Killer cytotoxicity and promotes spleen colonization. *Cancer Res.* **56**, 412–418

Yoshimura, M., Nishikawa, A., Ihara, Y., Taniguchi, S., and Taniguchi, N. (1995) Suppression of lung metastasis of B16 mouse melanoma by N-acetylglucosaminyltranferase III gene transfection. *Proc. Natl. Acad. Sci., USA* **92**, 8754–8758

Yoshimura, M., Ihara, Y., Matsuzawa, Y., Taniugchi, N. (1996) Aberrant glycosylation of E-cadherin enhances cell-cell binding to suppress metastasis. *J. Biol. Chem.* **272**, 13811–13815, 1996

Yousefi, S., Higgins, E., Daoling, Z., Pollex-Kruger, A., Hinds-gaul, O., and Dennis, J. (1991) Increased UDP-GlcNAc: Gal β1-3 GalNAc-R(GlcNAc to GlcNAc) β-1-6-N-acetylglucosaminyltransferase activity in metastatic murine tumor cell lines. Control of polylactosamine synthesis. *J. Biol. Chem.* **266**, 1772–1782

B2. β1,6N-acetylglucosaminyltransferase V is a Determinant of Cancer Growth and Metastasis

James W. Dennis and Maria Granovsky

β1,6GlcNAc-branched N-glycans are up-regulated in oncogene transformed cells, and in human carcinomas where expression correlates with disease stage and prognosis. Forced expression of β1,6N-acetylglucosaminyltransferaseV (GlcNAc-TV) in cell lines induces transformation and tumor formation in mice. *Mgat5* gene expression is induced through Ras signaling pathways, and the branched N-glycan products are required for oncogene-dependent tumor growth and metastasis *in vivo*. In this regard, breast cancer growth and metastasis induced by a polyomavirus middle T transgene in mice is markedly suppress in $Mgat5^{-/-}$ mice compared to their $Mgat5^{+/+}$ littermates. The β1,6GlcNAc-branched N-glycans destabilise integrin receptor clustering, thereby enhancing focal adhesion turnover, associated intracellular growth signals and cell motility.

1. TUMOR PROGRESSION

Multiple genetic changes combined with selection pressure of the surrounding tissue environment are the primary features of tumor formation and progression. Gene mutations that promote growth and survival are maintained as tumors expand and evolve (Cairns, 1981). Tumor evolution or progression produces phenotypic and molecular heterogeneity in the cell population, and this can complicate the analysis of the malignant phenotypes at the level of biochemistry and cell biology (Poste & Fidler, 1980). Cancer mutations result in either loss-of-function in a gene product, designated "tumor suppressor proteins" (e.g. p53, APC, WT1), or missense mutations which is gain-of function or activate of "proto-oncogenes" (e.g. H-Ras). Cancer genes have also been described as "caretaker" and "gatekeeper" genes, the latter regulate aspects of cellular proliferation, and the former control genomic integrity (reviewed in Kinzler & Vogelstein, 1997). The retinoblastoma gene Rb-1 is a gatekeeper gene, where loss-of function mutations promote entry into the cell cycle and tumorigenesis in retinal epithelial cells. The mismatch DNA repair enzymes MSH2 and MLH1 are caretaker genes, their inactivation in colon cancer leads to mutations in other genes that can enhance growth, such as the TGF-β receptor type II gene (Grady *et al.*, 1999). P53 and other genes that ensure faithful segregation of chromosomes during mitotic also serve a caretaker function. Mutations in genes that block cell death by apoptosis (e.g. Bax) contribute to cell survival in the hypoxic environment of tumors (Dang & Semenza, 1999). Many cancer mutations have been engineered in mice, and shown to increase the incidence of pre-malignant lesion and tumors, thereby providing evidence of cause and effect (reviewed in (Jacks, 1996; Webster & Muller, 1994). In addition, familial or heritable mutations in tumor suppressor genes such as P53, Rb-1, and APC increase the risk of cancer in certain tissues.

Ras proto-oncogenes sustain activating mutations in approximately 20% of all human tumors. In addition, Ras signaling is induced by other common mutations such as amplification of Neu/ErbB-2 in breast cancer. Ras is a GTPase and activator of Raf kinase leading to the activation of AP1 (i.e. c-Fos/c-Jun dimers) and Ets transcription factors (Wasylyk *et al.*, 1998). These transcription factors regulate expression of multiple genes involved in cell cycle progression (e.g. cyclin D), cell motility (Rho/Ccd42/Rac-1/Taim1), as well as metalloproteases (e.g. MMP-2, -3, -9), growth factors (e.g. VEGF and bFGF) and glycosyltransferases (e.g GlcNAc-TV). Genetic analysis in mice has validated the relationship between RAS activation and downstream gene transcription required for tumor growth. Expression of activated RAS in transgenic mice results in invasive skin tumors following application of skin carcinogens, but on a c-Fos deficient genetic background, matrix metalloproteases (MMPs) and vascular endothelial growth factor (VEGF) transcripts are suppressed, and only hyperkeratinized benign tumors are observed (Saez *et al.*, 1995). VEGF induces microvasculature in tumors, thereby providing the necessary oxygen and nutrients to the expanding tumor (Risau, 1997). MMPs secreted by tumor cells digest extracellular matrix and facilitate tumor cell invasion through extracellular matrix that separates tissue compartments.

The search for somatic and heritable mutations responsible for cancer initiation or progression does not yet include genes in the protein glycosylation pathways. However, activating mutations in Ras up-regulate GlcNAc-TV and its products, β1,6GlcNAc-branched N-glycans, a positive regulator of tumor growth and metastasis.

2. GlcNAc-TV AND β1,6 GlcNAc-BRANCHED N-GLYCANS IN CANCER

Larger glycopeptides are observed in transformed cell compared to their non-transformed counter part, a phenomena first reported for chicken embryo fibroblasts and BHK cells transformed by Rous sarcomavirus (Warren *et al.*, 1972), and subsequently, polyomavirus transformed BHK cells (Pierce & Arango, 1986), and NIH-3T3 cells transformed with activated H-Ras (Easton *et al.*, 1991; Lu & Chaney, 1993). The transformed morphology and larger glycopeptides were concordant and reversible in cells infected with a temperature sensitive RSV virus, (Warren *et al.*, 1972). Transient expression of H-Ras in NIH-3T3 results in the appearance of the larger N-glycans 25h prior to morphological transformation of the cells (van Beek *et al.*, 1984). Comparison of N-glycan structures in polyomavirus transformed and non-transformed BHK cells suggested that increased β1,6GlcNAc- branching of the trimannosyl core is the basis for the larger sized N-glycans (Yamashita *et al.*, 1984). GlcNAc-TV activity increased 3–5 fold in oncogene transformed cells, while several other glycosyltransferases activities measured in these experiments showed little change (Yamashita *et al.*, 1985). Similar increases in GlcNAc-TV activity were observed in rat2 fibroblasts and SP1 epithelial cells transfected with activated Ras or the v-Fps oncogene (Dennis *et al.*, 1989). GlcNAc-TV transcript levels are increased in RSV transformed BHK cell (Buckhaults *et al.*, 1997), and in pre-malignant hepatitis, tumors and metastases of LEC rats, a strain which has a hereditary predisposition to hepatitis and hepatocarcinomas (Miyoshi *et al.*, 1993).

In human tumors of breast, colon, and skin, *β*1,6GlcNAc-branched N-glycan measured by L-PHA immunohistochemistry also increase compared to surrounding normal tissues. (Fernandes *et al.*, 1991). L-PHA lectin binds to Gal*β*1,4GlcNAc*β*1, 6(Gal*β*1,4GlcNAc*β*1,2) Man*α* portion of tri- and tetraantennary N-glycans dependent on the action of GlcNAc-TV (Cummings & Kornfeld, 1982). L-PHA staining in human colorectal carcinoma sections is associated with the presence of lymph node metastases, and provides an independent prognostic indicator of tumor recurrence and patient survival and (Seelentag *et al.*, 1998). L-PHA reactivity in esophageal carcinoma has been observed to be most intense at the invasive edge of (Takano *et al.*, 1990).

Rat GlcNAc-TV is a 740 amino acid type II transmembrane glycoprotein, typical of the Golgi glycan processing enzymes (Shoreibah *et al.*, 1993). The C-terminal portion comprising $S_{213-740}$ is essential for the catalytic activity (Korczak *et al.*, 2000). A missense mutation changing Leu to Arg at position 188, distal to the catalytic region causes mis-localization of the enzyme without affecting catalytic activity (Weinstein *et al.*, 1996). The mammalian (*Mgat5*) and *C. elegans* (*gly-2*) GlcNAc-TV genes share significant sequence homology, catalytic specificity and the Golgi localization signal defined by the region containing Leu188. The *C. elegans* gene is a functional homologue of the mammalian gene, as it complements the Lec4 mutation in CHO cells. (Warren *et al.*, 2000). A targeted null mutation of the Mgat5 gene in mice results in loss of detectable GlcNAc-TV enzyme activity, and L-PHA reactive N-glycan structures in brain, intestine, kidney and lymphoid cells, suggesting a single gene encodes GlcNAc-TV activity (Granosvsky *et al.*, 2000). The GlcNAc-TV transcript is widely expressed in the E9.5 mouse embryo but becomes restricted to neural epithelium and basal epithelial cell layers in many tissues of the E17.5 mouse embryo (Granovsky *et al.*, 1995).

Mgat5 gene transcription is up-regulated with Ras activation. Transcription factor binding sites in 5′ region upstream region of *Mgat5* have been found for AP1, AP2 LFA1, HNF1, HP1 and PEA3/ets (Saito *et al.*, 1995). The Ets binding sites are functional, as suggested by the observation that transformation with the *v-src* oncogene stimulates transcription from the GlcNAc-TV promoter, through activation of Raf-1 kinase and the Ets-2 transcription factor (Buckhaults *et al.*, 1997). Ras-Raf-Mapk signaling activates the Ets transcription factors, an important pathway regulating cell proliferation and transformation (Galang *et al.*, 1996) (Figure 2). Her-2/Neu/Erb transformation also activates the Ras signaling pathway and results in increased *Mgat5* transcript levels (Chen *et al.*, 1998). *Mgat5* transcripts also increase in pre-malignant hepatitis, tumors and metastases of LEC rats, a strain with a hereditary predisposition to hepatitis and hepatocarcinomas (Miyoshi *et al.*, 1993). Ras proto-oncogenes sustain activating mutations in approximately 20% of all human tumors. In addition, Ras signaling is induced by other common mutations such as amplification of Neu/ErbB-2 in breast cancer (Slamon *et al.*, 1987). These observations suggest that cancer-associated increases in GlcNAc-TV may commonly occur at the level of gene expression.

Splicing of the *Mgat5* pre-message is altered in tumors, and may have an additional unusual effect on tumor growth. An intron sequence of the human *Mgat5* gene encodes a transcript in tumor cells, but not normal tissues and results in expression of a tumor-associated antigen in a high proposition of cancers. This peptide sequence was recognized by tumor infiltrating lymphocytes in an HLA dependent manner and tumor cells displaying it were lysed by cytotoxic T lymphocytes *in vitro*. The antigen was expressed in fifty percent of human melanomas, and also observed in tumors of other tissues (Guilloux *et al.*, 1996). Tumor antigens may potentially be recognized by the host immune response and restrict tumor growth. Carcinomas do not induce clinically meaningful anti-tumor immune responses that result in spontaneous regression, even though "tumor antigens" may be present. Tumors produce cytokines such as TGF-β and IL-10 that act to suppress cell-mediated immunity.

3. GlcNAc-TV PRODUCTS ENHANCES MALIGNANCY

Mutant MDAY-D2 tumor cell lines selected for resistance to L-PHA and found to be deficient in GlcNAc-TV produced 95% fewer spontaneous metastases in liver, and solid tumor growth rate was ~50% slower compared to wild-type cells (Dennis *et al.*, 1987). Similarly, GlcNAc-TV deficient mutants of 168.1 mammary tumor cell line cells formed 2–10 times fewer colonies in the lungs following intravenous injection (Lu *et al.*, 1994). However, as the nature of the mutations in these cell lines remain uncharacterized, it may be argued that GlcNAc-TV depletion in the mutant cell lines is secondary to a mutation in another gene important to cancer progression.

To test for transforming activity of GlcNAc-TV, Mv1Lu cells, an immortalized lung epithelial cell line was transfected with a GlcNAc-TV expression vector (Demetriou *et al.*, 1995). Strikingly, the transfected cells were observed to form tumors when injected into nude mice. The GlcNAc-TV expressing cells in culture showed loss of contact inhibition of growth, enhanced cell motility, and appeared morphological transformed as suggested by layering and foci formation in culture. The cells were less adhesive on fibronectin and collagen type IV, suggesting that turnover of substratum adhesions may be enhanced. In addition, the GlcNAc-TV expressing cells showed accelerated apoptosis under high-density low serum conditions where the parental Mu1Lv cells become growth arrested. Forced-expression of GlcNAc-TV in mammary carcinoma cells resulted in an enhancement of lung metastasis by 4–40 fold when the cells were injected intravenously into mice (Seberger & Chaney, 1999). Forced expression of GlcNAc-TV in epithelial cells increased β1,6GlcNAc-branched glycans on integrin chains α_5, α_v and β_1, suggesting that enzyme activity is not saturating in epithelial cells prior to transformation (Demetriou *et al.*, 1995). However, cell-surface levels of $\alpha_5\beta_1$ or $\alpha_v\beta_3$ integrins were unchanged in transfected cells. The α and β chains of integrin receptors each have multiple N-glycosylation sites, and β1,6GlcNAc-branched glycans have previously been detected on $\alpha_5\beta_1$ fibronectin receptor (Asada *et al.*, 1991; Nakagawa *et al.*, 1996). Furthermore, studies on *Mgat5*$^{-/-}$ mice suggest that GlcNAc-TV dependent glycosylation reduce integrin-mediated adhesion, enhancing focal-adhesion turnover and accelerate cell motility and growth signals (Granovsky *et al.* 2000) (Figure 2). The presence of β1,6GlcNAc-branched oligosaccharides on cell adhesion receptors may enhance on-off rates of cell-substratum contacts, thereby accelerating cell motility. Indeed, the extracellular adhesive environment has been shown to be rate-limiting for cell motility. Optimal migration speed depends upon both ligand concentration (e.g. fibronectin) and integrin levels (Palecek *et al.*, 1997). Rates of both focal-adhesion turnover and cell motility exhibit bell-shaped responses to changing ligand or receptor levels.

Substitution of the trimannosyl core by GlcNAc-TIII renders glycan intermediates poor substrate for α-mannosidase II, GlcNAc-TII, GlcNAc-TIV and GlcNAc-TV, and thereby routes the pathway into hybrid glycans (Schachter, 1986) (Figure 1). B16 mouse melanoma cells transfected with a GlcNAc-TIII were observed to be poorly metastatic in syngeneic mice, and less invasive through an extracellular matrix barrier *in vitro* (Yoshimura *et al.*, 1995). The transfected cells showed the expected increase in hybrid-type and a concomitant reduction in β1,6GlcNAc-branched glycans. E-cadherin and CD44 (hyaluronate receptor) adhesion molecules were subject to GlcNAc-TIII-dependent glycosylation in transfected B16 cells (Yoshimura *et al.*, 1996) (Sheng *et al.*, 1997). Over-expression of either E-cadherin and CD44 in tumor cells has been shown to suppress cancer growth in experimental models (Schmits *et al.*, 1997a) (Birchmeier, 1995; Schmits

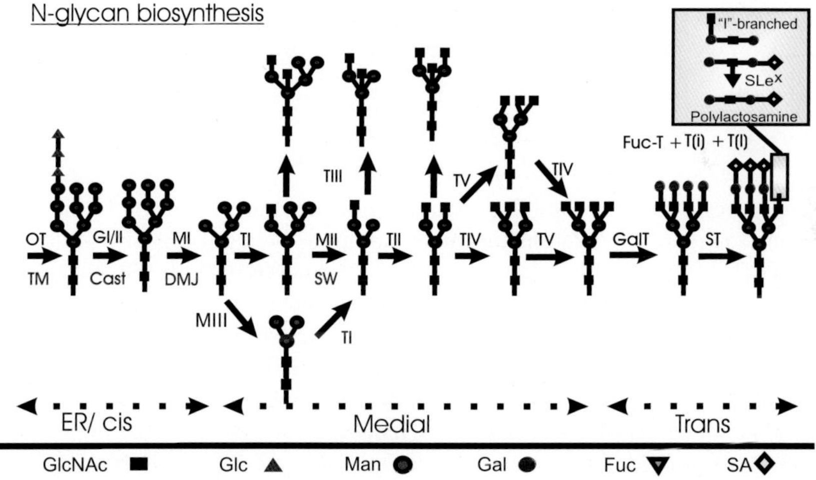

Figure 1: Schematic of N-linked glycan biosynthesis showing the Golgi compartments. Abbreviations used are oligosaccharyltransferase, OT; the α-glucosidases, GI, GII; the β-N-acetylglycosaminyltransferases, TI, TII, TIII, TIV, TV, T(i), T(I); the α1,2mannosidases, MI, α1,3/6mannosidases MII, MIII; β1-galactosyltransferases (Gal-T), α-fucosyltransferases (Fuc-T), α-sialyltransferases (ST), swainsonine (SW), tunicamycin (TM), castanospermine (Cast), deoxymannojirimycin (DMJ). The terminal sequences are added to both N- and O-linked glycans by β1 4Gal-T, β1,3GlcNAc-T(i), β1,6GlcNAc-T(I), α2 3ST, α1,3Fuc-T. The lactosamine antenna initiated by GlcNAc-TV is preferentially elongated with polylactosamine and Lewis antigens as indicated by the gray box. Note that GlcNAc-TIII re-directs the pathway into "bisected glycans" rather than "complex-type".

et al., 1997b). E-cadherin homotypic adhesion in epithelial cells is critical for maintenance of cell-cell contacts and growth control (Christofori & Semb, 1999). Germline mutations in the E-cadherin gene have been found in cases of familial gastric cancer. In sporadic carcinomas of most tissues, either E-cadherin expression or its function is suppressed. E-cadherin expression as well as homotypic cell adhesion was enhanced in GlcNAc-TIII transfected cells, suggesting E-cadherin activity in tumor cells may also be regulated by glycosylation. GlcNAc-TIII over-expression in B16 melanoma cells also enhanced CD44 mediating cell adhesion (Sheng *et al.*, 1997). CD44 activity is glycosylation-dependent, as enzymatic removal of sialic acid and galactose from complex-type N-glycans of CD44 enhanced binding to hyaluronate (Skelton *et al.*, 1998). EGF receptor and Trk/NGF receptor were also observed to be substrates of GlcNAc-TIII in transfected cell lines (Ihara *et al.*, 1997; Rebbaa *et al.*, 1997). EGF receptor levels on the cell surface and signaling activity was decreased in GlcNAc-TIII transfected U373 glioma cells. Similarly, dimerization and phosporylation of the Trk receptor was reduced in GlcNAc-TIII transfected PC12 cells. These studies suggest that the levels of GlcNAc-TIII and GlcNAc-TV in tumor cells have opposing effects on the function of several cell adhesion receptors.

Mice treated with the carcinogen diethylnitrosamine develop hepatocarinomas that grow more slowly in a *Mgat3*$^{-/-}$ genetic background that in wild type littermates (Bhaumik *et al.*, 1998). However, the mouse hepatocarinomas did not express GlcNAc-TIII transcripts. This result suggests that a host paracrine effect promoting tumor growth is dependent upon GlcNAc-TIII. For example, it is possible that host-derived growth factor acting on the tumor may require glycosylation by GlcNAc-TIII. These results show that glycoslyation may regulated host paracrine as well as tumor cell autonomous phenotypes.

4. POLYLACTOSAMINE

GlcNAc residues of both N- and O-glycan intermediates are extension in the *trans*-Golgi compartment (Figure 1). Polylactosamine addition is regulated by multiple factors including GlcNAc-branching in the medial Golgi, expression of β1,3GlcNAc-T(i) activity (Holmes *et al.*, 1987), glycoprotein transit time in the *trans*-Golgi (Wang *et al.*, 1991), and competition by chain-terminating enzymes including α1,2Fuc-T and α2,6SA-T (Prieto *et al.*, 1997). The antenna initiated by GlcNAc-TV is preferred for extension by β1,3GlcNAc-T(i) in the biosynthesis of polylactosamine (van den Eijnden *et al.*, 1988). GlcNAc-TV deficient BW5147-PHA^{R}2.1 lymphoma cells are depleted in polylactosmine (Cummings & Kornfeld, 1984). Similarly, a 3–5 fold decrease in GlcNAc-TV activity in the MDAY-D2 mutants results in severe loss of polylactosamine in N-glycans, while O-linked polylactosmaine is unaffected (Yousefi *et al.*, 1991).

O-glycans in cancer cells also show increased β1,6GlcNAc branching and polylactosamine due to increased expression of core 2 GlcNAc-T(L) (Li *et al.*, 1996) (Shimodaira *et al.*, 1997). MDAY-D2 mutants are deficient in UDP-Gal transporter, lack Gal in both O- and N-glycans, and show a most extreme attenuation of tumor growth and metastasis (Dennis *et al.*, 1981). When injected subcutaneously into mice, the mutant cells grow very slowly for 0–30 days following injection. However, the mutant tumor cells fuse at low frequency with host cells, an event which suppresses the recessive UDP-Gal transporter mutation and results in progressive dominance by the hybrid cells over the period of 30–70 days (Kerbel *et al.*, 1983). Only hybrid cells are detected in metastases at the 60–70 days after injection. Revertants of the UDP-Gal transporter mutation selected *in vitro* regain the malignant phenotype confirming that suppression of the mutation could restored malignant potential (Dennis & Laferte, 1986). A UDP-Gal transporter mutant of the human melanoma cell line MeWo shown a similar loss of metastatic potential in athymic nude mice (Ishikawa *et al.*, 1988). As mutants affecting specifically O-glycans are not available, the relative contribution of polylactosamine in O- and N- glycans to cancer progression remains to be explored. Gene knockout mice for GlcNAc-TV and core 2 GlcNAc-T, provide a means of exploring this question.

Galectin1 and 3 expressed on the surface of B16 melanoma, the UV-2237 fibrosarcoma and the K-1735 melanoma cells have previously been shown to facilitate organ colonization and metastasis by blood-borne tumor cells (Ohannesian *et al.*, 1995). Intravenous infusion of galactose or arabinogalactan has been shown to inhibit liver colonization by murine tumor cells, presumably by blocking their retention to the microvasculature (Beuth *et al.*, 1987). Restoring β1,4Gal to the surface of UDP-Gal transporter mutants using bovine β1,4Gal-T and injecting the cells iv into mice increased tumor cell adhesion to non-activated endothelial and produced 30 fold more liver metastases (Cornil *et al.*, 1990). These observations suggest that galectins on enothelium and β1,4Gal on tumor cells could potentially play a role in tumor cell adhesion and metastasis. Alternatively, galectins acting in *cis* on the cell surface may affect the distribution and aggregation of signaling receptors with effects on other aspects of tumor cell biology as suggested by the model in Figure 2.

5. SIALYLATION AND LEWIS ANTIGENS

The branch extensions and sialylation of glycan structures also contribute to cancer growth and metastasis. An MDAY-D2 mutant that over-expresses α2,6SA-T (ie. 40 fold) due to a retroviral insertion into the gene promoter, showed 3–10 fold fewer metastases and 60% slower tumor growth (Neng-Wen Lo *et al.*, 1999). The mutant cells expressed predominantly α2,6SA rather than the wild-type α2,3SA on surface glycoproteins (Takano *et al.*, 1994). Suppression of sialylation in the Wa4 mutant of B16 melanoma due to the over-express α1,3Fuc-T is associated with loss of metastatic potential (Finne *et al.*, 1982). CMP-NeuNAc-hydroxylase expressing mutants of MDAY-D2 remained metastatic but grew more slowly as solid tumors (Dennis, 1986; Shaw *et al.*, 1991). In contrast to the MDAY-D2 model, α2,6-sialylation of *N*-acetyllactosamine in *Ras*-transformed rat fibroblasts correlates positively with invasive potential (Le Marer & Stéhelin, 1995).

The Lewis carbohydrate antigens Le[x], Sialyl-Le[x] and Le[y], SLe[a] are often over-expressed in human carcinomas (Itzkowitz *et al.*, 1986), and the dimeric and extended chain forms correlate with poor prognosis in colon cancers (Hoff *et al.*, 1989). Polylactosamine forms the backbone of the dimeric Le[x] and Le[y] antigens. SLe[x] sequences on O-glycans structures of PSGL-1 in leukocytes, bind to L- and P- selectins on endothelial cells, and mediate rolling over endothelial cells of inflamed tissues. Mice lacking α1,3Fuc-TVII are deficient in leukocyte extravasation, a phenotype similar to P- and E- selectin deficient mice and to LADII in humans (Maly *et al.*, 1996). Selectin binding is also dependent upon regulated expression of core 2 GlcNAc-T in lymphoid cells, as the initiation of the β1,6GlcNAc branch appears to be required for polylactosamine and polymeric Le[x] addition (Nakamura *et al.*, 1998). Since carcinoma cells do not generally express leukocyte cell surface glycoproteins (e.g. PSGL-1),

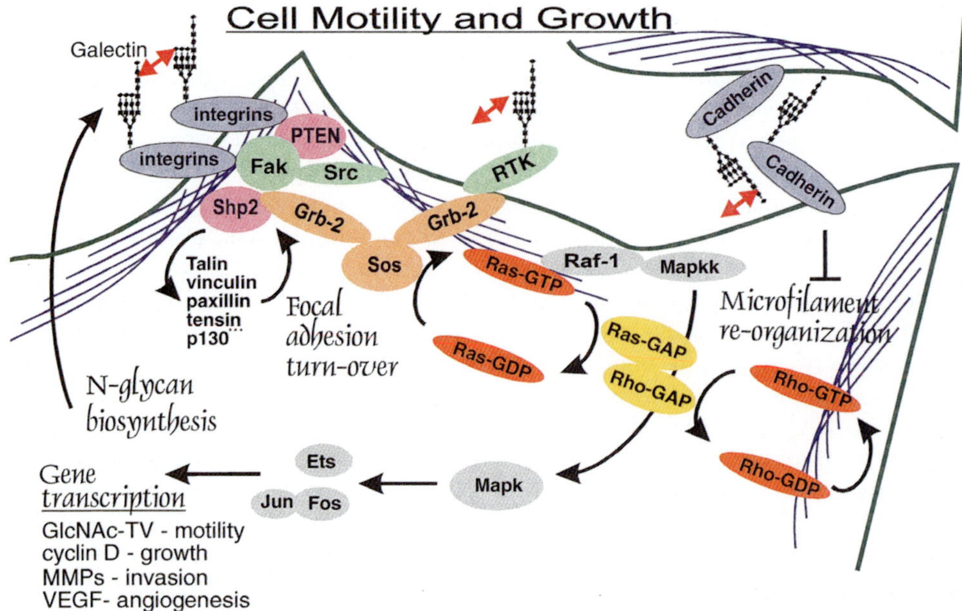

Figure 2: A model depicting up-regulation of GlcNAc-TV gene expression *via* activation of the Ras pathway, which increases β1,6GlcNAc-branching and polylactosamine glycans on cell surface receptors, leading to enhanced cell motility, with a positive feed-back amplifying the Ras-dependent signaling pathway. The red arrows represent galectins. Integrins and cadherins are adhesion receptors; RTK represents receptor tyrosine kinases; Shp2 and PTEN are phosphatases; Fak and Src are intracellular tyrosine protein kinases, Raf-1, Mapkk and MAPK are S/T protein kinases; Grb-2 is an adapter proteins; Sos is a GDP exchange protein; Ras-GAP and Rho-Gap are GTPase activating proteins; Ets, Jun, Fos are transcription factors. Proteins found in focal adhesion plaques include actin microfilaments, talin, vinculin, tensin, paxillin, and P130[cas], the latter two proteins are tyrosine phosphorylated and dephosphorylated during focal adhesion turnover.

the contribution of SLe[x] on tumor cells to organ colonization and metastasis *in vivo* is uncertain. Forced expression of E-selectin in liver of transgenic mice enhanced metastasis of carcinoma cells to liver rather than their usual destination of the lung (Biancone *et al.*, 1996). However P- and E-selectins expression are normally induced in inflamed endothelium, rather than constitutively expressed in hepatocytes. Furthermore, it is unclear that attachment of blood-borne tumor cells to endothelium is a rate-limiting step in clinical metastasis. For example, clinical studies on patients treated with peritoneal shunts to maintain their salt balance pushes millions of peritoneal tumor cells into the circulation, but does not significantly increase the number of metastases observed at autopsy (Jamjoom *et al.*, 1993). However, tumor growth and invasion of local tissues are rate-limiting events, and the evidence that glycosylation affects these phenotypes is compelling.

6. CPIs INHIBIT CANCER GROWTH AND METASTASIS

The glycosidase inhibitor swainsonine and castanospermine block tumor cell metastasis *in vivo*, and invasion through extracellular matrix *in vitro* (Humphries *et al.*, 1986; Yagel *et al.*, 1989). Swainsonine is a competitive inhibitor of Golgi α-mannosidaseII, which blocks the N-glycan biosynthetic pathway prior to β1,6GlcNAc-branching, and results in production of hybrid-type glycans (Figure 1). Swainsonine-treated cells showed increased transcription rates for tissue inhibitor of metalloproteinases (TIMP-1), a negative regulator of invasion (Korczak & Dennis, 1993). Swainsonine also suppresses MMP-2 expression in human tumor cells (Seftor *et al.*, 1991), a metalloproteinase associated with cancer progression in humans (Kanayama *et al.*, 1998).

In experiments assessing the anti-cancer activity of swainsonine in mice, toxicity was minimal and both slowing of solid tumor growth and >90% reduction in lung metastases have been observed (reviewed in (Goss *et al.*, 1995). In phase I clinical trials, swainsonine show both low toxicity and evidence of clinical responses. In the first trial, swainsonine was administered by continuous intravenous infusion over 5 days (Goss *et al.*, 1994; Goss *et al.*, 1997). Doses were 50–550 mg/kg/day administered to 19 patients with advanced cancers for a total of 31 courses of treatment. Common side effects included peripheral edema (n = 11/19), mild liver dysfunction in all patients (AST up to 4 fold normal) and a rise in serum amylase (n = 8/19). One patient with head and neck cancer showed >50% tumor shrinkage, and two patients with lymphangitis carcinomatosis on chest X-ray showed symptomatic improvement during the infusion of swainsonine and for a week thereafter. Clearance and serum half-life for swainsonine were determined to be approximately 2 ml/h·Kg, and 0.5 days, respectively. Chronic oral administration of swainsonine in 16 cancer patients was also well tolerated at 150 g/kg/day given twice weekly, and similar rises in liver enzymes as well as fatigue were also reported as toxicities. Phase II trials designed to measure efficacy of oral swainsonine treatment in renal and colon carcinoma are currently being done.

Swainsonine's therapeutic profile can be improved by creating analogues that lack activity against lysosomal α-mannosidases but retain potency for Golgi α-mannosidase II and III. This would reduce lysosomal storage and vacuolisation, which may be dose-limiting for swainsonine. Somatic tumor cell mutants with a deficiency in UDP-Gal transport activity show the most severe attenuation of malignancy, suggesting that a blocker of lactosamine extension in N- and O-glycans may be effective. Inhibitors of polylactosamine extension, β1,6GlcNAc-branching in the N- and O-glycosylation pathways, or a combination may have potent anti-cancer activity. Studies of cancer development in mutant mice lacking specific glycosyltransferase gene will increase our understanding of the relative importance of these glycan structures to, and should provide direction to anti-cancer drug development efforts.

7. GlcNAc-TV DEFICIENT MICE

Targeted gene disruption of the *Mgat5* gene in mice results in the loss of measurable GlcNAc-TV enzymatic activity, and of β1,6GlcNAc-branched N-glycans in all tissues examined (Granovsky *et al.*, 2000). *Mgat5*$^{-/-}$ mice are viable and fertile, but differ in response to extrinsic stimuli, including inflammatory stimuli and oncogenesis. *Mgat5*$^{-/-}$ mice were crossed with MMTV-PyMT mice to examine breast tumor induction, growth and metastasis in the absence of GlcNAc-TV. PyMT mice develop multifocal tumor in all 10 mammary fat pads, and also metastases in the lungs (Guy *et al.*, 1992). The PyMT oncogene binds and activates c-Src, Shc and PI3 kinase, which are all required for full transforming activity and carcinoma formation in the PyMT transgenic mice (Guy *et al.*, 1994). Tumor formation was delayed in PyMT $Mgat5^{-/-}$ mice, but occurred in 10/10 mammary fat pads. In addition to a delay in formation, tumor growth rate was suppressed 4 fold in the PyMT $Mgat5^{-/-}$ mice. The proliferating fraction of cells in hyperplasia, dysplasia and tumor tissues in $Mgat5^{-/-}$ mice was reduced compared to GlcNAc-TV expressing mice. A similar delay in tumor progression has also been reported for PyMT mice on a $Grb2^{+/-}$ genetic background, an adapter protein in the Ras signaling pathway (Cheng *et al.*, 1999). The incidence of lung metastases was reduced by 20 fold in GlcNAc-TV deficient mice, similar to that previously reported for somatic tumour cell mutants deficiency in GlcNAc-TV (Dennis *et al.*, 1987).

The PyMT $Mgat5^{-/-}$ breast tumour cells were examined for evidence of altered focal adhesion turnover. Both impaired membrane ruffling and activation of PI3 kinase/PKB pathways were reduced in PyMT $Mgat5^{-/-}$ cells compared to PyMT $Mgat5^{+/+}$ cells. The results suggest that GlcNAc-TV has a tumor cell autonomous effect on tumor growth (Granovsky *et al.* 2000). The $Mgat5^{-/-}$ mutation also affected cell attachment and phosphorylation of PKB in the absence of an oncogene. These studies suggest the intrinsic defect in $Mgat5^{-/-}$ cells is an inability to accelerate focal adhesion turnover and signaling via PI3 kinase/PKB as required for full transformation by PyMT. Indeed, the recruitment of leukocytes *in vivo* into inflamed sites was severely impaired in $Mgat5^{-/-}$ mice, and leukocytes were more adhesive on fibronectin as well. *Mgat5* gene expression is induced by T cell activation, and the glycans appear to both dampen T cell responsiveness and accelerate cell migration into inflammatory reaction (Demetriou *et al.* in preparation). The expression of β1,6GlcNAc-branched oligosaccharides present on integrins and other adhesion receptors may facilitate the turnover (on-off rates) of cell-cell and substratum contacts stimulating intracellular growth signals and cell motility (Figure 2). The bulky β1,6GlcNAc-branched glycan chains with polylactosamine extensions may slow receptor movement in the plane of the membrane, and parameters including residency time on the cell surface, ligand-dependent aggregation, internalization, or protease susceptibility.

The β1,6GlcNAc-branched carbohydrates are added to multiple glycoproteins, but those that affect cell motility and proliferation are not well defined. It is likely that regulation of other enzymes that contribute to lactosamine content affect receptor-mediated cell-cell interaction in a similar manner. In choriocarcinomas for example, β1,4GlcNAc-branching of the trimannosyl core is increased presumably due to an increased in GlcNAc-TIV activity (Kobata, 1988). Lactosamine antannae on glycoprotein receptors may bind multivalent lectins and form lattice structures on the cell surface, which acts as a barrier to spontaneous aggregation in the absence of ligand. The galectins, a widely expressed family of mammalian lectins are candidates for this role, as they bind to lactosamine antennae of N- and O-glycans (Perillo *et al.*, 1998). CD22, a transmembrane co-receptor on B cells is an example of a lectin that acts as a docking protein in *cis* on the cell surface. The extracellular domain of CD22 has affinity for the product of ST6Gal (ie. SAα2,3Galβ1,4GlcNAc). Following antigen stimulation, the cytosolic domain of CD22 is phosphorylated by Lyn, causing recruitment of Shp1 tyrosine phosphatase signaling complex, which dampens intracellular signaling (Tedder *et al.*, 1997). B cells of CD22 deficient mice are hypersensitive to antigens stimulation. Conversely, ST6Gal knocked-out mice show impaired B cell maturation and IgM production (Hennet *et al.*, 1998), suggesting that CD22 and SAα2,6Gal interact, and both are required for regulation of B cell signaling.

The conservation of GlcNAc-TV in mammals and C. elegans suggests the β1,6GlcNAc-branched glycans have a fundamental role in certain cellular processes. Although GlcNAc-TV is highly expressed in the developing embryo during organogenesis, it appears to be dispensable for normal development of the embryo. During normal adult life GlcNAc-TV appears to regulate immune cells migration, and may enhance the fitness of the immune systems. However, the advantage gained by regulating *Magt5* gene expression in leukocyte responses comes with the penalty of enhanced cancer growth and metastasis with expression of the enzyme in cancer cells.

ACKNOWLEDGMENTS

J.W. Dennis research was supported by grants from NCI of Canada, the Mizutani Foundation, the National Science and Engineering Research Council of Canada, and GlycoDesign Inc., Toronto.

REFERENCES

Asada, M., Furukawa, K., Kantor, C., Gahmberg, C.G. and Kobata, A. (1991) Structural study of the sugar chains of human leukocyte cell adhesion molecules CD11/CD18. *Biochemistry* **30**, 1561–1571.

Beuth, J., Ko, H.L., Oette, K., Pulverer, G., Roszkowski, K. and Uhlenbruck, G. (1987) Inhibition of liver metastasis in mice by blocking hepatocyte lectins with arabinogalactan infusions and D-galactose. *J. Cancer Res. Clin. Oncol.* **113**, 51–55.

Bhaumik, M., Harris, T., Sundaram, S., Johnson, L., Guttenplan, J., Rogler, C. and Stanley, P. (1998) Progresson of hapatic neoplasms in severely retarded in mice lacking the bisecting N-acetylglucosamine on N-glycans: evidence for a glycoprotein factor that facilitates hepatic tumor progression. *Cancer Res.* **58**, 2881–2887.

Biancone, L., Araki, M., Araki, K., Vassalli, P. and Stamenkovic, I. (1996) Redirection of tumor metastasis by expression of E-selectin in vivo. *J. Exp. Med.* **183**, 581–587.

Birchmeier, W. (1995) E-cadherin as a tumor (invasion) suppressor gene. *BioEssays* **17**, 97–99.

Buckhaults, P., Chen, L., Fregien, N. and Pierce, M. (1997) Transcriptional regulation of N-acetylglucosaminyl-transferase V by the *src* Oncogene. *J. Biol. Chem.* **272**, 19575–19581.

Cairns, J. (1981) The origin of human cancers. *Nature* **289**, 353–358.

Chen, L., Zhang, W., Fregien, N. and Pierce, M. (1998) The her-2/neu oncogene stimulates the transcription of N-acetylglucosaminyltransferase V and expression of its cell surface oligosaccharide products. *Oncogene* **17**, 2087–2093.

Cheng, A.M., Saxton, T.M., Sakai, R., Kulkarni, S., Mbamalu, G., Vogel, W., Tortorice, C.G., Cardiff, R.D., Cross, J.C., Muller, W.J. and Pawson, T. (1999) Mammalian Grb2 regulates multiple steps in embryonic development and malignant transformation. *Cell* **95**, 793–803.

Christofori, G. and Semb, H. (1999) The role of the cell-adhesion molecule E-cadherin as a tumour-suppressor gene. *Trends Biochem. Sci.* **24**, 73–76.

Cornil, I., Kerbel, R.S. and Dennis, J.W. (1990) Tumor cell surface β1-4 linked galactose binds to lectin(s) on microvascular endothelial cells and contributes to organ colonization. *J.Cell Biol.* **111**, 773–782.

Cummings, R.D. and Kornfeld, S. (1982) Characterization of the structural determinants required for the high affinity interaction of asparagine-linked oligosaccharides with immobilized *Phaseolus vulgaris leukoagglu-tinating and erythroagglutinating lectin. J. Biol. Chem.* **257**, 11230–11234.

Cummings, R.D. and Kornfeld, S. (1984) The distribution of repeating Gal β1-4GlcNAc β1-3 sequences in asparagine-linked oligosaccharides of the mouse lymphoma cell line BW5147 and PHAR 2.1. *J. Biol. Chem.* **259**, 6253–6260.

Dang, C.V. and Semenza, G.L. (1999) Oncogenic alterations of metabolism. *Tends Biochem. Sci.* **24**, 68–72.

Demetriou, M., Nabi, I.R., Coppolino, M., Dedhar, S. and Dennis, J.W. (1995) Reduced contact-inhibition and substratum adhesion in epithelial cells expressing GlcNAc-transferase V. *J. Cell Biol.* **130**, 383–392.

Dennis, J.W. (1986) Different metastatic phenotypes in two genetic classes of WGA-resistant tumor cell mutants. *Cancer Res.* **46**, 4594–4600.

Dennis, J.W., Donaghue, T., Florian, M. and Kerbel, R.S. (1981) Apparent reversion of stable in vivo genetic markers detected in tumor cells from spontaneous metastases. *Nature* **292**, 242–245.

Dennis, J.W., Kosh, K., Bryce, D.-M. and Breitman, M.L. (1989). Oncogenes conferring metastatic potential induce increased branching of Asn-linked oligosaccharides in rat2 fibroblasts. *Oncogene* **4**, 853–860.

Dennis, J.W. and Laferte, S. (1986) Co-reversion of a lectin-resistant mutation and non-metastatic phenotype in murine tumor cells. *Int. J. Cancer* **38**, 445–450.

Dennis, J.W., Laferte, S., Waghorne, C., Breitman, M.L. and Kerbel, R.S. (1987) β1-6 branching of Asn-linked oligosaccharides is directly associated with metastasis. *Science* **236**, 582–585.

Easton, E.W., Bolscher, J.G.M. and van den Eijnden, D.H. (1991) Enzymatic amplification involving glycosyltransferases forms the basis for the increased size of asparagine-linked glycans at the surface of NIH 3T3 cells expressing the N-*ras* proto-oncogene. *J. Biol. Chem.* **266**, 21674–21680.

Fernandes, B., Sagman, U., Auger, M., Demetriou, M. and Dennis, J.W. (1991) β1-6 branched oligosaccharides as a marker of tumor progression in human breast and colon neoplasia. *Cancer Res.* **51**, 718–723.

Finne, J., Burger, M.M. and Prieels, J.P. (1982) Enzymatic basis for a lectin resistant phenotype: increase in a fucosyltransferase in mouse melanoma cells. *J. Cell Biol.* **92**, 277–282.

Galang, C.K., Garcia-Ramirez, J., Solski, P.A., Westwick, J.K., Der, C.J., Neznanov, N.N., Oshima, R.G. and Hauser, C.A. (1996) Oncogenic Neu/ErbB-2 increases ets, AP-1, and NF-kappaB-dependent gene expression, and inhibiting ets activation blocks Neu-mediated cellular transformation. *J. Biol. Chem.* **271**, 7992–7998.

Goss, P.E., Baker, M.A., Carver, J.P. and Dennis, J.W. (1995) Inhibitors of carbohydrate processing: A new class of anticancer agents. *Clin. Cancer Res.* **1**, 935–944.

Goss, P.E., Baptiste, J., Fernandes, B., Baker, M. and Dennis, J.W. (1994) A study of swainsonine in patients with advanced malignancies. *Cancer Res.* **54**, 1450–1457.

Goss, P.E., Reid, C.L., Bailey, D. and Dennis, J.W. (1997) Phase IB clinical trial of the oligosaccharide processing inhibitor swainsonine in patients with advanced malignancies. *Clinical Cancer Res.* **3**, 1077–1086.

Grady, W.M., Myeroff, L.L., Swinler, S.E., Rajput, A., Thiagalingam, S., Lutterbaugh, J.D., Neumann, A., Brattain, M.G., Chang, J.k.S.J., Kinzler, K.W., Vogelstein, B., Wilson, J.K. and Markowitz, S. (1999) Mutational inactivation of transforming growth factor b receptor type II in microsatellite stable colon cancers. *Cancer Res.* **59**, 320–324.

Granovsky, M., Fode, C., Warren, C.E., Campbell, R.M., Marth, J.D., Pierce, M., Fregien, N. and Dennis, J.W. (1995) GlcNAc-transferase V and core 2 GlcNAc-transferase expression in the developing mouse embryo. *Glycobiology* **5**, 797–806.

Granovsky, M., Fata, J., Pawling, J., Muller, W.J., Khokha, R. & Dennis J.W. (2000) Suppression of tumor growth and metastasis in MGAT5 deficient mice. *Nature Medicine* **6**, 306–312

Guilloux, Y., Lucas, S., Brichard, V.G., Van Pel, A., Viret, C., De Plaen, E., Brasseur, F., Lethe, B., Jotereau, F. and Boon, T. (1996) A peptide recognized by human cytolytic T lymphocytes on HLA-A2 melanomas is incoded by an intron sequence of the N-acetylglucosaminyltransferase V gene. *J.Exp.Med.* **183**, 1173–1183.

Guy, C.T., Cardiff, R.D. and Muller, W.J. (1992) Induction of mammary tumors by expression of polyomavirus middle T oncogene: a transgenic mouse model for metastatic disease. *Mol.Cell.Biol.* **12**, 954–961.

Guy, C.T., Muthuswamy, S.K., Cardiff, R.D., Soriano, P. and Muller, W.J. (1994) Activation of the c-Src tyrosine kinase is required for the induction of mammary tumors in transgenic mice. *Genes Dev.* **1**, 23–32.

Hennet, T., Chui, D., Paulson, J.C. and Marth, J.D. (1998) Immune regulation by the ST6Gal sialytransferase. *Proc. Natl. Acad. Sci. USA* **95**, 4504–4509.

Hoff, S.D., Matsushita, Y., Ota, D.M., Cleary, K.R., Yamori, T., Hakomori, S.-I. and Irimura, T. (1989) Increased expression of sialyl-dimeric Le^x antigen in liver metastases of human colorectal carcinoma. *Cancer Res.* **49**, 6883–6888.

Holmes, E.H., Hakomori, S.-I. and Ostrander, G.K. (1987) Synthesis of type 1 and 2 lacto series glycolipid antigens in human colonic adenocarcinoma and derived cell lines is due to activation of a normally unexpressed
β1-3N-acetylglucosaminyltransferase. *J. Biol. Chem.* **262**, 15649–15657.

Humphries, M.J., Matsumoto, K., White, S.L. and Olden, K. (1986) Oligosaccharide modification by swainsonine treatment inhibits pulmonary colonization by B16-F10 murine melanoma cells. *Proc. Natl. Acad. Sci. USA* **83**, 1752–1756.

Ihara, Y., Sakamoto, Y., Mihara, M., Shimizu, K. and Taniguchi, N. (1997) Overexpression of N-acetylglucosaminyltransferase III disrupts the tyrosine phosphorylation of Trk with resultant signaling dysfunction in PC12 cells treated with nerve growth factor. *J. Biol. Chem.* **272**, 9629–9634.

Ishikawa, M., Dennis, J.W. and Kerbel, R.S. (1988) Isolation and characterization of spontaneous wheat germ agglutinin-resistant human melanoma mutants displaying remarkable different metastatic profiles in nude mice. *Cancer Res.* **48**, 665–670.

Itzkowitz, S.H., Yuan, M., Fukushi, Y., Palekar, A., Phelphs, P.C., Shamsuddin, A.M., Trump, B.F., Hakomori, S. and Kim, Y.S. (1986) Lewis^x and sialylated Lewis^x-related antigen expression in human malignant and non-malignant colonic tissue. *Cancer Res.* **46**, 2627–2632.

Jacks, T. (1996) Tumor suppressor gene mutations in mice. *Annu. Rev. Genet.* **30**, 603–636.

Jamjoom, Z.A., Jamjoom, A.B., Sulaiman, A.H. and Naim-Ur-Rahman, a.R.A. (1993) Systemic metastasis of medulloblastoma through vetriculoperitoneal shunt report of a case and critical analysis of the literature. *Surg. Neurol.* **40**, 403–410.

Kanayama, H., Yokota, K., Kurokawa, Y., Murakami, Y., Nishitani, M. and Kagawa, S. (1998) Prognostic values of matrix metalloproteinase-2 and tissue inhibitor of metalloproteinase-2 expression in bladder cancer. *Cancer* **82**, 1359–1366.

Kerbel, R.S., Lagarde, A.E., Dennis, J.W. and Donaghue, T.P. (1983) Spontaneous fusion *in vivo* between normal host and tumor cells: possible contribution to tumor progresion and metastasis studied with a lectin-resistant mutant tumor. *Mol. Cell. Biol.* **3**, 523–538.

Kinzler, K.W. and Vogelstein, B. (1997) Cancer-susceptibilty genes: Gatekeepers and caretakers. *Nature* **386**, 761–763.

Kobata, A. (1988) Structures, function, and transformational changes of the sugar chains of glycohormones. *J. Cell Biochem.* **37**, 79–90.

Korczak, B. and Dennis, J.W. (1993) Inhibition of N-linked oligosaccharide processing in tumor cells is associated with enhanced tissue inhibitor of metalloproteinases (TIMP) gene expression. *Int. J. Cancer* **53**, 634–639.

Korczak, B., Le, T., Elowe, S. Donavan, R., Datti A. & Dennis, J.W. (2000) Minimal catalytic domain of N-Acetylglucosaminyltransferase V. *Glycobiology* **10**, 595–599.

Le Marer, N. and Stéhelin, D. (1995) High alpha-2,6-sialylation of *N*-acetyllactosamine sequences in *ras*-transformed rat fibroblasts correlates with high invasive potential. *Glycobiology* **5**, 219–226.

Li, F., Wilkins, P.P., Crawley, S., Weinstein, J., Cummings, R.D. and McEver, R.P. (1996) Post-translational modifications of recombinant P-selectin glycoprotein ligand-1 required for binding to P- and E-selectin. *J. Biol. Chem.* **271**, 3255–3264.

Lu, Y. and Chaney, W. (1993) Induction of N-acetylglucosaminyltransferase V by elevated expression of activated or proto- Ha-ras oncogenes. *Mol.Cell.Biochem.* **122**, 85–92.

Lu, Y., Pelling, J.C. and Chaney, W.G. (1994) Tumor cell surface beta 1-6 branched oligosaccharides and lung metastasis. *Clin. Exp. Metastasis* **12**, 47–54.

Maly, P., Thall, A.D., Petryniak, B., Rogers, C.E., Smith, P.L., Marks, R.M., Kelly, R.J., Gersten, K.M., Cheng, G., Saunders, T.L., Camper, S.A., Camphausen, R.T., Sullivan, F.X., Isogai, Y., Hindsgaul, O., von Andrian, U.H. and Lowe, J.B. (1996) The α (1,3)fucosyltransferase fuc-TVII controls leukocyte trafficking through an essential role in L-, E-, and P-selectin ligand biosynthesis. *Cell* **86**, 643–653.

Miyoshi, E., Nishikawa, A., Ihara, Y., Gu, J., Sugiyama, T., Hayashi, N., Fusamoto, H., Kamada, T. and Taniguchi, N. (1993) N-acetylglucosaminyltransferase III and V messenger RNA levels in LEC rats during hepatocarcinogenesis. *Cancer Res.* **53**, 3899–3902.

Nakagawa, H., Zheng, M., Hakomori, S.-I., Tsukamoto, Y., Kawamura, Y. and Takahashi, N. (1996) Detailed oligosaccharide structures of human integrin $\alpha5\beta1$ analyzed by a three-dimensional mapping technique. *Eur.J.Biochem.* **237**, 76–85.

Nakamura, M., Kudo, T., Narimatsu, H., Furukawa, Y., Kikuchi, J.A.S., Yang, W., Iwase, S., Hatake, K. and Miura, Y. (1998) Single glycosyltransferase, core 2 β1-6-*N*-acetylglucosaminyltransferase, regulates cell surface sialyl-le^{x}expression level in human pre-B lymphocytic leukemia cell line KM3 treated with phorbolester. *J. Biol.Chem.* **273**, 26779–26789.

Neng-Wen Lo, Dennis, J.W. and Lau, J.T.Y. (1999) Over-expression of the α2,6-sialyltransferase, ST6Gal 1, in a low metastatic variant of a murine lymphoblastoid cell line is associated with Appearance of a Unique ST6Gal 1 mRNA. *Biochem. Biophys. Res. Commun.* **264**, 619–621.

Ohannesian, D.W., Lotan, D., Thomas, P., Jessup, J.M., Fukuda, M., Gabius, H.-J. and Lotan, R. (1995) Carcioembryonic antigen and other glycoconjugates act as ligands for galectin-3 in human colon carcinoma cells. *Cancer Res.* **55**, 2191–2199.

Palecek, S.P., Loftus, J.C., Ginsberg, M.H., Lauffenburger, D.A. and Horwitz, A.F. (1997) Integrin-ligand binding properties govern cell migration speed through cell-substratum adhesiveness. *Nature* **385**, 537–540.

Perillo, N.L., Marcus, M.E. and Baum, L.G. (1998) Galectins versatile modulators of cell adhesion, cell proliferation, and cell death. *J.Mol.Med.* **76**, 402–412.

Pierce, M. and Arango, J. (1986). Rous sarcoma virus-transformed baby hamster kidney cells express higher levels of asparagine-linked tri- and tetraantennary glycopeptides containing [GlcNAc-β(1,6)Man-α(1,6)Man] and poly-N-acetyllactosamine sequences than baby hamster kidney cells. *J. Biol. Chem.* **261**, 10772–10777.

Poste, G. and Fidler, I.J. (1980) The pathogenesis of cancer metastasis. *Nature* **283**, 139–146.

Prieto, P.A., Larsen, R.D., Cho, M., Rivera, H.N., Shilatifard, A., Lowe, J.B., Cummings, R.D. and Smith, D.F. (1997) Expression of human H-type α1,2-fucosyltransferase encoding for blood group H (0) antigen in Chinese hamster ovary cells. *J.Biol.Chem.* **272**, 2089–2097.

Rebbaa, A., Yamamoto, H., Saito, T., Meuillet, E., Kim, P., Kersey, D.S., Bremer, E.G., Taniguchi, N. and Moskal, J.R. (1997) Gene transfection-mediated overexpression of beta 1,4-N-acetylglucosamine bisecting oligosaccharides in glioma cell line U373 MG inhibits epidermal growth factor receptor function. *J. Biol. Chem.* **272**, 9275–9279.

Risau, W. (1997) Mechanisms of angiogenesis *Nature* **386**, 671–674.

Saez, E., Rutberg, S.E., Mueller, E., Oppenheim, H., Smoluk, J., Yuspa, S.H. and Spiegelman, B.M. (1995) c-*fos* is required for malignant progression of skin tumors. *Cell* **82**, 721–732.

Saito, H., Gu, J., Nishikawa, A., Ihara, Y., Fujii, J., Kohgo, Y. and Taniguchi, N. (1995) Organization of the human N-acetylglucosaminyltransferase V gene. *Eur.J. Biochem.* **233**, 18–26.

Schachter, H. (1986) Biosynthetic controls that determine the branching and microheterogeneity of protein-bound oligosaccharides *Biochem. Cell Biol.* **64**, 163–181.

Schmits, R., Filmus, J.G.N., Senaldi, G., Kiefer, F., Kundig, T., Wakeham, A., Shahinian, A., Catzavelos, C., Rak, J., Furlonger, C., Zakarian, A., Simard, J.J.L., Ohashi, P.S., Paige, C.J., Guiterrez-Ramos, J.C. and Mak, T.W. (1997a) CD44 regulates hematopoietic progenitor distribution, granuloma formation, and tumorigenicity. *Blood*, **90**, 2217–2233.

Seberger, P.J. and Chaney, W.G. (1999) Control of metastasis by asn-linked, β1-6 branched oligosaccharides in mouse mammary cancer cells. *Glycobiology*, **9**, 235–241.

Seelentag, W.K., Li, W.P., Schmitz, S.F., Metzger, U., Aeberhard, P., Heitz, P.U. and Roth, J. (1998) Prognostic value of b1,6-branched oligosaccharides in human colorectal carcinoma. *Cancer Res.*, **58**, 5559–5564.

Seftor, R.E.B., Seftor, E.A., Grimes, W.J., Liotta, L.A., Stetler-Stevenson, W.G., Welch, D.R. and Hendrix, M.J.C. (1991) Human melanoma cell invasion is inhibited *in vitro* by swainsonine and deoxymannojirimycin with a concomitant decrease in collagenase IV expression. *Melanoma Res.* **1**, 43–54.

Shaw, L., Yousefi, S., Dennis, J.W. and Schauer, R. (1991) CMP-*N*-acetylneuraminic acid hydroxylase activity determines the wheat germ agglutinin-binding phenotype in two mutants of the lymphoma cell line MDAY-D2. *Glycoconj. J.* **8**, 434–441.

Sheng, Y., Yoshimura, M., Inoue, S., Oritani, K., Nishiura, T., Yoshida, H., Ogawa, M., Okajima, Y., Matsuzawa, Y. and Taniguchi, N. (1997) Remodeling of glycoconjugates on CD44 enhances cell adhesion to hyaluronate, tumor growth and metastasis in B16 melanoma cells expressing β1,4-N-acetylglucosaminyltransferase III. *Int. J. Cancer* **73**, 850–858.

Shimodaira, K., Nakayama, J., Nakamura, N., Hasebe, O., Katsuyama, T. and Fukuda, M. (1997) Carcinoma-associated expression of core 2 β-1,6-N-Acetylglucosaminyltransferase gene in human colorectal cancer: Role of *O*-glycans in tumor progression. *Cancer Res.* **57**, 5201–5206.

Shoreibah, M., Perng, G.-S., Adler, B., Weinstein, J., Basu, R., Cupples, R., Wen, D., Browne, J.K., Buckhaults, P., Fregien, N. and Pierce, M. (1993) Isolation, characterization, and expression of cDNA encoding *N*-Acetylglucosaminyltransferase V. *J. Biol. Chem.* **268**, 15381–15385.

Skelton, T.P., Zeng, C., Nocks, A. and Stamenkovic, I. (1998) Glycosylation provides both stimulatory and inhibitory effects on cell surface and soluble CD44 binding to hyaluronan. *J. Cell Biol.* **140**, 431–446.

Slamon, D.J., Clark, G.M., Wong, S.G., Levin, W.J., Ulrich, A. and McGuire, W.L. (1987) Human breast cancer: correlation of relapse and survival with amplification of the Her-2/neu oncogene. *Science* **235**, 177–182.

Takano, R., Muchmore, E.A. and Dennis, J.W. (1994) Sialylation and malignant potential in tumor cell glycosylation mutants. *Glycobiology* **4**, 665–674.

Takano, R., Nose, M., Nishihira, I. and Kyogoku, M. (1990) Increase of β1-6-branched oligosaccharides in human esophageal carcinomas invasive against surrounding tissue *in vivo* and *in vitro*. *Am. J. Pathol.* **137**, 1007–1011.

Tedder, T.F., Tuscano, J., Sato, S. and Kehrl.J.H. (1997) CD22, a B lymphocyte-specific adhesion molecule that regulates antigen receptor signaling. *Annu. Rev. Immunol.* **15**, 481–504.

van Beek, W., Tulp, A., Bolscher, J., Blanken, G., Roozendaal, K. and Egbers, M. (1984) Transient versus permanent expression of cancer-related glycopeptides on normal versus leukemic myeloid cells coinciding with marrow egress. *Blood* **63**, 170–176.

van den Eijnden, D.H., Koenderman, A.H.L. and Schiphorst, W.E.C.M. (1988) Biosynthesis of blood group i-active polylactosaminoglycans. *J. Biol. Chem.* **263**, 12461–12465.

Wang, W.C., Lee, N., Aoki, D., Fukuda, M.N. and Fukuda, M. (1991) The poly-N-acetyllactosamines attached to lysosomal membrane glycoproteins are increased by the prolonged association with the Golgi complex. *J. Biol. Chem* **266**, 23185–23190.

Warren, L., Critchley, D. and Macpherson, I. (1972) Surface glycoproteins and glycolipids of chicken embryo cells transformed by a temperature-sensitive mutant of rous sarcoma virus. *Nature* **235**, 275–278.

Wasylyk, B., Hagman, J. and Gutierrez-Hartmann, A. (1998) Ets transcription factors: nuclear effectors of the Ras-MAP-kinase signaling pathways. *Trends Biochem. Sci* **23**, 213–216.

Webster, M.A. and Muller, W.J. (1994) Mammary tumorigenesis and metastasis in transgenic mice. *Semin. Cancer Biol.* **5**, 69–76.

Weinstein, J., Sundaram, S., Wang, X., Delgado, D., Basu, R. and Stanley, P. (1996) A point mutation causes mistargeting of golgi GlcNAc-TV in the Lec4A Chinese hamster ovary glycosylation mutant. *J. Biol. Chem.* **271**, 27462–27469.

Yagel, S., Feinmesser, R., Waghorne, C., Lala, P.K., Breitman, M.L. and Dennis, J.W. (1989) Evidence that β1-6 branched Asn-linked oligosaccharides on metastatic tumor cells facilitate invasion of basement membranes. *Int. J. Cancer*, **44**, 685–690.

Yamashita, K., Ohkura, T., Tachibana, Y., Takasaki, S. and Kobata, A. (1984) Comparative study of the oligosaccharides released from baby hamster kidney cells and their polyoma transformant by hydrazinolysis. *J. Biol. Chem.* **259**, 10834–10840.

Yamashita, K., Tachibana, Y., Ohkura, T. and Kobata, A. (1985) Enzymatic basis for the structural changes of asparagine-linked sugar chains of membrane glycoproteins of baby hamster kidney cells induced by polyoma transformation. *J.Biol.Chem.* **260**, 3963–3969.

Yoshimura, M., Ihara, Y., Matsuzawa, Y. and Taniguchi, N. (1996) Aberrant glycosylation of E-cadherin enhances cell-cell binding to suppress metastasis. *J. Biol. Chem.* **271**, 13811–13815.

Yoshimura, M., Nishikawa, A., Ihara, Y., Taniguchi, S. and Taniguchi, N. (1995) Suppression of lung metastasis of B16 mouse melanoma by N-acetylglucosaminyltransferase III gene transfection. *Proc. Natl. Acad. Sci. USA* **92**, 8753–8758.

Yousefi, S., Higgins, E., Doaling, Z., Hindsgaul, O., Pollex-Kruger, A. and Dennis, J.W. (1991) Increased UDP-GlcNAc:Gal β1-3GalNAc-R (GlcNAc to GalNAc) β1-6 N-acetylglucosaminyltransferase activity in transformed and metastatic murine tumor cell lines: control of polylactosamine-synthesis. *J. Biol. Chem.* **266**, 1772–1783.

B3. β(1-6)-*N*-Acetylglucosamine-Branched *N*-Glycans in Normal and Pathological Lymphocyte Behavior

Christian Derappe

1. INTRODUCTION

Lymphocytes, like all eukaryote cells, possess at their surface carbohydrate chains linked to either proteins or lipids to form glycoproteins and glycolipids, respectively. These chains exhibit a wide variety of structures due to the multiple possibilities of linkages for glycan residues. Different biological roles have been proposed for carbohydrate chains (for example, see, Allen and Kisailus 1992): mainly, protection against hydrolases, involvement in protein conformation or as ligands for lectin-like proteins to mediate cellular interactions. Structures of carbohydrate chains are able to change as a function of various biological events, such as embryogenesis or cell differentiation, and disease processes, for example cancer (Hakomori 1996). Consequently, specific carbohydrate structures have been associated with several pathophysiological states and have been proposed as glycosidic markers.

One of them, the β(1-6)-*N*-acetylglucosamine-branched *N*-glycan results from the linkage of an *N*-acetylglucosamine (GlcNAc) to the C-6 position of an α-mannose, himself linked on C-6 position of α β-mannose (Fig. 1). The potential importance of this structure emerged twelve years ago, when the Canadian group of Dennis *et al.* (1987) showed that these β(1-6)-linked glycans were associated with the metastatic character of several tumoral murine cell lines. This structure gives rise to a particular conformation characterized by the presence of two sucessive (1-6)-linkages (Fig. 1) in which, the substituting glycosyl residue is not directly attached to the pyranosyl ring, like for other *O*-glycosidic linkages, but by the intermediate of a methylene bridge. Such linkages might allow for the glycosyl residue to fold back towards the protein backbone, where it could affect protein conformation and/or activity. Moreover, it was shown that the elongation or ramification of the carbohydrate chain occurred preferentially on these β(1-6)-linked antennae (Yousefi *et al.* 1991), allowing attachment of numerous sialic acid residues. This structure could provide better protection against degradation but also mediate specific biological properties · in relationship with changes of glycoproteins conformation. Moreover, β(1-6)-linked structures could constitute a specific determinant for lectins (for example, for the leukoagglutinin of *Phaseolus vulgaris*) or possibly for some unknown lectin-like proteins.

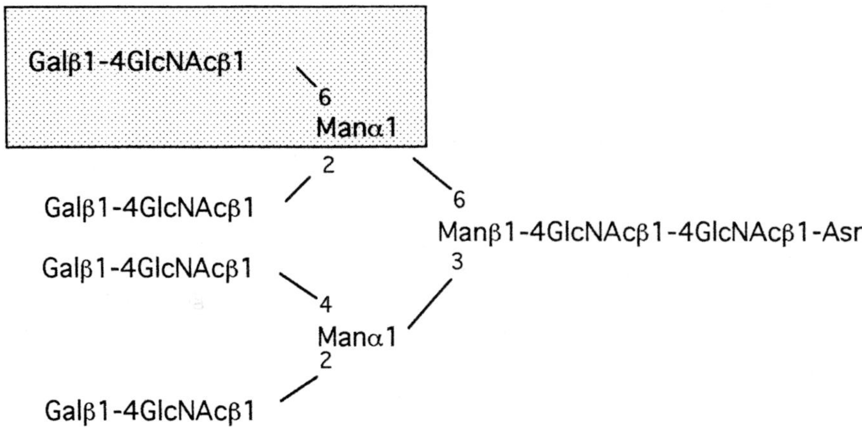

Figure 1: *β*(**1-6**)-*N*-**acetylglucosamine-branched** *N*-**glycans.**

More recently, the *β*(1-6)-*N*-acetylglucosamine-branched *N*-glycans have been associated with other pathophysiological events: activation of normal human T lymphocytes and Sézary syndrome. These two aspects are developed in this present chapter.

2. BIOSYNTHESIS OF *β*(1-6)-*N*-ACETYLGLUCOSAMINE-BRANCHED *N*-GLYCANS

Carbohydrate chains are synthesized in the endoplasmic reticulum and Golgi apparatus and require many glycosyltransferases and glycosidases. Over the past ten years, many studies have been particularly devoted to glycosyltransferases. Numerous reviews have been published (for example, see the basic review by Kornfeld and Kornfeld (1986) and several chapters in this book and, for glycosyltransferases the review by Kleene and Berger (1993). For most of these glycosyltransferases, specific antibodies and cDNA probes are available, thereby allowing their expression to be investigated.

As regards the *β*(1-6)-*N*-acetylglucosamine-branched *N*-glycans, some recent studies focused on the molecular biology of four families of glycosyltransferases, particularly those involved in its biosynthesis: *β*-*N*-acetylglucosaminyltransferases (GnT), *β*-galactosyltransferases (GalT), sialyltransferases (SialT) and fucosyltransferases (FucT). Regulation of their expressions and activities is highly complex, especially with the involvement of cellular factors, such cytokines, kinases, transcriptional factors, and receptors, such as integrins, operating at transcriptional and/or translational levels. Moreover, it was observed that for a given enzymatic activity, several isoenzymes or isoforms could be involved (Chang *et al.* 1995, Lo *et al.* 1998). Thus, all changes in the expression of these glycosyltransferases could be associated with a given pathology or favor of the development of a certain pathology.

After *β*(1-2)-*N*-acetylglucosaminyltransferases I (EC 2.4.1.101) and II (EC 2.4.1.143) generate biantennary structures (Fig. 2a), another GlcNAc residue could be transferred onto one of the following sites:

– the C-4 position of the β(1-4)-linked mannose (bisecting position) by the β(1-4)-*N*-acetylglucosaminyltransferase III (GnT III) (EC 2.4.1.144). No other GlcNAc residue can be transferred onto the resulting structure.

– the C-6 position of the a(1-6)-linked mannose, by the β(1-6)-N-acetylglucosaminyltransferase V (GnT V) (EC 2.4.1.155). This GlcNAc residue can also transferred onto the triantennary structure (Fig. 2b). It constitutes the first step in the biosynthesis of the α(1-6)-branched antennae.

GnT III and GnT V can use the same substrate to generate two types of glycans (Fig. 3) with different biological properties and the analyse of the factors controling the expression of these two glycosyltransferases, constitutes an important challenge for the understanding of the processes of tumorigenesis and metastasis (Yoshimura *et al.* 1995).

Rat and human GnT V have been cloned, respectively by Shoreiba *et al.* (1993) and Saito *et al.* (1994). The human *GnT* V gene was mapped to chromosome 2q21; it comprises 17 exons, spanning 155 kb. This transferase can be activated by membrane protein kinase C or A, indirectly or directly, *via* phosphorylation of Ser/Thr residues (Ju *et al.* 1995).

After linkage of the GlcNAc residue to the C-6 position, antennae are elongated by the addition of a galactose residue in presence of the GalT (EC 2.4.1.38). GalT is a *trans*-Golgi resident type II membrane-bound glycoprotein, which is responsible for the biosynthesis of *N*- and *O*-linked carbohydrate chains by transferring galactose to the acceptor sugar GlcNAc. Bovine GalT was cloned in 1986 (Shaper *et al.* 1986), then

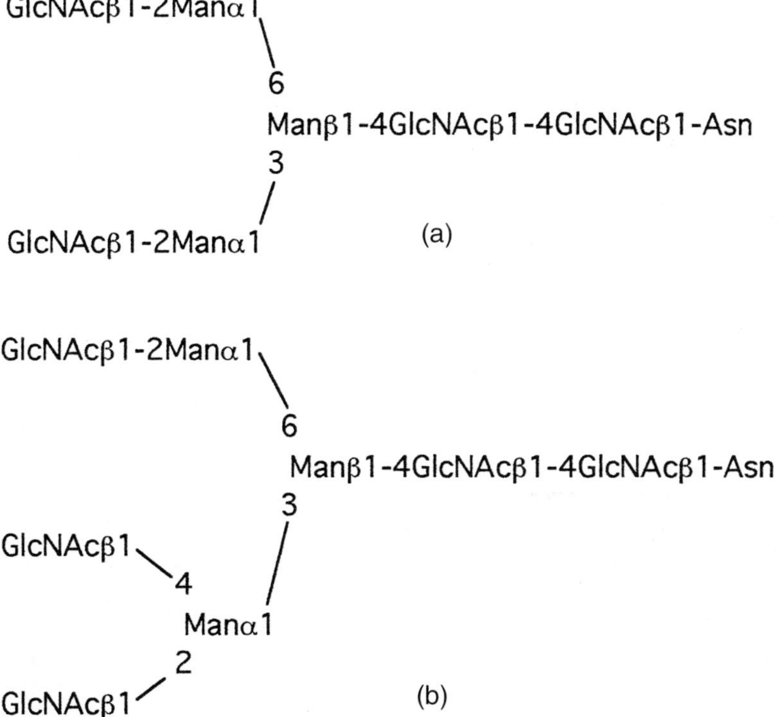

Figure 2: Biantennary (a) and triantennary (b) *N*-linked glycans.

                                        Christian Derappe

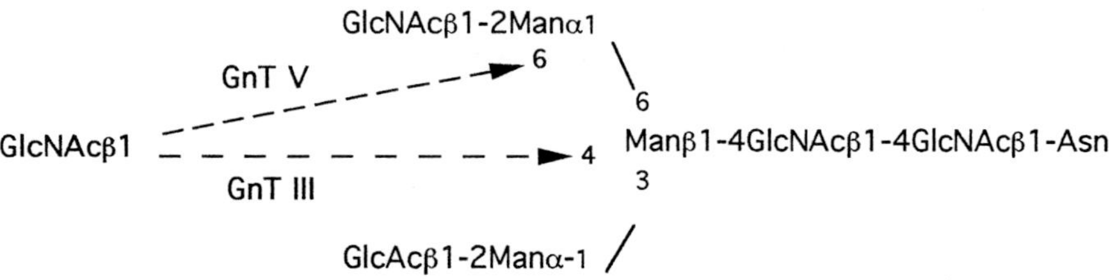

Figure 3: Specificities of *N*-acetylglucosaminyltransferase-III (GnT-III) and -V (GnT-V).

murine GalT cDNA (Shaper *et al.* 1988) and human GalT cDNA (Masri *et al.* 1988) were obtained. Two biological functions, both involving the transfer of a galactose residue were associated with GalT: biosynthesis of *N*- and *O*-linked carbohydrate chains and, together with lactalbumin, biosynthesis of lactose in the mammary gland. This human GalT was mapped to chromosome 9p13. Recently, the GalT family was expanded by five new GalT (Almeida *et al.* 1997, Lo *et al.* 1998, Sato *et al.* 1998): GalTII to GalTVI were mapped to human chromosomes 1p33-34, 1q21-23, 3q13, 11 and 18q11, respectively. These GalT exhibit different patterns of tissular expression.

On the non-reducing terminal galactose, several other glycosyltransferases can add glycosyl residues. Sialyltransferases can transfer sialic acid in either an α(2-3)- or α(2-6)-linkage and enhanced α(2-6) activity was observed in metastasizing human colorectal tumor tissue (Gessner *et al.* 1993) and in ras-transformed rat fibroblasts (Le Marer and Stéhelin 1995). Several fucosyltransferases can attache fucose residues via an α(1-2), α(1-3) or α(1-4) linkage and GlcNAc residues can also be transferred onto the C-6 position of galactose to give branched antennae.

3. INVOLVEMENT OF β(1-6)-*N*-ACETYLGLUCOSAMINE-BRANCHED *N*-GLYCANS IN T-CELL ACTIVATION

One of the most characteristic properties of lymphocytes is their ability to migrate from the bloodstream through tissues to lymphoid organs and back to the blood (Yednock and Rosen 1989) (Fig. 4). Thus, the possibility for lymphocytes to recirculate ensures efficient protection of tissues against potential pathogens and facilitates the quick accumulation of immune cells at the level of the inflamed or injured sites. During recirculation, lymphocytes are subjected to various and complex cellular interactions with other circulating cells and the vascular endothelium or with proteins of the extracellular matrix, the basement membrane and finally tissue cells (Springer 1994). Different classes of adhesive receptors are involved in these interactions (Picker and Butcher 1992). The first step in the interactions between endothelial cells and leukocytes involves three selectins (Bevilacqua and Nelson 1993, McEver 1997): **L-selectin**, expressed on lymphocytes and granulocytes, which binds to constitutively expressed ligands on high endothelial venules of peripheral lymph nodes, to inducible ligands on endothelium at sites of inflammation, and to ligands on other leukocytes, **E-selectin** expressed on activated endothelium, and **P-selectin**, expressed on activated platelets and endothelium. Specific glycan structures, sialylLewis[x] and sialylLewis[a] (Fig. 5), have been proposed as ligands for these selectins, but the affinity of selectins for isolated sialylLewis[x/a]-related oligosaccharides is very low. Selectins bind with higher affinity several sialylated and fucosylated glycoproteins present on leukocytes or endothelium: for example, glycosylated cell adhesion molecule-1 (GlyCAM-1), mucosal addressin cell adhesion molecule-1 (MAdCAM-1) or P-selectin glycoprotein ligand-1 (PSGL-1). These interactions between selectins and their ligands allows to trigger the tethering and the rolling of leukocytes on the vessel wall and subsequent steps in the immune response.

The following step consists of adhesion of leukocytes to the vascular endothelium. This adhesion involves other adhesion molecules, the integrins. Several authors (for example, Diamonds *et al.* 1991) suggested that carbohydrate chains could modulated the properties of β_2 integrin family, but the possibility of involvement of carbohydrates in the modulation of adhesive properties of other cell adhesion molecules, like β_1 integrins, cadherins or other proteins (for example, lectins exhibiting cell-adhesive properties) has been also proposed (Bowman *et al.* 1998, Hennet *et al.* 1998). After their adhesion on the vessel wall, leukocytes are able to pass throught the vessel wall (extravasation) and to migrate in tissues (Fig. 4).

The key step in the immune response is antigenic presentation, which occurs between an antigen-presenting cell and a T lymphocyte, and involves several cell-surface glycoproteins. Afterwards, the T lymphocyte changes from a resting cell into an activated one characterized by the expression of new cell-surface glycoproteins (for example, interleukine-2 receptor), the modification of migration behavior and the acquisition of mitogenic properties (Janeway and Goldstein 1993). Together with these modifications, important changes in cell glycosylation were also observed (Lemaire *et al.* 1994). Using affinity chromatography on columns of immobilized lectins, it was shown that, in vitro activation of normal human CD4[+] and CD8[+] T lymphocytes, induced numerous changes of glycoproteins, the most intriguing being the expression of high levels of $\beta(1\text{-}6)$-GlcNAc-branched *N*-glycans: 44 % for CD 4[+] and 33% for CD 8[+] T lymphocytes, whereas in corresponding resting T lymphocytes only 14% and 1% respectively, were observed. These results were confirmed by measurement of GnT V activity in both T-lymphocyte populations before and after *in vitro* activation.

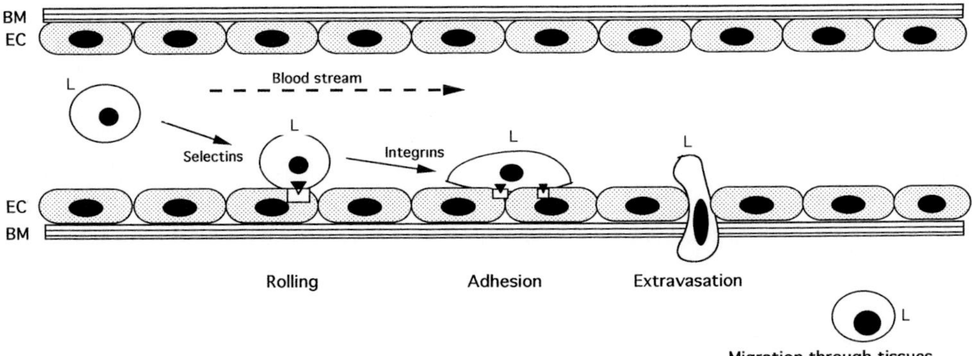

Figure 4: Interactions between vascular endothelial cells (EC) and leukocytes (L).
BM: Basement membrane.

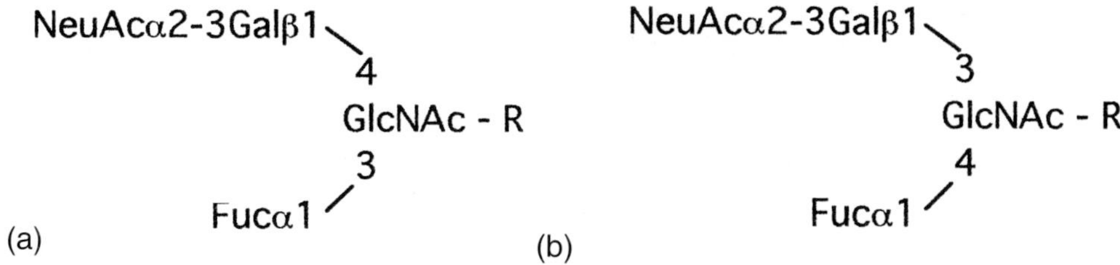

Figure 5: SialylLewis[a] (a) and sialylLewis[x] (b) glycan determinants.

Thus, the same β(1-6)-linked carbohydrate structure was observed in both metastatic tumor cells and normal activated T lymphocytes. This finding suggests a possible relationship between glycan structure and the adhesive properties of the cells. As indicated above, these properties be attributable to the spatial configuration resulting from the β(1-6) linkage. It should be noted that similar results were previously obtained with O-linked carbohydrate chains (Piller *et al.* 1988): activation of normal human T lymphocytes give rise to the expression of *O*-glycans having an antenna with an α(1-6)-linked GlcNAc residue. These observations are in agreement with the possibility of an important functional role for the β(1-6)-linked carbohydrates, such as the modulation of adhesive properties. Moreover, this hypothesis was supported by results obtained with pathological lymphocytes from patients with Sézary syndrome (Derappe *et al.*, 1996).

4. INVOLVEMENT OF β(1-6)-*N*-ACETYLGLUCOSAMINE-BRANCHED *N*-GLYCANS IN SÉZARY SYNDROME

Sézary syndrome is a cutaneous T-cell lymphoma characterized by slowly progressing clonal proliferation of $CD4^+$, CD 45RO$^+$ T cells, involving skin, lymph nodes, blood and visceral organs (Edelson 1980). Generally, patients are over 60 years old and present with generalized erythroderma associated with lymphadenopathy. It was observed in this pathology abnormal cellular interactions between malignant lymphocytes and cells of the skin and several authors postulated that expression of integrins of the β_1 and β_2 families were involved in the development of this pathology (Nickoloff *et al.* 1989, Savoia *et al.* 1992). Nevertheless, evaluation of adhesive properties of Sézary cells on resting or interferon-γ-activated human keratinocytes showed that the abnormal adhesion exhibited by Sézary cells cannot be explained only in terms of integrin expression (Braut-Boucher *et al.* 1998). That study and a previous one (Derappe *et al.*, 1996) on the role of carbohydrates as factors modulating adhesive capacity, suggest that glycosylation of cell glycoproteins could be involved in the disease. Analysis of the glycosylation of Sézary-cell glycoproteins led to the hypothesis of a relationship between the expression of β(1-6)-GlcNAc-branched *N*-glycans and the ability of Sézary cells to adhere to keratinocytes. Two groups of patients can be defined: in the first group, Sézary cells express β(1-6)-linked glycans and adhere poorly to keratinocytes; in the second group, pathological lymphocytes adhere onto keratinocytes and express only low levels of β(1-6)-linked glycans. Moreover, only the β_1 integrins from the lymphocytes of the first group of Sézary patients contained β(1-6)-linked glycans. This difference at the molecular level corresponds to observed clinical differences, since patients from the second group have different clinical features: they are younger and their lymphocytes do not respond to mitogenic stimulation.

5. PERSPECTIVES

More than ten years ago, the involvement of β(1-6)-GlcNAc-branched *N*-glycans in the metastatic properties of several murine tumor cell lines was postulated by Dennis *et al.* (1987). Since then, important advances have been obtained by the molecular biological analyses of glycosyltransferases involved in the biosynthesis of this carbohydrate structure. However, the exact molecular mechanisms of cellular interactions involving this carbohydrate structure remain hypothetical. In the future, progress could be made by developing physicochemistry technology (nuclear magnetic resonance, X-ray diffraction, . . .). On the other hand, utilization of new models (*in vivo* ?), giving a more global view of these interactions, might be able to provide new informations. Another point remains unexplored, the role of glycosidases' in the expression of β(1-6)-GlcNAc-branched *N*-glycans, indeed comparison with protein biology where the role of proteases is well established, suggests that, for carbohydrate biology, similar participation might be essential to the understanding of mechanisms mediating the expression of β(1-6)-linked carbohydrate structures.

REFERENCES

Almeida, R., Amado, M., David, L., Levery, S.B., Holmes, E.H., Merkx, G., Vankessel, A.G., Rygaard, E., Hassan, H., Bennett, E., Clausen, H. (1997) A family of human beta-4-galactosyltransferases – cloning and expression of two novel UDP-galactose-beta-N-acetylglucosamine beta-1,4-galactosyltransferases, beta-4Gal-T2 and beta-4Gal-T3. *J. Biol. Chem.*, **272**, 31979–31991.

Allen, H.J., Kisailus, E.C. (1992) Glycoconjugates: composition, structure and function, Marcel Dekker, NewYork.

Bevilacqua, M.P., Nelson, R.M. (1993) Selectins. *J. Clin. Invest.* **91**, 379–387.

Bowman, K.G., Hemmerich, S., Bhakta, S., Singer, M.S., Bistrup, A., Rosen, S.D., Bertozzi, C.R. (1998) Identification of an N-acetylglucosamine-6-O-sulfotransferase activity specific to lymphoid tissue—An enzyme with a possible role in lymphocyte homing. *Chem. Biol.*, **5**, 447–460.

Braut-Boucher, F., Font, J., Pichon, J., Paulin, Y., Boukhélifa, M., Aubery, M., Derappe, C. (1998). T lymphocytes from Sézary syndrome patients express β_1 integrins whose β(1-6)-branched *N*-linked oligosaccharides reflect their adhesive capacity. *Leuk. Res.*, **22**, 947–952.

Chang, M.L., Eddy, R.L., Shows, T.B., Lau, J.T.Y. (1995) Three genes that encode human β-galactoside α2,3-sialyltransferases. Structural analysis and chromosomal mapping studies. *Glycobiology*, **5**, 319–325.

Dennis, J.W., Laferté, S., Waghorne, C., Breitman, M.L., Kerbel, R. (1987) β1-6 branching of Asn-linked oligosaccharides is directly associated with metastasis. *Science*, **236**, 582–585.

Derappe, C., Haentjens, G., Lemaire, S., Feugeas, J.P., Lebbe, C., Pasqualetto, V., Bussel, A., Aubery, M., Néel, D. (1996) Circulating malignant lymphocytes from Sézary syndrome express high levels of glycoproteins carrying β(1-6)N-acetylglucosamine-branched N-linked oligosaccharides. Leukemia, **10**, 138–141.

Diamonds, M.S., Staunton, D.E., Marlin, S.D., Springer, T.A. (1991) Binding of the integrin Mac1 (CD11b/CD18) to the third immunoglobulin-like domain of ICAM-1 (CD54) and its regulation by glycosylation. *Cell*, **65**, 961–971.

Edelson, R.L. (1980) Cutaneous T cell lymphoma: mycosis fungoides, Sézary syndrome, and other variants. *J. Am. Acad. Dermatol.*, **2**, 89–106.

Gessner, P., Riedl, S., Quentmaier, A., Kemmner, W. (1993) Enhanced activity of CMP-NeuAc:Galβ1-4GlcNAc:α2,6-sialyltransferase in metastasizing human colorectal tumor tissue and serum of tumor patients. *Cancer Lett.*, **75**, 143–149.

Hakomori, S.I. (1996) Tumor malignancy defined by aberrant glycosylation and sphingo(glyco)lipid metabolism. *Cancer Res.*, **56**, 5309–5318.

Henet, T., Chui, D., Paulson, J.C., Marth, J.D. (1998) Immune regulation by the ST6Gal sialyltransferase. *Proc. Natl. Acad. Sci. USA*, **95**, 4504–4509.

Hynes, R.O. (1987) Integrins: a family of cell surface receptors. *Cell*, **48**, 549–554.

Janaway Jr, C.A., Goldstein, P. (1993) Lymphocyte activation and effector functions. The role of cell surface molecules. *Current Opinion in Immunology*, **5**, 313–323.

Jasulionis, M.G., Chammas, R., Ventura, A.M., Travassos, L.R., Brentani, R.R. (1996). $\alpha_6\beta_1$-integrin, a major cell surface carrier of β1-6-branched oligosaccharides, mediates migration of EJ-*ras*-transformed fibroblasts on laminin-1 independently of its glycosylation state. *Cancer Res.*, **56**, 1682–1689.

Ju, T.Z., Chen, H.L., Gu, J.X., Qin, H. (1995). Regulation of *N*-acetylglucosaminyltransferase V by protein kinases. *Glycobiology*, **5**, 767–772.

Kleene, R., Berger, E.G. (1993) The molecular and cell biology of glycosyltransferases. *Biochim. Biophys. Acta*, **1154**, 283–325.

Kornfeld, R., Kornfeld, S. Assembly of Asparagine-linked oligosaccharides. *Annu. Rev. Biochem.*, **54**, 631-664.

Lemaire, S. , Derappe, C., Michalski, J.C, Aubery, M., Néel, D. (1994) Expression of β1-6-branched *N*-linked oligosaccharides is associated with activation in human T4 and T8 cell populations. *J. Biol. Chem.*, **269**, 8069–8074.

Le Marer, N., Stéhelin, D. (1995) High alpha-2-6 sialylation of *N*-acetyllactosamine sequences in ras-transformed rat fibroblasts correlates with high invasive potential. *Glycobiology*, **5**, 219–226.

Lo, N.W., Shaper, J.H., Pevsner, J., Shaper, N.L. (1998) The expanding β4-galactosyltransferase gene family: messages from the databanks. *Glycobiology*, **8**, 517–526.

McEver, R.P. (1997) Selectin-carbohydrate interactions during inflammation and metastasis. *Glycoconjugate J.*, **14**, 585–591.

Masri, K.A., Appert, H.E., Fukuda, M.N. (1988) Identification of the full-length coding sequence for human galactosyltransferase (β-*N*-acetylglucosamide:β1,4-galactosyltransferase). *Biochem. Biophys. Res. Commun.*, **157**, 657-663.

Nickoloff, B.J., Griffiths, C.E.M., Baadsgaard, O., Voorthees, J.J., Hanson, C.A., Cooper, K.D. (1989) Markedly diminished epidermal keratinocyte expression of Intercellular Adhesion Molecule-1 (ICAM-1) in Sézary syndrome. *Jama*, **261**, 2217-2221.

Picker, L.J., Butcher, E.C. (1992) Physiological and molecular mechanisms of lymphocyte homing. *Annu. Rev. Immunol.*, **10**, 561-591.

Piller, F., Piller, V., Fox, R.I., Fukuda, M. (1988) Human T-lymphocyte activation is associated with changes in *O*-glycan biosynthesis. *J. Biol. Chem.*, **263**, 15146-15150.

Saito, H., Nishikawa, A., Gu, J., Thara, Y., Soejima, H., Wada, Y., Sekiya, C., Niikawa, N., Taniguchi, N. (1994) cDNA cloning and chromosomal mapping of human *N*-acetylglucosaminyltransferase V. *Biochem. Biophys. Res. Commun..* **198**, 318–327.

Sato, T., Furukawa, K., Bakker, H., Van den Eijnden, D.H., Van Die, I. (1998) Molecular cloning of a human cDNA encoding beta-1,4-galactosyltransferase with 37% identity to mammalian UDP-Gal:GlcNAc beta-1,4-galactosyltransferase. *Proc. Natl. Acad. Sci. USA*, **95**, 472–477.

Savoia, P., Novelli, M., Fierro, M.T., Cremona, O., Marchisio, P.C., Bernengo, M.G. (1992) Expression and role of integrin receptors in Sézary syndrome. *J. Invest. Dermatol.*, **99**, 151–159.

Shaper, N.L., Shaper, J.H., Meuth, J.L., Fox, J.L., Chang, H., Kirsch, I.R., Hollis, G.F. (1986) Bovine galactosyltransferase: identification of a clone by direct immunological screening of a cDNA expression library. *Proc. Natl. Acad. Sci. USA*, **83**, 1573–1577.

Shaper, N.L., Hollis, G.F., Douglas, J.G., Kirsch, I.R., Shaper, J.H. (1988) Characterization of the full length cDNA for murine β-1-4-galactosyltransferase. Novel features at the 5'-end predict two translational start sites at two in-frame AUGs. *J. Biol. Chem.*, **263**, 10420–10428.

Shoreiba, M., Perng, G.S., Adler, B., Weinstein, J., Basu, R., Cupples, R., Wen, D., Browne, J.K., Buckhaults, P., Fregien, N., Pierce, M. (1993) Isolation, characterization and expression of a cDNA encoding *N*-acetylglucosaminyltransferase V. *J. Biol. Chem.*, **268**, 15381–15385.

Springer, T.A. (1994) Traffic signals for lymphocyte recirculation and leukocyte emigration: the multistep paradigm. *Cell*, **76**, 301–314.

Yednock, T.A., Rosen, S.D. (1989) Lymphocyte homing. *Adv. Immunol.*, **44**, 313-378.

Yoshimura, M., Nishikawa, A., Ihara, Y., Taniguchi, S.I., Taniguchi, N. (1995) Suppression of lung metastasis of B16 mouse melanoma by *N*-acetylglucosaminyltransferase III gene transfection. *Proc. Natl. Acad. Sci. USA*, **92**, 8754–8758.

Yousefi, S., Higgins, E., Daoling, Z., Pollex-Krüger, A., Hindsgaul, O., Dennis, J.W. (1991) Increased UDP-GlcNAc:Galβ 1-3GalNAc-R (GlcNAc to GalNAc) β-1,6-*N*-acetylglucosaminyltransferase activity in metastatic murine tumor cell lines. Control of polylactosamine synthesis. *J. Biol. Chem.*, **266**, 1772–1782.

B4. Roles of Glycans in Bacterial Infections: Interaction Host-Mycobacteria

Martine Gilleron, Michel Rivière and Germain Puzo

1. INTRODUCTION

Over the past 10 years, a significant and always growing literature suggests that the glycoconjugates of the mycobacterial envelope play a crucial role in the immuno-pathogenesis of tuberculosis infection and also in protective immunity. Like for many other cases of intracellular bacterial pathogens, the glycoconjugates of the envelope constitute a primary line of interactions between the infecting mycobacteria and their host-cells. Most of the recent findings indicate that the recognition by specific cell receptor of these complex glycoconjugates such as the lipoarabinomannans (LAM), may determine the correct addressing and the adhesion of the pathogenic mycobacteria to their target cells: the alveolar macrophages. Moreover, several mycobacterial glycoconjugates are also known to modulate both the microbicidal activity and the cytokine secretion of the phagocytes supporting the idea that they may be involved in the intramacrophagic survival of the bacteria and therefore in the bacterial virulence. In addition, mycobacterial glycoconjugates were found to stimulate human $\alpha\beta$ T-cells suggesting their putative involvement in the protective immunity against tuberculosis.

The most pathogenic strains, such as *Mycobacterium tuberculosis*, are ordinarily intracellular parasites of host mononuclear phagocytes. So, the first step of the primary infection is the invasion of the target cells and great efforts were made to delineate the molecular basis of the bacilli adhesion to the host cells. As for most intracellular pathogens of mammalian cells, the entrance of *M. tuberculosis* and other pathogenic mycobacteria into human mononuclear phagocytes occurs through a conventional receptor-mediated phagocytosis which depends on the type of receptor present on the host cell surface. There are now evidences that several glycoconjugates are involved in the different interaction mechanisms adopted by mycobacteria to invade target cells. These different entry routes are likely to determine the fate of the bacilli into the phagocytes. They are mediated by specific carbohydrate recognition through several different opsonic or nonopsonic pathways including the complement, the collectins and the mannose receptor pathways.

Recently, important findings reveal that non-peptide antigens, assigned to mycobacterial lipoglycans and glycolipids are recognized by human CD1-restricted $\alpha\beta$ T-cells highlighting the possibility of elaboration of molecular vaccines based on glycolipids.

In this review, we will focus on the structure of the mycobacterial glycoconjugates and their role in the binding of mycobacteria to phagocytes and in the T-cells activation.

2. THE MYCOBACTERIAL LIPOGLYCANS: PURIFICATION AND STRUCTURE

2.1. Introduction

Both the architecture and the glycoconjugate structure of the mycobacterial envelope are unique in the bacterial field (Puzo, 1990; Brennan *et al.*, 1995). Among these glycoconjugates, the lipoarabinomannans (LAM), are known to play a key role in the control of the host immune response. So, this chapter summarizes our current knowledge of the structure of the lipoarabinomannans and structurally related molecules: the lipomannans (LM), the arabinomannans (AM) and the phosphatidyl-inositol-mannosides (PIM).

The structural models of LAM (Figure 1) show that they are composed by a carbohydrate backbone constituted by two homopolysaccharides, the mannan core and the arabinan domain. At its reducing end, the mannan core is terminated by the phosphatidylinositol mannoside anchor. The arabinan domain was shown to be capped by mannosyl residues or phosphoinositides. This structural difference allowed to classify the LAM into two classes: the ''ManLAM'', characterized by the presence of mannosyl residues and the ''AraLAM'', containing phosphoinositide caps. Beside the LAM, other cell wall components share similar structural motifs: the LM which roughly correspond to arabinose-free LAM, the AM, described as LAM devoid of phosphatidyl-inositol anchor and finally the PIM, consisting of phosphatidyl-inositol bearing from two (PIM$_2$) to six (PIM$_6$) mannosyl units.

All these molecules are heterogeneous in size. This was first suggested by SDS-PAGE analysis of the LAM and LM which migrate as broad bands around 30–40 kDa and 15–20 kDa respectively (Venisse *et al.*, 1993). This heterogeneity was further

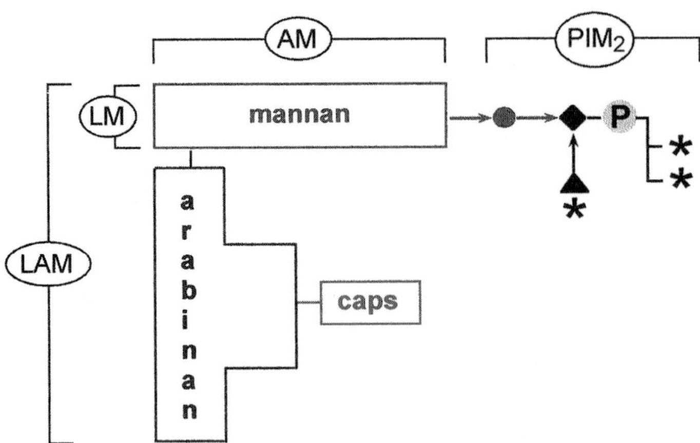

Figure 1: Schematic representation of the structural relationship of PIM$_2$, LM, AM and LAM. LAM and LM are presented as putative extension of PIM$_2$.

supported by gel permeation analysis. In these experiments, the deacylated LAM (dLAM) and LM (dLM) were purified, from a BioGel P100, as broad peaks despite use of denaturing buffers to disrupt molecular aggregates. In these conditions, the LAM molecular weight was estimated to be around 10 to 15 kDa. Finally, MALDI-TOF mass spectrometry allowed to access to more precise molecular weights of native and deacylated LAM (Venisse *et al.*, 1993). The native LAM of *Mycobacterium bovis* BCG and *M. tuberculosis* gave a broad peak centered at 17.4 kDa (16.7 kDa after deacylation) with a size distribution of, at least, 4 kDa. Once again, the shape of the signal reflected the molecular heterogeneity of the LAM fraction. Recently, this MS analytical approach applied to deacylated and permethylated LAM from *Mycobacterium smegmatis* revealed a molecular weight around 13 kDa ± 3 kDa (Khoo *et al.*, 1996).

2.2. LAM Purification

The global purification procedures of the native LAM and LM developed by Brennan's and Puzo's groups are both based on the same general features, including i) solvent extractions, ii) enzyme digestion and dialysis to remove DNA, RNA, proteins and glucans and iii) separation as individual entities using size exclusion column chromatography (Sephacryl 200 or BioGel P100), checked by SDS-PAGE. However, they differ by a major point concerning the extraction mode. Brennan's group uses disrupted delipidated cells to prepare an unique lipoglycans fraction. In contrast, our strategy (Figure 2) results in the obtention of two distinct manno-glycoconjugates (AM / LAM / LM) fractions namely "parietal" and "cellular". The parietal manno-glycoconjugates are obtained by ethanol/water extraction of delipidated cells. Afterwards, these cells are disrupted and extracted again with an ethanol/water mixture leading to the cellular manno-glycoconjugates. This dual extraction was applied to *M. bovis* BCG (Venisse *et al.*, 1993; Delmas *et al.*, 1997; Nigou *et al.*, 1997), *M. smegmatis* (Gilleron *et al.*, 1997) and *M. tuberculosis* H37Rv (Gilleron *et al.*, 2000).

The obtention of lipoglycans fraction (LAM and LM) devoid of AM is a critical step. Separation of AM from LAM and LM was first achieved by phenol-water biphasic wash of the mycobacterial organic solvent extract (Brennan *et al.*, 1995). In order to improve this separation, a Triton X-114 phase separation step was introduced prior to gel filtration. By this way, the lipoglycans (LAM and LM) were recovered from the detergent rich phase while the glycans, mainly AM, remained in the aqueous phase (Nigou *et al.*, 1997).

Another way of purification, based on the hydrophobic interaction of the lipoglycans with the octyl-Sepharose matrix, was used to retain the lipoglycans (LAM, LM and PIM) using low concentration of propanol, while the nucleic acid and the remaining AM were eluted (Leopold *et al.*, 1993). By increasing the content of propanol in water, LAM, LM and PIM were shown to be fractionated according to the decreasing size difference of the glycan moiety and increasing size of the lipidic part. This strategy was applied by Khoo *et al.* (1996) to recover LAM and LM containing fraction from non-acylated polysaccharides and to separate the mannan core from free oligosaccharides after endoarabinanase digestion of LAM. Likewise, use of affinity chromatography on ConA-Sepharose matrix was reported to prepare a LAM fraction from a pellet of an ultracentrifugated phenol/water extract of *Mycobacterium paratubercutosis* (Sugden *et al.*, 1987).

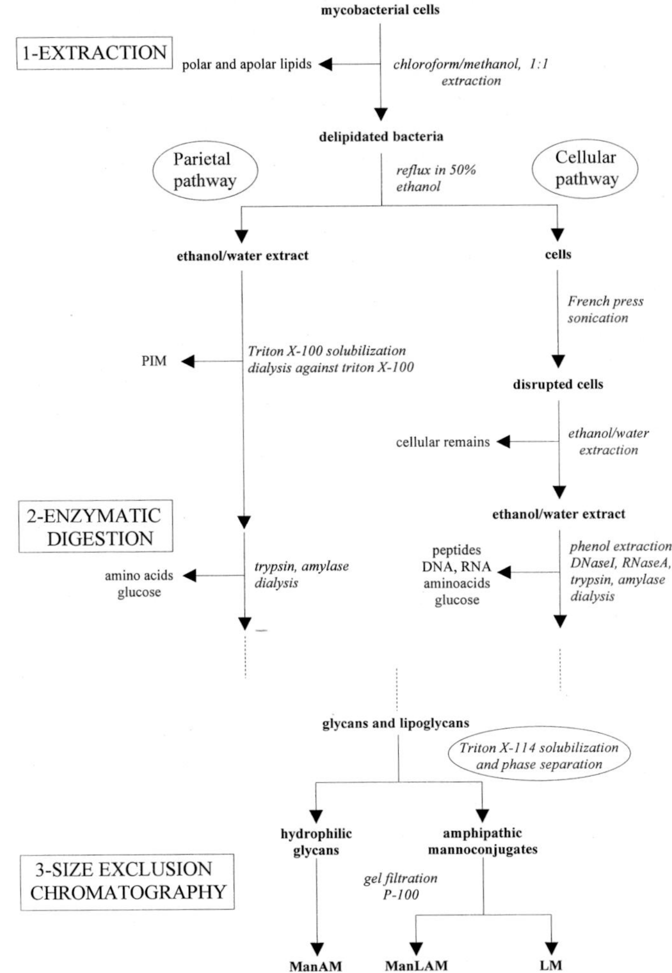

Figure 2: Purification scheme of the parietal and cellular LAM and related compounds (AM and LM) from the *M. bovis* BCG Pasteur strain (from Nigou *et al.*, 1997).
The purification steps in boxes represent routine strategy while those in circle were newly developed.

2.3. Carbohydrate Backbone

2.3.1. Mannan core

The mannan core of LAM from different mycobacteria strains has been studied: *M. tuberculosis* Erdman (Chatterjee *et al.*, 1993), *M. bovis* BCG (Venisse *et al.*, 1995b) (Figure 3C) and *M. smegmatis* (Khoo *et al.*, 1996). They share a common structural feature assigned to a linear α-(1→6)-Man*p* backbone with single α-(1→2)-Man*p* side chains (as described in Misaki *et al.*, 1977). However, they differ by the level of branching which was found higher for the LAM of *M. tuberculosis* Erdman (Chatterjee *et al.*, 1993) compared to those of *M. bovis* BCG parietal (Venisse *et al.*, 1995b) and *M. smegmatis* (Khoo *et al.*, 1996). Likewise, in *M. bovis* BCG, it was established that the mannan core of the parietal AM (Figure 3D) was more branched than that of the parietal LAM. Also, these cores are composed by a variable number of mannosyl residues contributing to the LAM heterogeneity. The major homologous of the mannan core from the *M. smegmatis* LAM is composed by 26 residues.

2.3.2. Arabinan domain

The structure of the arabinan domain was provided by Chatterjee *et al.* (1991) studying the "AraLAM" from a fast growing *Mycobacterium* sp. strain, which was considered at that time to be a *M. tuberculosis* H37Ra variant. They demonstrated that the arabinosyl residues are furanosides by ^{13}C NMR analysis of the native molecule and by

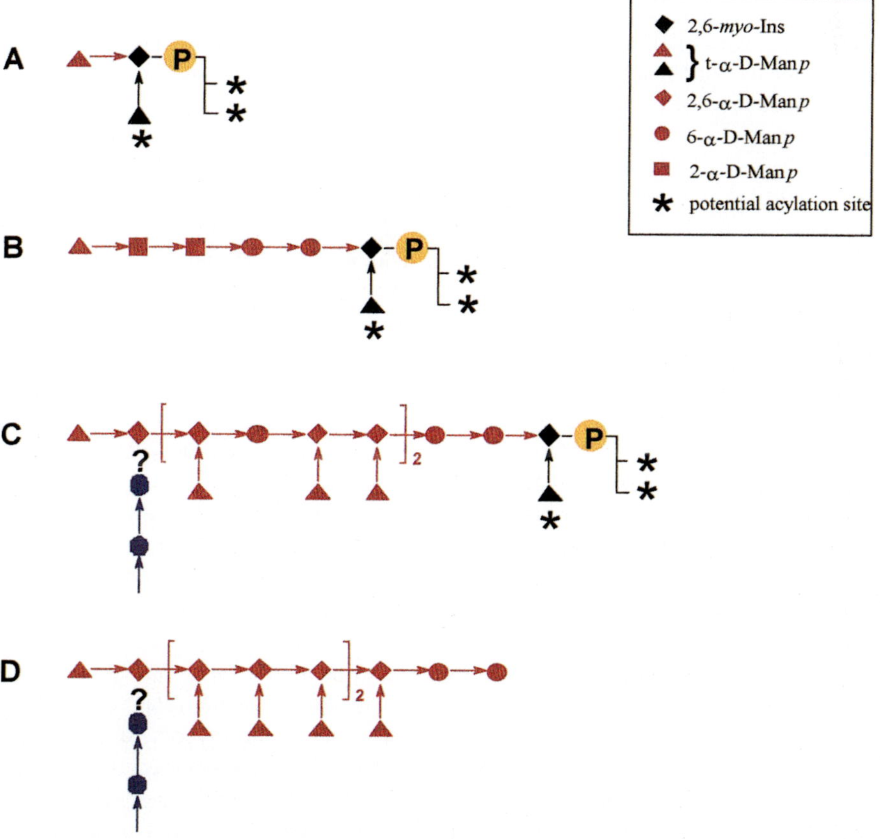

Figure 3: Structural models of PIM$_2$ (A), PIM$_6$ (B), and of mannan cores from parietal *M. bovis* BCG LAM (C) and AM (D).

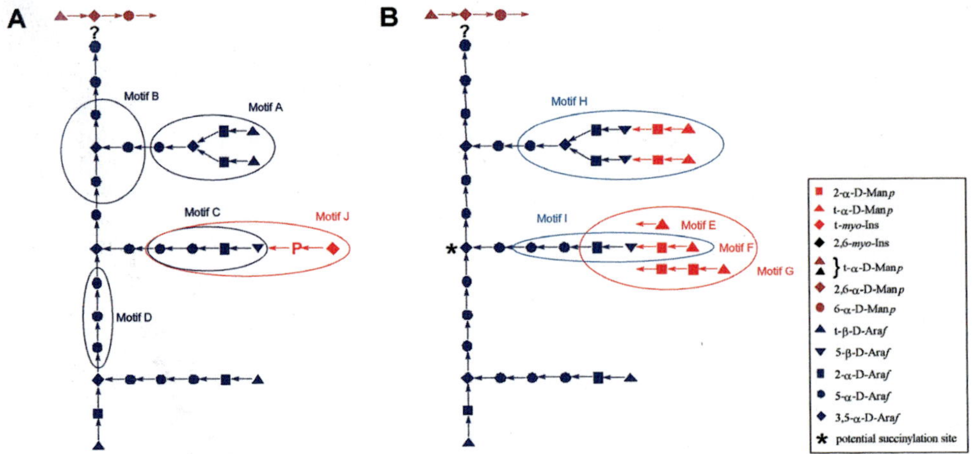

Figure 4: Structure of the arabinan domains of the *M. smegmatis* LAM (A) and *M. bovis* BCG LAM (B).
The major motifs structurally elucidated (A to J) are surrounded. Details about these motifs are given in the text. The arabinan linkage to the mannan core is not yet established. The proposed model does not account for the total number of Ara*f* residues, the branching point and the number of lateral chains and the number of residues on each lateral chains.

GC/MS of alditol acetates derived from sophisticated per-O-alkylated LAM including per-O-methylation and per-O-ethylation. Four oligoarabinofuranoside structural motifs (motifs A to D) were then characterized by a sequence of degradations, derivatizations, separations and GC/MS analysis (Figure 4A). From this study, a model was proposed showing that the arabinan domain is composed by i) a linear oligosaccharide of 5-α-Ara*f* (motif D), ii) side chains composed by 5-α-Ara*f*, attached to the C3 of the linear oligosaccharide (motif B) and iii) non-reducing terminal segments which consist of either linear Ara_4 (motif C) or biantennary Ara_6 (motif A). The anomeric configuration of the arabinosyl residues was deduced from the 1D ^{1}H NMR analysis of the different fragments, highlighting 2-α-Ara*f*, 5-α-Ara*f*, 3,5-α-Ara*f* and t-β-Ara*f*.

The motifs A and C were also obtained by enzymatic digestion of deacylated LAM with an endoarabinanase from a soil *Cellulomonas* species. Their precise structure was further determined by methylation (McNeil *et al.*, 1994) and MS/MS analysis (Khoo *et al.*, 1995a). A protocol based on digestion by this enzyme and HPAEC purification of the digestion products coupled to FAB/MS analysis was developed by Khoo *et al.* (1996) in order to quantify both motifs in LAM. The structural model proposed for the arabinan domain appears similar in all the LAM investigated. However, the relative abundance of the motifs described above could be different according to the mycobacterial strain.

The presence of succinyl residues in LAM fraction was first reported by Hunter *et al.* (1986). Recently, working on LAM from *M. bovis* BCG, we proved the presence of such succinyl residues which were localized on the arabinan domain (Delmas *et al.*, 1997). By 2D NMR experiments performed on the native molecules, they were shown to esterify the OH-2 of 3,5-α-Ara*f* units. Moreover, the average number of succinyl groups was estimated between 1 and 4 per LAM molecule depending on the *M. bovis* BCG strain.

2.4. Capping Motifs of the Arabinan Domain

2.4.1. Mannose capping in ManLAM

The concept of mannose capping the Ara*f*-containing oligoarabinofuranosides was first proposed by Ohashi (1970). Indeed, among the fragments arising from partial acid hydrolysis of arabinomannan, the two trisaccharides Man-Man-Ara and Man-Ara-Ara were identified. It was proposed that they represent a part of the non-reducing terminus. Similar motifs were found at the non-reducing end of the arabinan side chains of the LAM from the virulent *M. tuberculosis* Erdman strain (Chatterjee *et al.*, 1991). Beside the previously observed arabinosyl motifs (A to D), new motifs containing mannose residues were found (Figure 4B), i.e. either a single α-D-Man*p* (motif E), a dimannoside (α-D-Man*p*-(1→2)-α-D-Man*p*) (motif F), or a trimannoside (α-D-Man*p*-(1→2)-α-D-Man-*p*(1→2)-α-D-Man*p*) (motif G). These mannooligosaccharides were localized at the nonreducing end of the biantennary and linear arabinan side chains according to the following strategy. The LAM was hydrolyzed with the endoarabinanase from *Cellulomonas*, and the resulting fragments were purified by BioGel P6. After routine chemical reactions, the derivatized oligoglycosyl alditols were analyzed by FAB/MS. The following structures, Manp_{2or3}Araf_4 (Figure 4B, motif I: Manp_2Araf_4) and Manp_{2to5}Araf_6 (Figure 4B, motif H: Manp_4Araf_6), were determined as the major structural motifs, typifying the presence of Man*p* residues at the extremities of both linear and biantennary arabinan side chains (Chatterjee *et al.*, 1993). Likewise, the presence of mannose caps were observed for the LAM of all the *M. tuberculosis* strains examined (Erdman, H37Rv, H37Ra). However, it was found that the mannose capping is not restricted to the *M. tuberculosis* LAM since it was also evidenced in LAM of *M. leprae* (Khoo *et al.*, 1995a) as well as in the parietal (Venisse *et al.*, 1993) and the global (Prinzis *et al.*, 1993) LAM of *M. bovis* BCG.

Studying the structure of the parietal LAM of BCG, Venisse *et al.* (1993) have proposed two powerful analytical strategies in order to reveal and to characterize the mannooligosaccharide caps. First, the presence of mannose capping was directly established by 1D ^{13}C and 2D ^{1}H-^{13}C HMQC NMR analysis of the native LAM. Indeed, 2-Man*p* typify the Man*p*-capping while 2,6-Man*p* belong to the mannan core (Figure 3B). Then, the presence of two sorts of 2-O-linked Man*p* (2,6-Man*p* of the mannan core and 2-Man*p* of the capping) proves that the analyzed LAM belong to the ManLAM class, whereas the presence of only one type of 2-O-linked Man*p* (2,6-Man*p* of the mannan core) defines them as AraLAM. This approach was then applied for ''LAM typing'' to demonstrate that the parietal and cellular LAM from four vaccine strains of *M. bovis* BCG (Pasteur, Glaxo, Copenhaguen, and Japanese strains) belong to the ManLAM class (Delmas *et al.*, 1997). The second analytical approach developed concerns the structural determination and the quantification of the mannooligosaccharide caps. Taking advantage of the acido lability of the interglycosidic Ara*f* linkages, the dLAM were hydrolyzed in mild acidic conditions (0.1 N HCl, 90 °C, 15 min) yielding to a mixture containing Ara*f* units, Man*p*-caps and the mannan core. The Man*p*-caps and the Ara*f* units were recovered by BioGel filtration and tagged with the paraaminobenzoate ethyl ester (ABBE) via a reductive amination reaction. Then, the ABBE derivatives were separated by reverse phase HPLC monitored by UV-detection and analyzed by FAB/MS and FAB/MS/MS. Among the following ABEE derivatives characterized, Ara*f*-Man*p*, Ara*f*-Man*p*-Man*p*, and Ara*f*-Man*p*-Man*p*-Man*p*, the trisaccharide was the most abundant (Venisse *et al.*, 1993).

Recently, a new strategy based on capillary electrophoresis (CE) analysis of fluorescent amino-pyrene-trisulfonate carbohydrate derivatives was successfully applied to the analysis of the mannose capping of the BCG LAM (Nigou *et al.*, 1997). The combined use of laser induced fluorescence detection of the APTS derivatives and CE allow high chromatography resolution as well as a 100 fold increase in sensitivity compared to the HPLC analysis. Indeed, starting from 1 μg of cellular BCG LAM (~50 pmoles), we were able to identify and to determine the relative amount of the APTS cap derivatives: Ara*f*-Man*p* (16%), Ara*f*-Man*p*-Man*p* (77%), and Ara*f*-Man*p*-Man*p*-Man*p* (7%). This approach applied to the parietal LAM from *M. bovis* BCG afforded similar relative abundance values. In contrast, the degree of capping was found higher for the parietal LAM from *M. bovis* BCG (76%) than for the cellular ones (48%). It was also different for the LAM from the *M. tuberculosis* strains Erdman (71%), H37Rv (44%), H37Ra (43%) and less than 30% of capping was observed for the LAM from *M. leprae* (Khoo *et al.*, 1995a). From all these studies, it seems reasonable to think that the mannose capping is a characteristic of the LAM from slow-growing mycobacterial strains.

2.4.2. *Phosphoinositide caps in AraLAM—characterization of PI-GAM*

The presence of Ins-P-Ara motifs capping the LAM of a fast growing *Mycobacterium sp.* was first evidenced by Khoo *et al.* (1995a). The LAM was hydrolyzed with trifluoroacetic acid under mild conditions (40 mM, 100 °C, 20 min). The resulting fragments were purified and perdeuteroacetylated for FAB/MS analysis. The negative FAB mass spectrum revealed mainly Ins-P-Ara$_1$ motif beside Ins-P-Ara$_{2-8}$ motifs. Upon digestion of the dLAM by the *Cellulomonas* endoarabinanase, mainly Ins-P-Ara$_4$ motif (Figure 4A, motif J) was retrieved with low amount of more glycosylated fragments containing from 5 to 8 Ara*f* units. These results indicate that the phosphoinositides cap a part of the linear arabinan chains (Figure 4A). These motifs were also found in the parietal LAM isolated from *M. smegmatis* (Gilleron *et al.*, 1997). Their structure was unambiguously established with a strategy based on both 1D ^{31}P and 2D ^{1}H-^{31}P HMQC and HMQC-HOHAHA NMR experiments applied to the native LAM. Indeed, the 1D ^{31}P NMR spectrum unambiguously reveals the presence of two types of phosphodiester groups (called Pl and P2) with an integration ratio of one to four. P2 was found esterified by a terminal *myo*-Ins through the *myo*-Ins OH-1 and a β-D-Ara*f* through the Ara*f* OH-5 typifying these phosphoinositide capping motifs. P1 was assigned to the phosphatidyl *myo*-Ins anchor phosphate. From the integration value, the presence of four phosphoinositide residues per LAM molecule was proposed.

Among these four phosphoinositide caps, three were found alkali-labile and one alkali-stable. Indeed, the 1D ^{31}P NMR spectrum of the *M. smegmatis* parietal LAM treated in mild alkaline conditions (0.1 NaOH, 30 min, 37 °C) shows a decrease of the P2 signal intensity leading to a P1/P2 ratio of 1. These two mild alkali-stable phosphodiester groups were attributed to the expected *myo*-Ins-1-P-Gro of the anchor, and to the {(t-*myo*-Ins-1)-P-(5-β-D-Ara*f*)} cap respectively. It is noteworthy that all P2 have the same substituents though they have different behavior toward mild alkaline hydrolysis. From a chemical point of view, the P2 alkali-labile hydrolysis mechanism remains an open question. Nevertheless, the presence of alkaline stable and labile inositol phosphates is in agreement with the previous observation showing that alkaline treatment of LAM, radiolabelled with ^{3}H *myo*-Ins, gives rise to free phospho-*myo*-Ins (Hunter *et al.*, 1990).

It is interesting to note that the phosphoinositides are localized on the OH-5 of the β-Ara$f$ replacing the Man$_1$, Man$_2$ or Man$_3$ units that typify ManLAM (Figure 4A). The parietal LAM of *M. smegmatis* are thus characterized by the presence of phosphoinositide caps and also by the absence of fatty acids on the anchor (see below) and so were named PI-GAM for PhospholInositol-GlyceroArabinoMannans. As they correspond to parietal, i.e. LAM obtained from extraction of delipidated cells, they do not represent the totality of the *M. smegmatis* LAM. But the cellular LAM, under investigation, appear to be acylated (unpublished results), clarifying the Chatterjee's comments (1998).

2.5. Phosphatidylinositol Mannoside Anchor

It has been proposed that LAM and LM are multiglycosylated forms of the phosphatidyl-inositol-mannoside molecules (PIM). The PIM are composed of phosphoinositol di-, tri-, tetra-, penta and hexamannosides (PIM$_{2 \text{ to } 6}$) (Ballou *et al.*, 1963; Lee *et al.*, 1965). PIM$_2$, the most abundant component, is composed by a phosphatidyl unit linked to *myo*-Ins glycosylated by two mannose units at the position 2 and 6 respectively (Figure 3A). PIM$_4$ is described as an extension of PIM$_2$ by addition of the disaccharide unit t-α-Man$p$(1→6)-α-Man$p$(1→6) on the Man$p$ linked to C-6 of the *myo*-Ins. In turn, PIM$_6$ results from the elongation of PIM$_4$ by the disaccharide unit t-α-Man$p$(1→2)-α-Man$p$(1→2) (Figure 3B). LM are described as polymannosylated PIM$_2$ (Figure 3C) while LAM seem to correspond to LM on which is attached the arabinan domain.

The presence of *myo*-Ins-1-P, as putative anchor motif of the LAM from *M. tuberculosis* was firstly claimed by Hunter *et al.* (1986). This assignment was based on the characterization, by GC/MS, of phospho-*myo*-Ins arising from LAM alkalinolysis (0.1N NaOH, 100 °C, 24 h). However, this assumption that *myo*-Ins-P typified the anchor was erroneous according to the two following points: i) the PIM phosphodiester linkage is known to be resistant to alkalinolysis and when the LM are treated in the same conditions, no significant release of *myo*-Ins-P can be detected, ii) the *M. tuberculosis* strain was probably a fast growing mycobacteria strain containing LAM capped with phosphoinositides. Indeed, by selective LAM radio-labeling (^{3}H *myo*-Ins), the presence of both alkali-stable and labile *myo*-Ins-l-P was demonstrated (Hunter *et al.*, 1990). The presence of the Ins-P-Gro structure was never evidenced, but the glycerol unit was found after acidic hydrolysis of the alkali-stable fraction. Moreover, the observation of palmitic and tuberculostearic fatty acids agrees with the idea that taken together, these motifs can be combined to form the phosphatidyl-*myo*-Ins anchor. Finally, this assumption was supported by the substitution pattern of the inositol at the C2 and C6 by Man$p$ units and at the Cl by the phosphate established by methylation analysis of the LAM before and after dephosphorylation (Chatterjee *et al.*, 1992a). Moreover, the presence of this anchor was evidenced in the dLM by digestion with an exo and endo α1 → 6 mannosidase which lead to the {1-(*sn*-glycerol-3-phospho)-D-*myo*-Ins-2,6-bis-α-D-Man$p$} fragment, indistinguishable from that derived from the PIM$_2$.

The presence and the structure of the phosphatidyl-*myo*-inositol anchor were definitively proved by 2D NMR analysis of the mannan core obtained by mild acidic hydrolysis of the parietal LAM from *M. bovis* BCG, and purified by gel filtration and anion exchange chromatography (Venisse *et al.*, 1995b). The phosphate was shown esterified by the OH-l of the *myo*-Ins unit and the OH-3 of the glycerol unit, resulting in the {(*myo*-Ins-1)-P-(3-Gro)} structure. Moreover, the characteristic chemical shifts of the H-2 and H-6 of the *myo*-Ins indicate that this residue is glycosylated on both C-2 and C-6 positions. Finally, this NMR analytical approach was successfully applied to characterize the presence and to establish the structure of the anchor from the native parietal and the deacylated cellular LAM of BCG (Nigou *et al.*, 1997) and from the *M. smegmatis* LAM (Gilleron *et al.*, 1997).

Moreover, recognition of key fragments (obtained after permethylation, partial acid hydrolysis, reduction, ethylation), analyzed by GC/MS demonstrates that only a single α-Man*p* is present on the C-2 of the inositol and that the extended mannan emanates from C-6 (Chatterjee *et al.*, 1992a). None of the fragments observed from PIM$_6$, containing linear 2-linked Man*p* residues were retrieved in LAM, but two fragments of probable structure {→6Man*p*-1→Man*p*-1→6Man*p*-1→6Ins-2←} and {→2,6Man*p*-1→6Man*p*-1→6Man*p*-l→6Ins-2←} were advanced leading to the anchor model described in Figure 3C.

An important variability of the acylation state of the LAM, LM and PIM phosphatidylinositol anchor has been described. Leopold *et al.* (1993) have recently purified, by octylsepharose chromatography, LAM and LM from *M. tuberculosis* into species bearing from two to four fatty acids. The presence of tri- and tetra-acylated forms of PIM was reported by several authors working on different mycobacterial strains (Pangborn *et al.*, 1966; Brennan *et al.*, 1967; Khuller *et al.*, 1968). In these different studies, the nature of the acyl groups was never rigorously characterized and their sites of attachment were not established. In order to get a better insight in the possible role of PIM as biosynthetic precursors of LM and LAM, the structure of the acylated moiety of PIM was recently revised by Khoo *et al.* (1995b). Using FAB/MS analysis, they presented evidences for tri-acylated PIM as a common lipid anchor for both *M. tuberculosis* Erdman strain and *M. leprae* LM and LAM, and postulated that acylation of the PIM$_2$ may constitute a key regulatory step in their biosynthesis. The fatty acids (essentially C$_{16}$, C$_{18}$ and C$_{19}$, beside C$_{14}$, C$_{16:1}$, C$_{16:2}$) were located on the PIM$_2$ on both position of the glycerol and on the C-6 position of mannose (Figure 3A). However, the parietal LAM of *M. smegmatis* (PI-GAM) have been described devoid of fatty acids on the glycerol moiety (Gilleron *et al.*, 1997) while the parietal and the cellular LAM from *M. bovis* BCG differ drastically in regard of the lipidic moiety (Nigou *et al.*, 1997). Indeed, the glycerol moiety of the cellular ManLAM showed a large heterogeneity due to a combination of palmitic and tuberculostearic acid while the glycerol moiety of the parietal ManLAM was esterified on its C-l by a novel fatty acid assigned to a 12-O-(methoxypropanoyl)-12-hydroxystearic acid.

3. GLYCANS MEDIATE MYCOBACTERIA BINDING TO PHAGOCYTES

3.1. Introduction

M. tuberculosis is an intracellular bacterial pathogen of mononuclear phagocytes. The first step of the infection consists in the binding of the bacteria to the macrophage and is triggered by specific receptor-ligand interactions. In the last few years, significant efforts have been made to define both the ligands of the mycobacteria envelope and their corresponding receptors on the macrophage. The following chapter summarizes our current knowledge of the role the mycobacterial envelope glycoconjugates in the binding of virulent and non-virulent mycobacteria to the macrophages.

3.2. Carbohydrate Interactions with the Complement Cascade Components

It is now well established that both the classical and the alternative complement pathways play an important role in the binding of several mycobacteria to mononuclear phagocytes (Swartz *et al.*, 1988) through their heat-labile serum components (C3, C3b and C3bi) and their specific receptors (CR1, CR3 and CR4) present at the surface of the phagocytes (Schlesinger *et al.*, 1990a; 1990b; 1991a) (Figure 5). However, very few is known regarding the mycobacterial cell wall acceptor of the soluble complement molecules C3b and C3bi. The *M. tuberculosis* trehalose dimycolate (Cord Factor: CF) activates the alternative complement pathway suggesting it might be a potential ligand for the C3 (Ramanathan *et al.*, 1980). Likewise, the purified phenolic glycolipid-I (PGL-I) from *M. leprae* was shown to bind the complement component C3 in solid phase assay suggesting that it could be a major mediator of the opsonic binding of the leprosy bacillus to phagocytes (Schlesinger *et al.*, 1991b).

Recently, the complement receptor CR3 has been found involved in the nonopsonic binding of *M. tuberculosis* to CR3 transfected Chinese hamster ovary (CHO) cells (Cywes *et al.*, 1996). CR3 is a $\beta2$ integrin present in the membrane of myeloid and lymphoid cells, and is mainly involved in phagocytosis of C3bi coated particles. It is composed of two polypeptides: CD11b and CD18. There are now convincing evidences that the CD11b chain presents a lectin site at its C-terminal domain, which may correspond to the previously reported phagocyte β-glucan receptor (Thornton *et al.*, 1996; Ross *et al.*, 1985). Inhibition by monoclonal antibodies directed against different moieties of the CR3

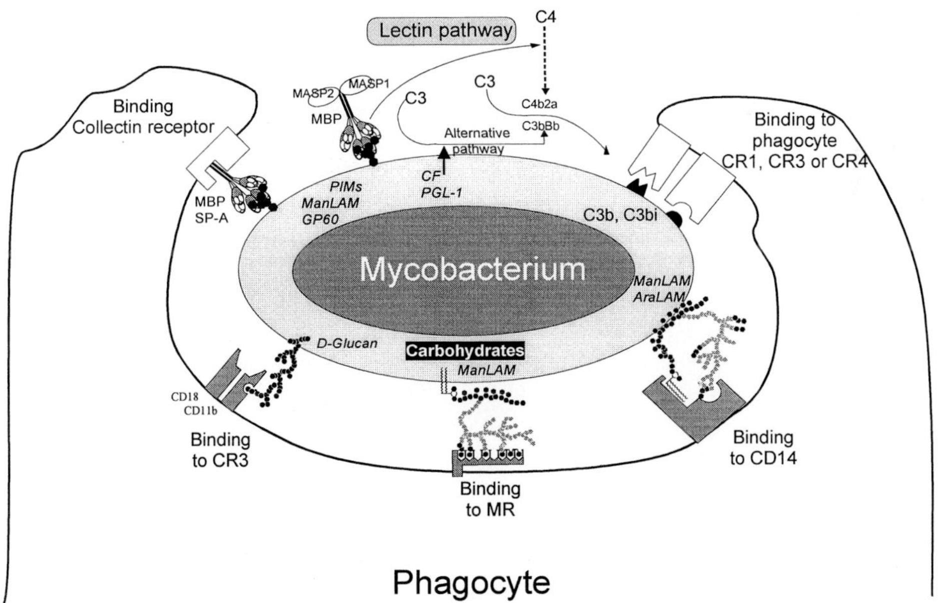

Figure 5: Carbohydrate mediated binding of mycobacterium to phagocyte.
CF: cord factor, GP-60: 60 kDa glycoprotein from *M. tuberculosis*, MBP: mannose binding protein, MR: mannose receptor, PGL-l: phenolic glycolipid 1 from *M. leprae*, SP-A: pulmonary surfactant protein A.

receptor revealed that *M. tuberculosis* interacts at a different site than C3bi binding site and most probably at the level of the β-glucan lectin domain (Cywes *et al.*, 1997). This was supported by the fact that the interaction is inhibited by laminarin, a ligand of the β-glucan receptor. Further investigations led to the conclusion that the nonopsonic binding of *M. tuberculosis* to the CR3 is mediated by the D-glucan, a major mycobacterial capsular polysaccharide.

It is noteworthy that CR3 (or CR4) mediated phagocytosis may present a particular interest for mycobacteria intracellular survival (Marra *et al.*, 1996; Isberg *et al.*, 1994) since this entry route does not activate the phagocyte microbicidal respiratory burst (Berton *et al.*, 1992; Wright *et al.*, 1983; Aderem *et al.*, 1985; Relman *et al.*, 1990).

3.3. Mycobacterial Glycoconjugate Ligands of Host Collectins

3.3.1. Mannose binding protein

Direct recognition of pathogen cell surface carbohydrates by collectin, and in particular by MBP and SP-A, is likely to be an important event in the first line host defense against a variety of microorganisms. However, only few data are available regarding the exact structure of their glycoconjugate ligands. Several studies provide evidences that the binding of collectins to gram-negative bacteria is mediated by their external membrane lipopolysaccharides (since presence of a capsule significantly impairs the interactions (van Emmerik *et al.*, 1994)) while interaction with viruses occurs mainly through their envelope glycoproteins. Regarding mycobacteria, it has been reported that sonicates of *M. tuberculosis* or *M. leprae* bind strongly MBP (Garred *et al.*, 1994). Likewise, previous results demonstrated that binding of the ManLAM from *M. bovis* BCG to phagocytes is enhanced by heat inactivated serum suggesting that MBP or other serum lectin may be involved in the adhesion (Venisse *et al.*, 1995a). Furthermore, the binding of MBP to *M. avium* increases by two-fold the uptake of the bacilli by human neutrophils in nonopsonic conditions. In order to determine the putative ligand recognized by MBP, several mycobacterial envelope components were tested by solid phase binding assay. ManLAM from *M. tuberculosis* were found to be the best ligand of MBP, exhibiting two times more binding activity compared to the AraLAM, PIM or LM (Polotsky *et al.*, 1997). In contrast, further studies indicate that MBP mediated binding of *M. tuberculosis* to Chinese hamster ovary (CHO) fibroblasts is strongly inhibited by monoclonal antibodies directed against PIM (Hoppe *et al.*, 1997). This result suggests that PIM$_t$ are the major MBP ligands on the mycobacterial envelope. The discrepancy between the observations made on purified molecules or on whole cells may result from a differential exposure of ManLAM and PIM at the surface of the bacilli and from the much higher amount of PIM present in the mycobacterial envelope.

3.3.2. Pulmonary surfactant protein

Recently, the pulmonary surfactant protein A (SP-A) (McCormack, 1997) involved in the phagocytosis of several pulmonary pathogens (Tino *et al.*, 1996) has been shown to bind to *M. tuberculosis* (Pasula *et al.*, 1997) or *M. bovis* BCG (Weikert *et al.*, 1997). Inhibition of these interactions by EDTA or carbohydrates supports the idea that the binding is mediated by specific carbohydrate-lectin interactions. Binding of SP-A to *M. bovis* BCG cells is inhibited by galactose but not by mannan or mannose suggesting, according to the authors, that the arabino-galactan could be the major *M. bovis* BCG cell surface ligand of SP-A (Weikert *et al.*, 1997). However, this proposal seems surprising, in regard to the generally admitted internal location of this constituent in the mycobacterial cell wall (Puzo, 1990).

In contrast, Pasula *et al.* (1997) concluded that SP-A interaction with *M. tuberculosis* may be mediated by a mycobacterial envelope mannose rich glycoprotein. Indeed, they observed that either addition of mannosylated BSA (Man-BSA) or pretreatment of mycobacterial cells by trypsin, significantly reduced the adhesion of SP-A to *M. tuberculosis*. The ligand was assigned to a glycoprotein with an apparent molecular weight of 60 kDa (GP 60). This was achieved by direct binding of SP-A to *M. tuberculosis* crude protein extract resolved by SDS-PAGE. Deglycosylation of SP-A completely abolished the binding to the bacilli, suggesting that the N-linked carbohydrate moiety of the collectin plays a determinant role in the interaction with the mycobacterial cell. Thus, the mechanism by which the SP-A binds to *M. tuberculosis* cells is still puzzling. Does the SP-A recognize the mannose moiety of the GP 60 or does the mycobacterial GP 60 bind the carbohydrate domain of SP-A? This point still remains an open question. It can not be excluded that the N-linked oligosaccharides of SP-A only act by maintaining the structural and conformational integrity of the molecule necessary for the functionality of the lectin domains (CRDs).

Although SP-A has been found to enhance the phagocytosis of pathogenic mycobacteria by macrophages, a full understanding of this activity awaits further investigations. Gaynor *et al.* (1995) reported that preincubation of macrophages with SP-A results in the increase of the uptake of *M. tuberculosis* by human phagocytes. However, the effect is still observed after removal of the SP-A by washing the macrophages prior addition of mycobacteria. This finding suggested that the SP-A is not directly involved in the binding of the infectious particle to phagocytic cells. Indeed, further data demonstrated that the adhesion of *M. tuberculosis* was mediated by the mannose receptor. In addition, the enhancement of the level of phagocytosis was shown to result from an upregulation of the mannose receptor induced by a direct interaction of SP-A with the phagocytes. Then, SP-A was considered as an important regulatory factor involved in the tuning of important biological functions of the macrophage (van Iwaarden *et al.*, 1990; Kalina *et al.*, 1995). In contrast, Weikert *et al.* (1997) provided evidences for a direct binding of the SP-A coated *M. bovis* BCG cells to phagocytes. Addition of type V collagen abolished the adhesion of the coated mycobacteria while it has no effect on the binding of the collectin to the bacilli. It was then assumed that the interaction of SP-A coated particles with macrophages is mediated by the collagen moiety of SP-A. Moreover, the specific SP-A receptor termed SPR2lO was found to be responsible of the attachment of the mycobacteria opsonised by SP-A to the macrophage. These results demonstrate that, in addition to its effect on the regulation of expression of the MR, SP-A may also act as an opsonin by performing a bridge between the infectious particle and the phagocyte.

3.3.3. Role of collectins

The consensus that emerges from recent finding is that the collectins are likely to bind to surface glycoconjugates of a wide range of microorganisms *via* their lectin domains and to activate the host antimicrobial immune defense by interacting with specialized cells or complement cascade activators (Figure 5). The precise physiological significance of this pathogen clearance pathway is not fully understood though it has been observed that young children presenting MBP deficiency suffer of severe recurrent infections due to defective opsonisation (Summerfield *et al.*, 1997). Kuhlman *et al.* (1989) have shown that MBP has a direct opsonizing activity by binding the mannose rich O-polysaccharide of *Salmonella montevideo* resulting in an enhanced phagocytosis by human monocytes. However, the mechanism by which the collectins participate to the killing of intracellular pathogens is still unclear. Indeed, MBP was described mediating the killing of *E. coli* through the classical complement pathway (Kawasaki *et al.*, 1989) by activating the $C1r_2C1s_2$ complex without the involvement of the C1q (Ikeda *et al.*, 1987). On the other end, MBP mediated phagocytosis and killing of Salmonella has been shown to occur via the alternative complement pathway (Schweinle *et al.*, 1989). Furthermore, it has been reported that MBP binding to capsular polysaccharides of several Salmonella or *E. coli* strains provokes a C1s independent activation of the C4 complement component through a new MBP associated serine protease (MASP) (Figure 5) (Matsushita *et al.*, 1992). An alternative mechanism of opsonisation mediated by MBP, independent of the complement cascade components, emerged from the report that MBP as well as SP-A are able to bind directly the leukocytes C1q receptor through their collagen domains (Malhotra *et al.*, 1990). All these functions are likely to result from the structural homology between the C1q, the MBP and the SP-A. Thus, whatever the pathway (dependent or independent of the complement cascade), the specific binding of MBP or SP-A to mycobacterial cell surface glycoconjugates results in the opsonophagocytosis of the bacilli. Moreover, the SP-A, present in the pulmonary surfactant covering the surface of the pulmonary alveoli, is of special interest since it is probably the first component of the host defense line, encountered by an invading *M. tuberculosis* cell. It seems obvious that some of the mycobacterial envelope glycoconjugates are produced by the bacilli to take advantage of such entrance route to gain access to their niche, where the most virulent will survive and multiply.

3.4. LAM Mediated Mycobacterial Phagocytosis via Macrophage Mannose Receptor

The active form of the macrophage mannose receptor (MR) (Stahl *et al.*, 1980) is a monomeric 180 kDa type I transmembrane glycoprotein (Lennartz *et al.*, 1987; Taylor *et al.*, 1990; Ezekowitz *et al.*, 1990) which displays five distinct functional domains: an amino terminal cysteine-riche region (GalNac-4-SO_4 binding domain (Fiete *et al.*, 1998)), a fibronectin type II repeat (FN-II), a series of eight tandem lectin such as carbohydrate recognition domains (CRDs), a transmembrane domain and a COOH cytosolic tail. It belongs to the multilectin receptor family which also includes the phospholipase A2 receptor (Zvaritch *et al.*, 1996) (eight CRDs), the dendritic cell receptor (DEC-205) (Jiang *et al.*, 1995) (ten CRDs) and the newly described mouse endothelial cell lectin (MMU56734) (Wu *et al.*, 1996) (eight CRDs). Although first report suggested that the MR was restricted to tissue macrophages, it is now clear that its expression is regulated and that it is present in several cell types (Stahl *et al.*, 1998). Beside its role in the elimination of potentially harmful endogenous manno-glycoconjugates, the MR was found also involved in the clearance of several intracellular pathogens (Wilson *et al.*, 1986; Ezekowitz *et al.*, 1991). This major biological function has been shown to be mediated by the carbohydrate recognition domains and thus much efforts were made to delineate the molecular basis of the adhesion of MR to pathogenic microorganisms. The eight CRDs although similar (30%

of homology) are not equivalent. Sequence comparisons among several C-type lectins including MBP, SP-A and asialoglycoprotein, reveal that only CRDs 4 and 5 exhibit the consensus aminoacids sequence (. . .*EPN*. . . .*E*. . . .*WND*. . . . .) involved in the binding of Ca^{++} and carbohydrate residues (Mullin *et al.*, 1997). This was also supported by Taylor *et al.* (1992; 1993) by testing several deletions of MR purified from insect cells. This study reveals that the CRD 4 is the smallest portion that retains the ability to interact with sugar ($K_I \sim 10^{-3}$ M) although CRDs 4 to 8 must be present to achieve the same level of affinity than the intact receptor for natural ligands ($K_I \sim 10^{-9}$ M). This discrepancy between these values is likely to result from the clustering of the different CRDs allowing a high affinity multivalent binding of branched ligands. In contrast to these well documented studies on the carbohydrate binding mechanisms, very few is known concerning the glycoconjugate ligands recognized by MR at the surface of the pathogen bacilli.

In 1993, Schlesinger reported that the phagocytosis of virulent, but not attenuated, strains of *M. tuberculosis* was mediated by the macrophage mannose receptor (MR) in addition to the complement receptor. Indeed, the use of monoclonal antibodies against complement receptor molecules (CR1, CR3 and CR4) leads to a significant decrease in the uptake by monocyte-derived macrophages of the three *M. tuberculosis* strains studied (Erdman, H37Rv and H37Ra). In contrast, inhibition of the mannose receptor functions, by either specific antibodies or competition with mannan or mannose-BSA, affected only the adhesion of the two virulent strains Erdman and H37Rv (decrease in the level of binding of about 60%) (Schlesinger, 1993). This possible new entry route specific of pathogenic mycobacteria has gained much interest since, for the first time, it provided some molecular basis of mycobacteria pathogenesis. Obviously, the characteristic ManLAM, isolated from the virulent *M. tuberculosis* Erdman strain (Chatterjee *et al.*, 1992b) was then suspected to be the specific ligand responsible of the MR mediated adhesion of virulent mycobacteria. This was supported by the observation that microspheres coated with *M. tuberculosis* Erdman ManLAM exhibit a more than three fold increase in adherence to monocyte-derived macrophages compared with microspheres coated with AraLAM issued from a non virulent fast growing mycobacterial species. Removal of terminal mannose from the ManLAM by α-exomannosidase treatment abolished the adhesion supporting the determinant role of the terminal manno-oligosaccharide caps in the ManLAM recognition (Schlesinger *et al.*, 1994). Moreover, the adhesion to macrophages of the ManLAM coated microspheres was found significantly reduced by mannan-induced down-regulation of the mannose receptor supporting its involvement in the interaction. Likewise, by flow cytometry using a phycoerythrin-streptavidin conjugate, biotinylated ManLAM of *M. bovis* BCG was shown to interact directly with murine phagocytes (Venisse *et al.*, 1995a). This binding to macrophages and granulocytes was found to be temperature- and divalent ions-dependant and inhibited by yeast mannan, suggesting the involvement of the mannose receptor in the binding process. In contrast, in the same conditions, AraLAM bind in much less extent (Venisse *et al.*, unpublished data). Recently, it has been reported that the level of binding is slightly higher for Erdman ManLAM compared to H37Rv or H37Ra ones (Schlesinger *et al.*, 1996). This result which affords some clue to the first observation of Schlesinger (1993, cited above) highlights some probable subtle structural differences between these ManLAM molecules, which still remain to be characterized. Thus, this macrophage invasion pathway is likely to be specific of the most virulent strains of mycobacteria. However, to date, no direct relationship has been found between this entry route and the ability of the virulent bacilli to survive and multiply inside the hostile macrophage. Several hypothesis were put forward, but the most attractive is the assumption that the adhesion mediated by the MR allows the virulent strains to escape from the phagocyte anti-microbial response by confining the bacilli in some protected cellular compartments.

3.5. Mycobacterial Lipoglycans and CD14

Two others phagocyte receptors, the CD14 and the scavenger receptors, are suspected to be involved in the glycoconjugate mediated phagocytosis of mycobacteria (Peterson *et al.*, 1995; Zimmerli *et al.*, 1996). CD14, known as the gram-negative bacteria lipopolysaccharide (LPS) receptor is a 55 kDa glycosyl-phosphatidylinositol anchored membrane protein present on several phagocyte surfaces including microglial cells. It has been reported that phagocytosis of the virulent *M. tuberculosis* H37Rv by the microglial cells may be through the CD 14. Furthermore, in an independent study, AraLAM were shown to compete for the binding of LPS to soluble CD14 providing evidence that AraLAM directly bind to CD14 (Pugin *et al.*, 1994). In a same way, the release of TNF-α and IL-1β by human mononuclear phagocyte stimulated with LAM can be blocked with an anti-CD14 monoclonal antibody. Thus, the modulation by LAM of the production of those cytokines is likely to result from the interaction between LAM and the CD14. However, the mechanism of CD14-mediated induction of TNF-α release by LAM-stimulated human monocytes must be distinct from that of LPS. Indeed, it is not inhibited by polymixin B suggesting that the activation does not involve the lipidic moiety of the LAM. This is further supported by the recent finding that the *M. smegmatis* PI-GAM, a new class of LAM-like molecules (see above) devoid of fatty acids on their phosphoglycerol anchor, induce TNF-α release probably through the inositolphosphate caps (Gilleron *et al.*, 1997). However, in this latter study, the involvement of the CD14 has not been investigated. Further studies are thus necessary to clarify the mechanism of interaction of AraLAM with the CD14 and the subsequent induction of an antimicrobial response. To date, no direct evidences are available concerning the possible interaction of ManLAM to the CD14, although it has been observed that ManLAM are 100 fold less potent than AraLAM in eliciting the TNF-α production by stimulated macrophages. At present, this lack of activity of ManLAM is attributed to the absence of phosphoinositol caps at the extremity of the arabinan domain, though it does not exclude that other still undetermined structural differences could be also involved. However, it can be suspected that mannose capping, which is specific of the virulent mycobacteria LAM, constitutes a mean to inhibit or to escape to the phagocyte antimicrobial response.

3.6. Conclusion

Mycobacteria can adopt different entry routes to invade the phagocytes, by interacting with specific receptors in either opsonic or non opsonic conditions (Figure 5). Of these, the binding mechanisms resulting from recognition of carbohydrate ligands are of special interest because of the extraordinary abundance and diversity of the mycobacterial envelope glycoconjugates. At present, several mycobacterial cell wall glycoconjugates were shown to be involved in four important carbohydrate mediated adhesion routes:

GP 60 or PIM collectin recognition, non opsonic binding of capsular D-glucan to CR3, ligation of ManLAM to MR and the interaction of the CD14 with the AraLAM. The relative significance of these pathways is still undetermined and phagocytosis may not be due to a single receptor, but rather the result of simultaneous binding to different receptor types. Furthermore, the different ways of entry are probably not truly independent since the expression of certain of these receptors has been shown to be tightly interrelated (inverse regulation of MR expression and macrophage SP-A binding (Gaynor *et al.*, 1995; Chroneos *et al.*, 1995)). Thus, these distinct pathways may cooperate to optimize the binding and phagocytosis of mycobacteria and the resulting effect may determine the fate of the infection.

It is also obvious that glycoconjugate ligands present at the surface of the mycobacteria may determine the preeminent internalization route and thereby may greatly affect also the outcome of the host immunological response. Then, the mycobacterial LAM are of special interest for the wide spectrum of associated modulatory activities of the macrophage effector functions (for review see Fenton *et al.*, 1996; Schlesinger, 1996; Ernst, 1998). However, it is noteworthy that ManLAM from slow growing mycobacterial strains and AraLAM from fast growing mycobacteria exert most of the time opposite effects by probably activating differently distinct cell signaling pathways through different phagocyte receptors (Bernardo *et al.*, 1998; Knutson *et al.*, 1998). AraLAM are likely to trigger the activation of the macrophage microbicidal functions by eliciting the macrophage activation early response genes (c-fos, JE, KC) (Roach *et al.*, 1994), and the release of several important antimicrobial response cytokines and chemokines including TNF-α, IL-1, IL-6 (Dahl *et al.*, 1996), IL-8 (Riedel *et al.*, 1997), IL-10 (Roach *et al.*, 1995), IL-12 (Yoshida *et al.*, 1997). In contrast, ManLAM activate in a much less extent or not at all, these antimicrobial responses. Thus, ManLAM have been considered as factors of virulence since, in comparison to AraLAM, they allow the intracellular survival and growth of virulent mycobacteria. Obviously, these different effects have been attributed to the nature of the caps located at the non reducing end of the arabinan domain which define the two classes of LAM. However, it is clear that the acyl groups linked to the glycero-phospho-*myo*-inositol moiety of the ManLAM constitute the second critical functional domain of these molecules. Indeed, removal of the fatty acids by deacylation, abolished almost all of their biological properties among which the binding to the CD1 antigen presenting molecules (Sieling *et al.*, 1995) or to the mannose receptor (Venisse *et al.*, unpublished results). It can be assumed that the lipidic moiety of the ManLAM may be involved in the tridimensional structure or the efficient presentation of the polysaccharide. Moreover, recent results demonstrate that in a given strain (i.e. *M. bovis* BCG), the cellular and parietal ManLAM induce very different levels of IL-8 and TNF-α release by human dendritic cells (Nigou *et al.*, 1997). This surprising effect has been attributed to subtle structural variations in the number and the nature of the acyl groups associated to the phophatidylinositol anchor of the molecule. Although considered as critical, it is the first time that a differential activity of these multifunctional lipopolysaccharides is attributed to the lipidic moiety of the molecule. One can imagine that the metabolic control of the acylation state of these lipoglycans may be involved in the fine tuning of their activities. This emphasizes the need for further fine structural investigations of these complex molecules to gain insight their biological properties and their role in the immunopathology of the tuberculous and other mycobacterial infections.

4. LIPOGLYCANS, A NEW CLASS OF T-CELL ANTIGENS

4.1. Introduction

It is now well established that T-cells play an essential role in the protective immunity against mycobacterial infections. For more than one decade the central dogma in immunology has been that $\alpha\beta$ T-cell activation process occurs via the presentation of foreign peptides by the MHC class I and II molecules.

However, recent *in vitro* studies in the context of $\alpha\beta$ T-cell activation by mycobacterial antigens have revealed that $\alpha\beta$ T-cells could recognize nonpeptide antigens (Beckman *et al.*, 1994). These mycobacterial antigens were identified as lipids, glycolipids and lipoglycans. The ability of $\alpha\beta$ T-cells to recognize non-peptide antigens is independent of class I and II MHC molecules but involves the CD1 as antigen-presenting molecules (Figure 6). Here, we provide an overview of the recognition of mycobacterial lipoglycans by $\alpha\beta$ T-cells. We will emphasize the role of their carbohydrate part.

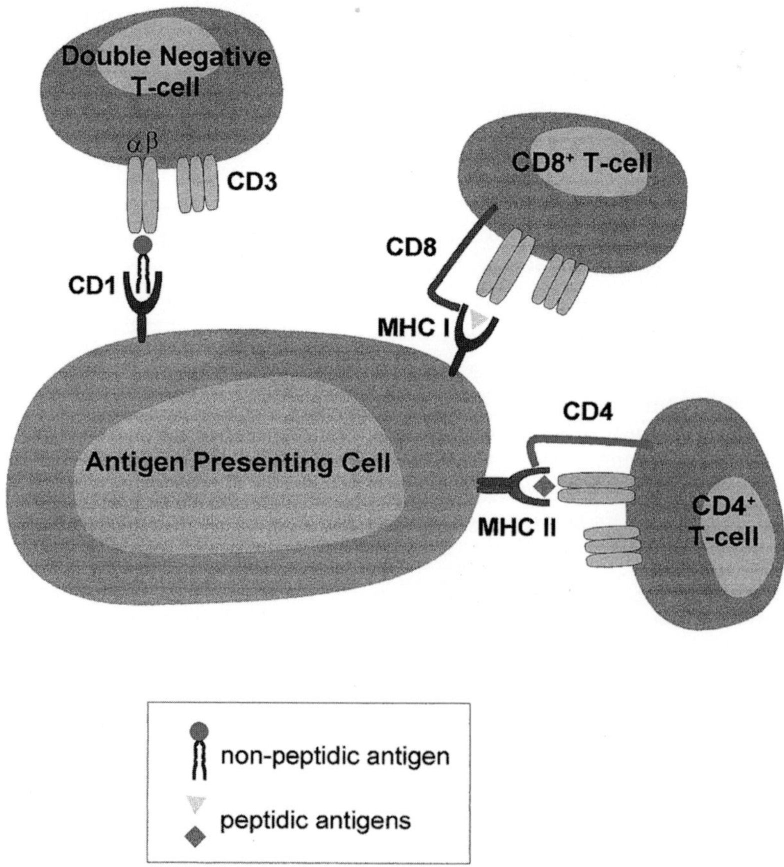

Figure 6: A schematic representation of T-cells recognition by peptide and non-peptide antigens.

4.2. The Human CD1 Membrane Glycoproteins

4.2.1. Expression

The CD1 molecules form a novel family of antigen-presenting molecules different from the MHC class I and II molecules (for review, see Porcelli, 1995). The CD1 genes are located on a different chromosome than the MHC genes and encode four isotypes which can be separated into two groups according to their aminoacid sequences. The classic or group I includes the CD1a, CD1b, and CD1c glycoproteins which are expressed on human lymphoid cells including the immature precursor of T-cells and a sub-population of B lymphocytes. These proteins are also expressed in dendritic cells which are widely distributed in lymphoid and non-lymphoid tissues, as well as in tissue macrophages at certain inflammatory sites. The group 2 includes the human CD1d and the mouse CD1 which are expressed in epithelial cells.

4.2.2. Function

There are now clear evidences that the CD1 molecules are involved in the specific recognition of self antigens (Joyce *et al.*, 1998) and microbial antigens, by a subset of T-cells characterized by the absence of expression of CD4 and CD8 co-receptor molecules. These novel T-cells, referred as double negative (DN $\alpha\beta$) (Porcelli, 1995), represent a quantitatively minor population, in comparison to the classic $\alpha\beta$ TCR expressing CD4$^+$ (helper/inducer) and CD8$^+$ (cytotoxic/suppressor) subsets. However, they act as components of efficient cell-mediated immunity against intracellular pathogens such as *M. tuberculosis*. So, a significant finding was the characterization, from the blood of normal human subjects or from granulomatous skin lesions of leprosy patients, of DN $\alpha\beta$ human T-cell clones that responded to *M. tuberculosis* antigens and that were dependent on CD1b expression.

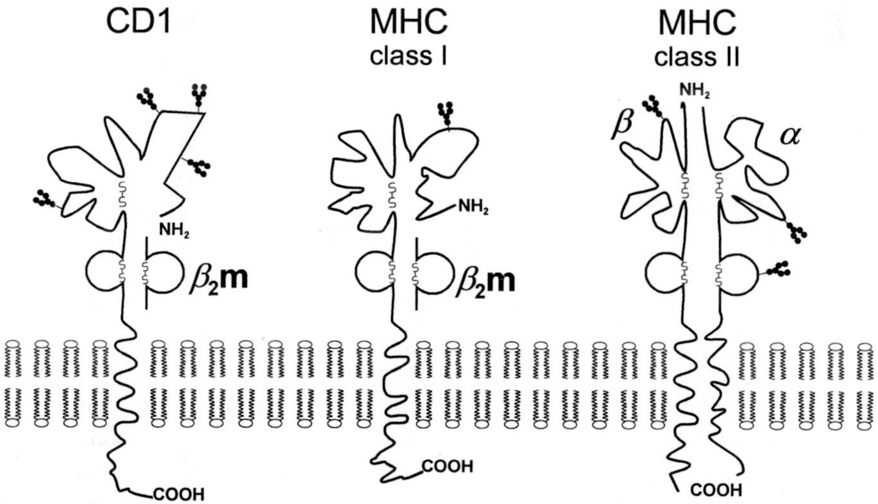

Figure 7: Drawing proposed by Porcelli *et al.* (Porcelli, 1995) of the CD1 glycoproteins organization in comparison to MHC class I and II glycoproteins.

4.2.3. Structure

CD1 is a family of non polymorphic glycoproteins composed by a heavy chain non covalently associated with β2-microglobulins (β2m), displaying structural features similar to the MHC class-1 molecules (Figure 7) (Wilson *et al.*, 1998; Melian *et al.*, 1996; Blumberg *et al.*, 1995). The CD1 heavy chain is a type I transmembrane glycoprotein with a short cytoplasmic tail (6 to 7 residues for CD1b), a transmembrane region and three extracellular domains namely α1, α2 and α3. The extracellular domains have a highly conserved length of approximately 90 aminoacids. The α3 domain shows a significant homology with the MHC molecules due to its association with the β-microglobulin (Figure 7). The α1 and α2 are less homologous with MHC proteins and are characterized by their high content of hydrophobic aminoacids. It was postulated that these hydrophobic aminoacids form an antigen-binding pocket allowing the anchoring of the amphipatic mycobacterial antigens.

The mCD1d corresponding to the mouse extracellular domain associated with β2-microglobulin chain has been crystallized in the absence of added antigen and its structure has been resolved at 2.8 Å of resolution (Zeng *et al.*, 1997). The secondary, tertiary and quaternary overall structures of mCD1d is similar to those of MHC class I proteins. The binding groove domain is formed by the α1 and α2 helical domains and an eight-stranded antiparallel β-pleated sheet platform. The relative positions of the helical segments and of the β-strands modulate the size of the cleft formed between the α1 and α2 helices (Figure 8). The entrance and the binding of the antigen in the groove could be mainly controlled by topographic parameters as, for instance, its size. In the case of mCD1d, the cleft is 14 Å wide and the CD1 α1 helix is 4 to 6 Å higher than in the MHC I molecules resulting in the formation of a deeper groove. The CD1d groove is formed by two pockets of different size labeled A' (18 to 20 Å long) and F' (10 to 12 Å long). The groove is closed at both ends and is only accessible through a narrow entrance of 6 to 7 Å wide. The A' pocket wall is made entirely by hydrophobic residues and the only polar residues are located at the surface and the entrance of the groove. From sequence homology, it was proposed that the pocket A' is probably part of all CD1 molecules, but significant differences can be noticed regarding the aminoacids composing the exposed surface and the carboxyl-terminus of the groove.

From this study, it can be proposed that the binding of lipid or glycolipid antigens is compatible with the hydrophobic character of the CD1 groove. Moreover, contrarily to specific fatty acid binding protein, no exposed basic residues are present on the CD1 groove floor, predicting that the putative ligand does not have amphipatic character. So, it can be deduced that the carbohydrates as well as all the hydrophilic residues of the lipoglycans are most likely located on the outside of the binding groove.

4.3. Mycobacterial Ligands: The Lipoarabinomannans

The unique known function of human CD1 molecules is the ability to present non-peptide antigens to T-cells. Indeed, it was found that human CD1b could present, to different human T-cell clones, several mycobacterial components which are mycolic acids (Moody *et al.*, 1997) and lipoglycans such as LAM (Sieling *et al.*, 1995). Human CD1c molecules have also been shown to present LAM (Beckman *et al.*, 1996).

Two DN $\alpha\beta$ T-cell lines, reactive to mycobacteria, were derived from the skin lesions of a leprosy patient (line LDN4) and from the peripheral blood mononuclear cells of a sane donor (line BDN2) (Sieling *et al.*, 1995). The antigen response of these two DN $\alpha\beta$ T-cell

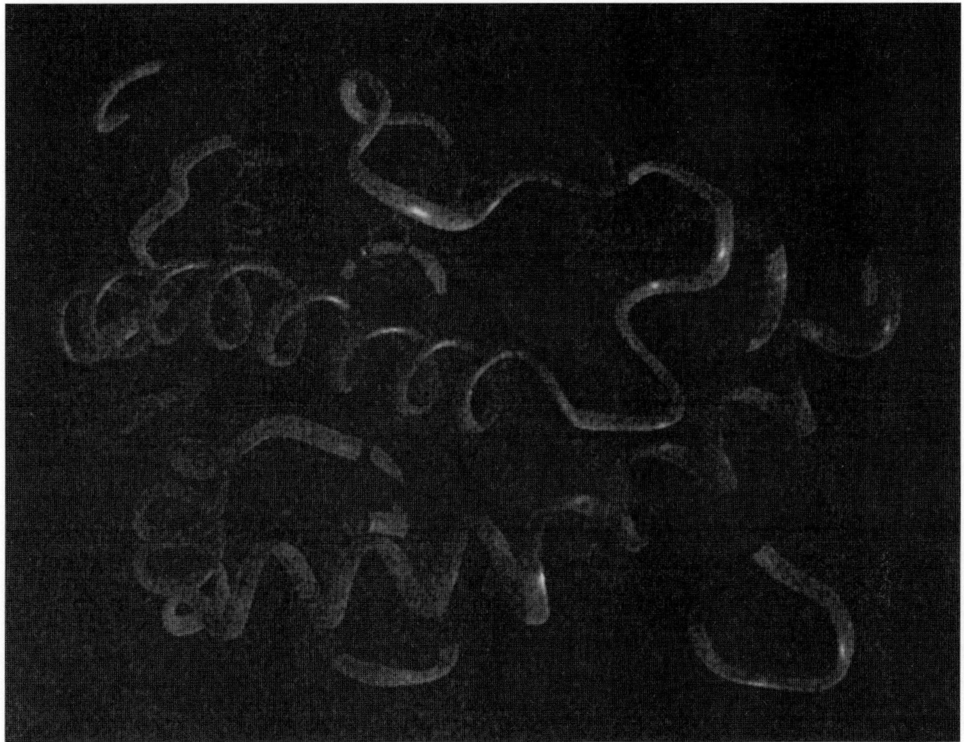

Figure 8: The mouse CD1d structure according to *Zeng et al.* (1997): top view of the ligand binding α1 and α2 domains.

lines was exclusively associated with components of the mycobacterial envelope which presentation was restricted by CD1b molecules. The major antigen of LDN4 was identified as the ManLAM of *M. leprae*, but LM were also found to present a low activity, approximately half compared to that of ManLAM. However, PIM, which represent the anchor structure of the ManLAM were not active. The majority of the activity of ManLAM and LM were alleviated either by digestion with the α-exomannosidase which provokes the loss of the α1→2 Man*p* caps, or by alkaline treatment which induces the loss of the anchor fatty acids. It was proposed that both the mannose capping and the native anchor motive are critical structural determinants for LDN4 T-cell responsiveness. Surprisingly, *M. tuberculosis* ManLAM, which also contains these two motives, were found inactive. It was suggested that this different behavior was due to the differences in the percentage of Man*p* capping in the ManLAM of *M. tuberculosis* and *M. leprae* (see above).

In contrast, the DN $\alpha\beta$ T-cell line BDN2 was stimulated by the ManLAM, the LM and the PIM from both *M. tuberculosis* and *M. leprae*. However, in this case, the T-cell proliferation, measured as the stimulation index, was considerably lower than that observed for the LDN4 T-cells. Again the activity was dramatically decreased after alkaline or mannosidase treatments, supporting the assumption that T-cell recognition requires both the Man*p* capping and the lipidic part of the anchor.

Another DN $\alpha\beta$ T-cell line (LDN5), specific for mycobacterial lipid antigens was derived from a skin biopsy of a cutaneous reaction to *M. leprae*. This T-cell line was found to proliferate via CD1b to a glycolipid fraction assigned to a 6-mycoloyl-D-Glc*p* (GMM) (Moody *et al.*, 1997). The T-cell response was not affected either by drastic variation in the chain length of mycolic acid residue or by different substitution in the meromycolic chain. For instance, a decrease of the chain length from C80 to C32 does not affect T-cell proliferation. However, it was clearly established that recognition of the GMM was dependent on both the α-branched chain and the β-hydroxy group which are typical of the mycoloyl acid structure. In order to investigate the role of the carbohydrate moiety in the specificity of the T-cell response, 6-mycoloyl-D-Man*p* and D-Gal1*p* were syntethized (Man*p* and Gal*p* are epimers of the Glc*p* at the 2 and 4 position of the pyranose ring respectively). Using these synthetic glycolipids, no T-cell proliferation was observed. These data reinforce the idea that the carbohydrate part of the glycolipid is directly involved in T-cell activation.

It was also shown that the human CD1c isotype can mediate antigen presentation to DN $\alpha\beta$ T-cell lines DN2, DN6 and BDN2e isolated from normal donor blood (Beckman *et al.*, 1996). It was established that the ManLAM from *M. leprae* but not from *M. tuberculosis* induced BDN2e T-cell proliferation. PIM$_2$ and PIM$_6$ were also recognized by both BDN2e and DN6 T-cells.

In summary, to date, two types of CD1 restricted mycobacterial glycoconjugate antigens have emerged. Firstly, the phosphoglycoconjugates composed by ManLAM, PIM and LM, and secondly, the glycolipids typified by the 6-mycoloyl-D-Glc*p*.

4.4. Putative Mechanism of Non-Peptide Antigen Presentation to CD1 Restricted T-cells

The exact mechanism of mycobacterial glycoconjugates presentation such as ManLAM, LM, PIM or GMM, is not yet well understood. This process includes at least two steps which consist in the antigen binding to the CD1 followed by the antigen presentation to the T-cell receptor by CD1 molecules.

4.4.1. *The antigen CD1 binding is monitored by the lipid structure of the phosphatidyl-myo-inositol anchor*

By analogy to the crystal structure of murine CD1d glycoprotein described above, it can be deduced that the antigen binding sites of the human CD1b and CD1c molecules are also composed of an hydrophobic groove with two pockets A$'$ and F$'$. In terms of size, shape, and electrostatic topography, the CD1b and CD1c binding grooves are ideally suited to interact with the fatty acid residues of the phosphatidyl-*myo*-inositol anchor shared by ManLAM, LM, PIM and with the mycoloyl moiety of the GMM. It is clear that lipoglycan and glycolipid antigens binding to the CD1 isotypes occurs through hydrophobic interactions between the fatty acid residues and the non polar aminoacids which compose the groove (Moody *et al.*, 1996). It seems possible that subtle differences in the phosphatidyl-*myo*-inositol anchor such as the fatty acid structure (see the "exotic structure" of the fatty acid from the BCG parietal ManLAM), the acylation degree (2 to 3 fatty acid residues) or the site of acylation (glycerol, Man*p*) could discriminate the antigen CD1 binding.

4.4.2. *The T-cell antigen recognition is controlled by the structure of the carbohydrate part*

This point was clearly demonstrated by the fact that T-cell proliferation was drastically modulated by changing the stereochemistry of the monosaccharide moiety of the GMM. In the case of the ManLAM, the mechanism is less well understood mainly due to the complexity and the molecular heterogeneity of these molecules. It is quite well established that the $\alpha(1\rightarrow2)$Man$p$ caps are involved in the recognition, and the frequency of capping could be one parameter involved in this process, as it is suggested by the different response with the ManLAM of *M. leprae* or *M. tuberculosis*. However, we do not know how these caps are exposed in the ManLAM molecules. Moreover, a discrimination can exist between the assortment of mono-, di- and trisaccharide Man$p$ which represent the cap structures.

In summary, the small differences in the carbohydrate part and lipid moiety must be the cause of the ability of CD1 restricted T-cells to distinguish between these LAM molecules.

4.5. Conclusion

The CD1 glycoproteins are expressed in immature human dendritic cells which also expressed the mannose receptor. The carbohydrate part of the mycobacterial antigens presented by CD1 have been shown to participate also in the uptake of these antigens by the mannose receptor. This point was demonstrated exclusively in the case of the ManLAM of *M. tuberculosis* and *M. bovis* BCG. Moreover, it was proposed that the mannooligosaccharide caps are involved in the ManLAM binding to the CRD of the mannose receptor. Dissociation of the ManLAM from the mannose receptor occurs in the endosome at low pH. The mannose receptor recycles to the cell surface and the ManLAM is delivered to the CD1 glycoprotein (Prigozy *et al.*, 1997). The ManLAM binding to the CD1 occurs in this case by the lipidic part of the phosphatidyl-*myo*-Ins unit. Most of the functional studies described above have been conducted *in vitro* and the *in vivo* relevance of the role of CD1 restricted T-cells during the immune response has not been yet precisely established (Banchereau *et al.*, 1998). However, it is likely that this process, involving ManLAM and glycolipids, could contribute to an increase of the protective immunity particularly against intracellular pathogens such as *M. tuberculosis*, which possesses an envelope abundantly composed of lipoglycans and glycolipids (Jullien *et al.*, 1997). So, since CD1 is a family of nonpolymorphic glycoproteins, lipoglycans and glycolipid molecules could be evaluated as putative vaccine candidates, with the advantage that the same antigen could elicit a CD1-restricted T cell response for all recipients.

ACKNOWLEDGEMENTS

We gratefully acknowledge Dr M. Eyrard for the figure of the mouse CD1d molecular model and Dr D. Zerbib for comments concerning the English writing.

APPENDIX—ABBREVIATIONS

ABEE: p-aminobenzoate ethyl ester
BCG: bacillus Calmette Guérin
CD: cluster of differentiation

CF: cord factor
CHO: Chinese hamster ovary
CRD: carbohydrate recognition domain
dLAM: deacylated LAM
dLM: deacylated LM
DN$\alpha\beta$: double negative–$\alpha\beta$
FAB/MS: fast atom bombardment mass spectrometry
GC/MS : gas chromatography/mass spectrometry
GMM: 6-mycoloyl-D-Glc*p*
HMQC: heteronuclear multiple quantum correlation spectroscopy
HOHAHA: homonuclear Hartmann-Hahn spectroscopy
HPAEC: high pH anion-exchange chromatography
HPLC: high performance liquid chromatography
LAM: lipoarabinomannans
MALDI-TOF: matrix assisted laser desorption ionization-time-of-flight
ManLAM: LAM with mannosyl extensions
MBP: mannose binding protein
MR: mannose receptor
LM: lipomannans
LPS: lipopolysaccharides
MHC: major histocompatibility complex
^{1}H, ^{13}C and ^{31}P NMR: proton, carbon and phosphore nuclear magnetic resonance
PGL-1: phenolic glycolipid-1 from *M. leprae*
PI anchor: phosphatidyl-*myo*-inositol anchor
PI-GAM: phosphoinositol-glyceroarabinomannans
PIM: phosphatidyl-*myo*-inositol mannosides
PIM$_2$: phosphatidyl-*myo*-inositol di-mannosides
P-Ins: inositol-1-phosphate
SDS-PAGE: sodium dodecyl sulfate-polyacrylamide gel electrophoresis
SP-A: pulmonary surfactant protein-A
TCR: T cells receptor
Ins: inositol; Man*p*: mannopyranose; Ara*f* arabinofuranose, Gro: glycerol

REFERENCES

Aderem, A.A., Wright, S.D., Silverstein, S.C. and Cohn, Z.A. (1985) Ligated complement receptors do not activate the arachidonic acid cascade in resident peritoneal macrophages. *J Exp Med*, **161**, 617–22.
Ballou, C.E., Vilkas, E. and Lederer, E. (1963) Structural studies on the myo-inositol phospholipids of Mycobacterium tuberculosis (var. bovis, strain BCG). *J Biol Chem*, **238**, 69–76.
Banchereau, J. and Steinman, R.M. (1998) Dendritic cells and the control of immunity. *Nature*, **392**, 245–52.
Beckman, E.M., Porcelli, S.A., Morita, C.T., Behar, S.M., Furlong, S.T. and Brenner, M.B. (1994) Recognition of a lipid antigen by CD1-restricted alpha beta$^+$ T cells. *Nature*, **372**, 691–4.
Beckman, E.M., Melian, A., Behar, S.M., Sieling, P.A., Chatterjee, D., Furlong, S.T., Matsumoto, R., Rosat, J.P., Modlin, R.L. and Porcelli, S.A. (1996) CD1c restricts responses of mycobacteriaspecific T cells. Evidence for antigen presentation by a second member of the human CD1 family. *J Immunol*, **157**, 2795–803.
Bernardo, J., Billingslea, A.M., Blumenthal, R.L., Seetoo, K.F., Simons, E.R. and Fenton, M.J. (1998) Differential responses of human mononuclear phagocytes to mycobacterial lipoarabinomannans: role of CD14 and the mannose receptor. *Infect Immun*, **66**, 28–35.

Berton, G., Laudanna, C., Sorio, C. and Rossi, F. (1992) Generation of signals activating neutrophil functions by leukocyte integrins: LFA-1 and gp150/95, but not CR3, are able to stimulate the respiratory burst of human neutrophils. *J Cell Biol*, **116**, 1007–17.

Blumberg, R.S., Gerdes, D., Chott, A., Porcelli, S.A. and Balk, S.P. (1995) Structure and function of the CD1 family of MHC-like cell surface proteins. *Immunol Rev*, **147**, 5–29.

Brennan, P. and Ballou, C.E. (1967) Biosynthesis of mannophosphoinositides by Mycobacterium phlei. The family of dimannophosphoinositides. *J Biol Chem*, **242**, 3046–56.

Brennan, P.J. and Nikaido, H. (1995) The envelope of mycobacteria. *Annu Rev Biochem*, **64**, 29–63.

Chatterjee, D., Bozic, C.M., McNeil, M. and Brennan, P.J. (1991) Structural features of the arabinan component of the lipoarabinomannan of Mycobacterium tuberculosis. *J Biol Chem*, **266**, 9652–60.

Chatterjee, D., Hunter, S.W., McNeil, M. and Brennan, P.J. (1992a) Lipoarabinomannan. Multiglycosylated form of the mycobacterial mannosylphosphatidylinositols. *J Biol Chem*, **267**, 6228–33.

Chatterjee, D., Lowell, K., Rivoire, B., McNeil, M.R. and Brennan, P.J. (1992b) Lipoarabinomannan of Mycobacterium tuberculosis. Capping with mannosyl residues in some strains. *J Biol Chem*, **267**, 6234–9.

Chatterjee, D., Khoo, K.H., McNeil, M.R., Dell, A., Morris, H.R. and Brennan, P.J. (1993) Structural definition of the non-reducing termini of mannose-capped LAM from Mycobacterium tuberculosis through selective enzymatic degradation and fast atom bombardment-mass spectrometry. *Glycobiology*, **3**, 497–506.

Chatterjee, D. and Khoo, K.H. (1998) Mycobacterial lipoarabinomannan: an extraordinary lipoheteroglycan with profound physiological effects. *Glycobiology*, **8**, 113–20.

Chroneos, Z. and Shepherd, V.L. (1995) Differential regulation of the mannose and SP-A receptors on macrophages. *Am J Physiol*, **269**, L721–6.

Cywes, C., Godenir, N.L., Hoppe, H.C., Scholle, R.R., Steyn, L.M., Kirsch, R.E. and Ehlers, M.R. (1996) Nonopsonic binding of Mycobacterium tuberculosis to human complement receptor type 3 expressed in Chinese hamster ovary cells. *Infect Immun*, **64**, 5373–83.

Cywes, C., Hoppe, H.C., Daffe, M. and Ehiers, M.R. (1997) Nonopsonic binding of Mycobacterium tuberculosis to complement receptor type 3 is mediated by capsular polysaccharides and is strain dependent. *Infect Immun*, **65**, 4258–66.

Dahl, K.E., Shiratsuchi, H., Hamilton, B.D., Elmer, J.J. and Toossi, Z. (1996) Selective induction of transforming growth factor beta in human monocytes by lipoarabinomannan of Mycobacterium tuberculosis. *Infect Immun*, **64**, 399–405.

Delmas, C., Gilleron, M., Brando, T., Vercellone, A., Gheorghui, M., Riviere, M. and Puzo, G. (1997) Comparative structural study of the mannosylated-lipoarabinomannans from Mycobacterium bovis BCG vaccine strains: characterization and localization of succinates. *Glycobiology*, **7**, 811–7.

Ernst, J.D. (1998) Macrophage receptores for Mycobacterium tuberculosis. *Infect Immun*, **66**, 1277–81.

Ezekowitz, R.A., Sastry, K., Bailly, P. and Warner, A. (1990) Molecular characterization of the human macrophage maninose receptor: demonstration of multiple carbohydrate recognition-like domains and phagocytosis of yeasts in Cos-1 cells. *J Exp Med*, **172**, 1785–94.

Ezekowitz, R.A., Williams, D.J., Koziel, H., Armstrong, M.Y., Warner, A., Richards, F.F. and Rose, R.M. (1991) Uptake of Pneumocystis carinii mediated by the macrophage mannose receptor. *Nature*, **351**, 155–8.

Fenton, M.J. and Vermeulen, M.W. (1996) Immunopathology of tuberculosis: roles of macrophages and monocytes. *Infect Immun*, **64**, 683–90.

Fiete, D.J., Beranek, M.C. and Baenziger, J.U. (1998) A cysteine-rich domain of the ''mannose'' receptor mediates GalNAc-4-SO4 binding. *Proc Natl Acad Sci USA*, **95**, 2089–93.

Garred, P., Harboe, M., Oettinger, T., Koch, C. and Svejgaard, A. (1994) Dual role of mannan-binding protein in infections: another case of heterosis? *Eur J Immunogenet*, **21**, 125–31.

Gaynor, C.D., McCormack, F.X., Voelker, D.R., McGowan, S.E. and Schlesinger, L.S. (1995) Pulmonary surfactant protein A mediates enhanced phagocytosis of Mycobacterium tuberculosis by a direct interaction with human macrophages. *J Immunol*, **155**, 5343–51.

Gilleron, M., Himoudi, N., Adam, O., Constant, P., Venisse, A., Riviere, M. and Puzo, G. (1997) Mycobacterium smegmatis phosphoinositols-glyceroarabinomannans. Structure and localization of alkali-labile and alkali-stable phosphoinositides. *J Biol Chem*, **272**, 117–24.

Gilleron M., Bala L., Brando T., Vercellone A. and Puzo G., (2000) *J. Biol. Chem.*, **275**, 677–684.

Hoppe, H.C., de Wet, B.J., Cywes, C., Daffe, M. and Ehlers, M.R. (1997) Identification of phosphatidylinositol mannoside as a mycobacterial adhesin mediating both direct and opsonic binding to nonphagocytic mammalian cells. *Infect Immun*, **65**, 3896–905.

Hunter, S.W., Gaylord, H. and Brennan, P.J. (1986) Structure and antigenicity of the phosphorylated lipopolysaccharide antigens from the leprosy and tubercie bacilli. *J Bio Chem*, **261**, 12345–51.

Hunter, S.W. and Brennan, P.J. (1990) Evidence for the presence of a phosphatidylinositol anchor on the lipoarabinomannan and lipomannan of Mycobacterium tuberculosis. *J Biol Chem*, **265**, 9272–9.

Ikeda, K., Sannoh, T., Kawasaki, N., Kawasaki, T. and Yamashina, I. (1987) Serum lectin with known structure activates complement through the classical pathway. *J Biol Chem*, **262**, 7451–4.

Isberg, R.R. and Tran Van Nhieu, G. (1994) Binding and internalization of microorganisms by integrin receptors. *Trends Microbiol*, **2**, 10–4.

Jiang, W., Swiggard, W.J., Heufler, C., Peng, M., Mirza, A., Steinman, R.M. and Nussenzweig, M.C. (1995) The receptor DEC-205 expressed by dendritic cells and thymic epithelial cells is involved in antigen processing. *Nature*, **375**, 151–5.

Joyce, S., Woods, A.S., Yewdell, J.W., Bennink, J.R., De Silva, A.D., Boesteanu, A., Balk, S.P., Cotter, R.J. and Brutkiewicz, R.R. (1998) Nature ligand of mouse CD1dd1: cellular glycosylphosphatidylinositol. *Science*, **279**, 1541–4.

Jullien, D., Stenger, S., Ernst, W.A. and Modlin, R.L. (1997) CD1 presentation of microbial nonpeptide antigens to T cells. *J Clin Invest*, **99**, 2071–4.

Kalina, M., Blau, H., Riklis, S. and Kravtsov, V. (1995) Interaction of surfactant protein A with bacterial lipopolysaccharide may affect some biological functions. *Am J Physiol*, **268**, L144–51.

Kawasaki, N., Kawasaki, T. and Yamashina, I. (1989) A serum lectin (mannan-binding protein) has complement-dependent bactericidal activity. *J Biochem (Tokyo)*, **106**, 483–9.

Khoo, K.H., Dell, A., Morris, H.R., Brennan, P.J. and Chatterjee, D. (1995a) Inositol phosphate capping of the nonreducing termini of lipoarabinomannan from rapidly growing strains of Mycobacterium. *J Biol Chem*, **270**, 12380–9.

Khoo, K.H., Dell, A., Morris, H.R., Brennan, P.J. and Chatterjee, D. (1995b) Structural definition of acylated phosphatidylinositol mannosides from Mycobacterium tuberculosis: definition of a common anchor for lipomannan and lipoarabinomannan. *Glycobiology*, **5**, 117–27.

Khoo, K.H., Douglas, E., Azadi, P., Inamine, J.M., Besra, G.S., Mikusova, K., Brennan, P.J. and Chatterjee, D. (1996) Truncated structural variants of lipoarabinomannan in ethambutol drug-resistant strains of Mycobacterium smegmatis. Inhibition of arabinan biosynthesis by ethambutol. *J Biol Chem*, **271**, 28682–90.

Khuller, G.K. and Subrabmanyam, D. (1968) On the mannophosphoinositides of Mycobacterium 607. *Experentia*, **24**, 851–52.

Knutson, K.L., Hmama, Z., Herrera-Velit, P., Rochford, R. and Reiner, N.E. (1998) Lipoarabinomannan of Mycobacterium tuberculosis promotes protein tyrosine dephosphorylation and inhibition of mitogen-activated protein kinase in human mononuclear phagocytes. Role of the Src homology 2 containing tyrosine phosphatase 1. *J Biol Chem*, **273**, 645–52.

Kuhlman, M., Joiner, K. and Ezekowitz, R.A. (1989) The human mannose-binding protein functions as an opsonin. *J Exp Med*, **169**, 1733–45.

Lee, Y.C. and Ballou, C.E. (1965) Complete structures of the glycophospholipids of Mycobacteria. *Biochemistry*, **4**, 1395–404.

Lennartz, M.R., Cole, F.S., Shepherd, V.L., Wileman, T.E. and Stahl, P.D. (1987) Isolation and characterization of a mannose-specific endocytosis receptor from human placenta. *J Biol Chem*, **262**, 9942–4.

Leopold, K. and Fischer, W. (1993) Molecular analysis of the lipoglycans of Mycobacterium tuberculosis. *Anal Biochem*, **208**, 57–64.

Malhotra, R., Thiel, S., Reid, K.B. and Sim, R.B. (1990) Human leukocyte C1q receptor binds other soluble proteins with collagen domains. *J Exp Med*, **172**, 955–9.

Marra, A. and Isberg, R.R. (1996) Common entry mechanisms. Bacterial pathogenesis. *Curr Biol*, **6**, 1084–6.

Matsushita, M. and Fujita, T. (1992) Activation of the classical complement pathway by mannose-binding protein in association with a novel C1s-like serine protease. *J Exp Med*, **176**, 1497–502.

McCormack, F. (1997) The structure and function of surfactant protein-A. *Chest*, **111**, 114S–9S.

McNeil, M.R., Robuck, K.G., Harter, M. and Brennan, P.J. (1994) Enzymatic evidence for the presence of a critical terminal hexa- arabinoside in the cell walls of Mycobacterium tuberculosis. *Glycobiology*, **4**, 165–73.

Melian, A., Beckman, E.M., Porcelli, S.A. and Brenner, M.B. (1996) Antigen presentation by CD1 and MHC-encoded class I-like molecules. *Curr Opin Immunol*, **8**, 82–8.

Misaki, A., Azuma, I. and Yamamura, Y. (1977) Structural and immunochemical studies on Darabino-D-mannans and D- mannans of Mycobacterium tuberculosis and other Mycobacterium species. *J Biochem (Tokyo)*, **82**, 1759–70.

Moody, D.B., Sugita, M., Peters, P.J., Brenner, M.B. and Porcelli, S.A. (1996) The CD1-restricted T-cell response to mycobacteria. *Res Immunol*, **147**, 550–9.

Moody, D.B., Reinhold, B.B., Guy, M.R., Beckman, E.M., Frederique, D.E., Furlong, S.T., Ye, S., Reinhold, V.N., Sieling, P.A., Modlin, R.L., Besra, G.S. and Porcelli, S.A. (1997) Structural requirements for glycolipid antigen recognition by CD1b- restricted T cells. *Science*, **278**, 283–6.

Mullin, N.P., Hitchen, P.G. and Taylor, M.E. (1997) Mechanism of Ca^{2+} and monosaccharide binding to a C-type carbohydrate- recognition domain of the macrophage mannose receptor. *J Biol Chem*, **272**, 5668–81.

Nigou, J., Gilleron, M., Cahuzac, B., Bounery, J.D., Herold, M., Thurnher, M. and Puzo, G. (1997) The phosphatidyl-myo-inositol anchor of the lipoarabinomannans from Mycobacterium bovis bacillus Calmette Guerin. Heterogeneity, structure, and role in the regulation of cytokine secretion. *J Biol Chem*, **272**, 23094–103.

Ohashi, M. (1970) Studies on the chemical structure of serologically active arabinomannan from Mycobacteria. *Japan. J. Exp. Med.*, **40**, 1–14.

Pangborn, M.C. and McKinney, J.A. (1966) Purification of serologically active phosphoinositides of Mycobacterium tuberculosis. *J. Lipid Res.*, **7**, 627–33.

Pasula, R., Downing, J.F., Wright, J.R., Kachel, D.L., Davis, T.E., Jr. and Martin, W.J., 2nd (1997) Surfactant protein A (SP-A) mediates attachment of Mycobacterium tuberculosis to murine alveolar macrophages. *Am J Respir Cell Mol Biol*, **17**, 209–17.

Peterson, P.K., Gekker, G., Hu, S., Sheng, W.S., Anderson, W.R., Ulevitch, R.J., Tobias, P.S., Gustafson, K.V., Molitor, T.W. and Chao, C.C. (1995) CD14 receptor-mediated uptake of nonopsonized Mycobacterium tuberculosis by human microglia. *Infect Immun*, **63**, 1598–602.

Polotsky, V.Y., Belisle, J.T., Mikusova, K., Ezekowitz, R.A. and Joiner, K.A. (1997) Interaction of human mannose-binding protein with Mycobacterium avium. *J Infect Dis*, **175**, 1159–68.

Porcelli, S.A. (1995) The CD1 family: a third lineage of antigen-presenting molecules. *Adv Immunol*, **59**, 1–98.

Prigozy, T.I., Sieling, P.A., Clemens, D., Stewart, P.L., Behar, S.M., Porcelli, S.A., Brenner, M.B., Modlin, R.L. and Kronenberg, M. (1997) The mannose receptor delivers lipoglycan antigens to endosomes for presentation to T cells by CD1b molecules. *Immunity*, **6**, 187–97.

Prinzis, S., Chatterjee, D. and Brennan, P.J. (1993) Structure and antigenicity of lipoarabinomannan from Mycobacterium bovis BCG. *J Gen Microbiol*, **139**, 2649–58.

Pugin, J., Heumann, I.D., Tomasz, A., Kravchenko, V.V., Akamatsu, Y., Nishijima, M., Glauser, M.P., Tobias, P.S. and Ulevitch, R.J. (1994) CD14 is a pattern recognition receptor. *Immunity*, **1**, 509–16.

Puzo, G. (1990) The carbohydrate- and lipid-containing cell wall of mycobacteria, phenolic glycolipids: structure and immunological properties. *Crit Rev Microbiol*, **17**, 305–27.

Ramanathan, V.D., Curtis, J. and Turk, J.L. (1980) Activation of the alternative pathway of complement by mycobacteria and cord factor. *Infect Immun*, **29**, 30–5.

Relman, D., Tuomanen, E., Falkow, S., Golenbock, D.T., Saukkonen, K. and Wright, S.D. (1990) Recognition of a bacterial adhesion by an integrin: macrophage CR3 (alpha M beta 2, CD11b/CD18) binds filamentous hemagglutinin of Bordetella pertussis. *Cell*, **61**, 1375–82.

Riedel, D.D. and Kaufmann, S.H. (1997) Chemokine secretion by human polymorphonuclear granulocytes after stimulation with Mycobacterium tuberculosis and lipoarabinomannan. *Infect Immun*, **65**, 4620–3.

Roach, T.I., Chatterjee, D. and Blackwell, J.M. (1994) Induction of early-response genes KC and JE by mycobacterial lipoarabinomannans: regulation of KC expression in murine macrophages by Lsh/Ity/Bcg (candidate Nramp). *Infect Immun*, **62**, 1176–84.

Roach, T.I., Barton, C.H., Chatterjee, D., Liew, F.Y. and Blackwell, J.M. (1995) Opposing effects of interferon-gamma on iNOS and interleukin-l0 expression in lipopolysaccharide- and mycobacterial lipoarabinomannan- stimulated macrophages. *Immunology*, **85**, 106–13.

Ross, G.D., Cain, J.A. and Lachmann, P.J. (1985) Membrane complement receptor type three (CR3) has lectin-like properties analogous to bovine conglutinin as functions as a receptor for zymosan and rabbit erythrocytes as well as a receptor for iC3b. *J Immunol*, **134**, 3307–15.

Schlesinger, L.S., Bellinger-Kawahara, C.G., Payne, N.R. and Horwitz, M.A. (1990a) Phagocytosis of Mycobacterium tuberculosis is mediated by human monocyte complement receptors and complement component C3. *J Immunol*, **144**, 2771–80.

Schlesinger, L.S. and Horwitz, M.A. (1990b) Phagocytosis of leprosy bacilli is mediated by complement receptors CR1 and CR3 on human monocytes and complement component C3 in serum. *J Clin Invest*, **85**, 1304–14.

Schlesinger, L.S. and Horwitz, M.A. (1991a) Phagocytosis of Mycobacterium leprae by human monocyte-derived macrophages is mediated by complement receptors CR1 (CD35), CR3 (CD11b/CD18), and CR4 (CD11c/CD18) and IFN-gamma activation inhibits complement receptor function and phagocytosis of this bacterium. *J Immunol*, **147**, 1983–94.

Schlesinger, L.S. and Horwitz, M.A. (1991b) Phenolic glycolipid-1 of Mycobacterium leprae binds complement component C3 in serum and mediates phagocytosis by human monocytes. *J Exp Med*, **174**, 1031–8.

Schlesinger, L.S. (1993) Macrophage phagocytosis of virulent but not attenuated strains of Mycobacterium tuberculosis is mediated by mannose receptors in addition to complement receptors. *J Immunol*, **150**, 2920–30.

Schiesinger, L.S., Hull, S.R. and Kaufinan, T.M. (1994) Binding of the terminal mannosyl units of lipoarabinomannan from a virulent strain of Mycobacterium tuberculosis to human macrophages. *J Immunol*, **152**, 4070–9.

Schlesinger, L.S. (1996) Entry of Mycobacterium tuberculosis into mononuclear phagocytes. *Curr Top Microbiol Immunol*, **215**, 71–96.

Schlesinger, L.S., Kaufinan, T.M., Iyer, S., Hull, S.R. and Marchiando, L.K. (1996) Differences in mannose receptor-mediated uptake of lipoarabinomannan from virulent and attenuated strains of Mycobacterium tuberculosis by human macrophages. *J Immunol*, **157**, 4568–75.

Schweinle, J.E., Ezekowitz, R.A., Tenner, A.J., Kuhlman, M. and Joiner, K.A. (1989) Human mannose-binding protein activates the alternative complement pathway and enhances serum bactericidal activity on a mannose-rich isolate of Salmonella. *J Clin Invest*, **84**, 1821–9.

Sieling, P.A., Chatterjee, D., Porcelli, S.A., Prigozy, T.I., Mazzaccaro, R.J., Soriano, T., Bloom, B.R., Brenner, M.B., Kronenberg, M., Brennan, P.J. and et al. (1995) CD1-restricted T cell recognition of microbial lipoglycan antigens. *Science*, **269**, 227–30.

Stahl, P., Schiesinger, P.H., Sigardson, E., Rodman, J.S. and Lee, Y.C. (1980) Receptor-mediated pinocytosis of mannose glycoconjugates by macrophages: characterization and evidence for receptor recycling. *Cell*, **19**, 207–15.

Stahl, P.D. and Ezekowitz, R.A. (1998) The mannose receptor is a pattern recognition receptor involved in host defense. *Curr Opin Immunol*, **10**, 50–5.

Sugden, E.A., Samagh, B.S., Bundle, D.R. and Duncan, J.R. (1987) Lipoarabinomannan and lipid-free arabinomannan antigens of Mycobacterium paratuberculosis. *Infect Immun*, **55**, 762–70.

Summerfield, J.A., Sumiya, M., Levin, M. and Turner, M.W. (1997) Association of mutations in mannose binding protein gene with childhood infection in consecutive hospital series. *Bmj*, **314**, 1229–32.

Swartz, R.P., Naai, D., Vogel, C.W. and Yeager, H., Jr. (1988) Differences in uptake of mycobacteria by human monocytes: a role for complement. *Infect Immun*, **56**, 2223–7.

Taylor, M.E., Conary, J.T., Lennartz, M.R., Stahl, P.D. and Drickamer, K. (1990) Primary structure of the mannose receptor contains multiple motifs resembling carbohydrate-recognition domains. *J Biol Chem*, **265**, 12156–62.

Taylor, M.E., Bezouska, K. and Drickamer, K. (1992) Contribution to ligand binding by multiple carbohydrate-recognition domains in the macrophage mannose receptor. *J Biol Chem*, **267**, 1719–26.

Taylor, M.E. and Drickamer, K. (1993) Structural requirements for high affinity binding of complex ligands by the macrophage mannose receptor. *J Biol Chem*, **268**, 399–404.

Thornton, B.P., Vetvicka, V., Pitman, M., Goldman, R.C. and Ross, G.D. (1996) Analysis of the sugar specificity and molecular location of the beta- glucan-binding lectin site of complement receptor type 3 (CD11b/CD18). *J Immunol*, **156**, 1235–46.

Tino, M.J. and Wright, J.R. (1996) Surfactant protein A stimulates phagocytosis of specific pulmonary pathogens by alveolar macrophages. *Am J Physiol*, **270**, L677–88.

van Emmerik, L.C., Kuijper, E.J., Fijen, C.A., Dankert, J. and Thiel, S. (1994) Binding of mannan-binding protein to various bacterial pathogens of meningitis. *Clin Exp Immunol*, **97**, 411–6.

van Iwaarden, F., Welmers, B., Verhoef, J., Haagsman, H.P. and van Golde, L.M. (1990) Pulmonary surfactant protein A enhances the host-defense mechanism of rat alveolar macrophages. *Am J Respir Cell Mol Biol*, **2**, 91–8.

Venisse, A., Berjeaud, J.M., Chaurand, P., Gilleron, M. and Puzo, G. (1993) Structural features of lipoarabinomannan from Mycobacterium bovis BCG. Determination of molecular mass by laser desorption mass spectrometry. *J Biol Chem*, **268**, 12401–11.

Venisse, A., Fournie, J.J. and Puzo, G. (1995a) Mannosylated lipoarabinomannan interacts with phagocytes. *Eur J Biochem*, **231**, 440–7.

Venisse, A., Riviere, M., Vercauteren, J. and Puzo, G. (1995b) Structural analysis of the mannan region of lipoarabinomannan from Mycobacterium bovis BCG. Heterogeneity in phosphorylation state. *J Biol Chem*, **270**, 15012–21.

Weikert, L.F., Edwards, K., Chroneos, Z.C., Hager, C., Hoffman, L. and Shepherd, V.L. (1997) SPA enhances uptake of bacillus Calmette-Guerin by macrophages through a specific SP-A receptor. *Am J Physiol*, **272**, L989–95.

Wilson, I.A. and Bjorkman, P.J. (1998) Unusual MHC-like molecules: CD1, Fc receptor, the hemochromatosis gene product, and viral homologs. *Curr Opin Immunol*, **10**, 67–73.

Wilson, M.E. and Pearson, R.D. (1986) Evidence that Leishmania donovani utilizes a mannose receptor on human mononuclear phagocytes to establish intracellular parasitism. *J Immunol*, **136**, 4681–8.

Wright, S.D. and Silverstein, S.C. (1983) Receptors for C3b and C3bi promote phagocytosis but not the release of toxic oxygen from human phagocytes. *J Exp Med*, **158**, 2016–23.

Wu, K., Yuan, J. and Lasky, L.A. (1996) Characterization of a novel member of the macrophage mannose receptor type C lectin family. *J Biol Chem*, **271**, 21323–30.

Yoshida, A. and Koide, Y. (1997) Arabinofuranosyl-terminated and mannosylated lipoarabinomannans from Mycobacterium tuberculosis induce different levels of interleukin-12 expression in murine macrophages. *Infect Immun*, **65**, 1953–5.

Zeng, Z., Casta, ntilde, o, A.R., Segelke, B.W., Stura, E.A., Peterson, P.A. and Wilson, I.A. (1997) Crystal Structure of Mouse CD1: An MHC-Like Fold with a Large Hydrophobic Binding Groove. *Science*, **277**, 339–45.

Zimmerli, S., Edwards, S. and Ernst, J.D. (1996) Selective receptor blockade during phagocytosis does not alter the survival and growth of Mycobacterium tuberculosis in human macrophages. *Am J Respir Cell Mol Biol*, **15**, 760–70.

Zvaritch, E., Lambeau, G. and Lazdunski, M. (1996) Endocytic properties of the M-type 1 80-kDa receptor for secretory phospholipases A2. *J Biol Chem*, **271**, 250–7.

B5. Neuropathologies Involving Endogenous Lectins and their Ligands

Jean-Pierre Zanetta and Jean-Claude Michalski

The interactions between endogenous lectins and their glycoconjugate ligands play essential roles in adhesion or recognition mechanisms as well as in the regulation of cell proliferation. Such interactions occur not only during the ontogenesis of the nervous tissue (myelinating cell proliferation, neuron migration, axon fasciculation, synaptogenesis and myelination), but also in maintaining cellular contacts in the adult (myelin compaction). These essential functions can be perturbed by anomalies (genetic or not) of the biosynthesis or the degradation of glycoconjugate constituents or by auto-immune attacks against glycans or lectins. This chapter will review the glycobiological basis of several neuropathologies, discussing these effects based on the knowledge on the morphological and functional features on the nervous tissue.

1. SURVEY ON THE STRUCTURES AND FUNCTIONS OF NERVOUS GLYCOCONJUGATES AND LECTINS

1.1. The Nervous Glycoconjugates

1.1.1. The nervous glycolipids

The chemical characteristic of the nervous tissue is the higher level of sialylated glycolipids, essentially gangliosides, but other compounds are also present, showing a cellular specificity. For example, myelin from both the central (CNS) and peripheral (PNS) nervous tissue contains high level of galactosyl-ceramide (Gal-Cer) and of its sulfated derivative on the 3 position of Gal, the sulfatide (SO4-Gal-Cer). In contrast, glucosyl-ceramide (Glc-Cer) is virtually absent in normal conditions. Myelin contains also sialylated compounds: GM_1 is the major myelin ganglioside in mammals, although its level is very low compared to that of neuronal membranes. Its actual localization in compact myelin is questioned, since its only clear localization is the areas of contacts between axons and myelin in the lateral loops of the nodes of Ranvier (Molander *et al.*, 1996). GM_4 (or G7; Neu5AcGal-Cer) is present in human myelin preparations (Mullin *et al.*, 1981) and in low amounts in other mammalian myelin fractions. Its actual localization in myelin is also questioned since immunohistochemical studies in the cerebellum indicated an exclusive localization in astrocytes present in the white matter and internal granular layer (Ozawa *et al.*, 1993). Consequently, this constituent is probably associated with the perinodal type-2 astrocytes (Miller and Raff, 1984), which insulate the Ranvier's nodes. Myelin (specially the human myelin) is also rich in glycolipids bearing the HNK-1 epitope (3-sulfated glucuronic acid; SO4-3-GlcA; Abo and Balch, 1981; Chou *et al.*, 1985). Astrocytic membranes seem to contain high levels of GM_3 and GD_3 (Cammer and Zhang, 1996a and b). In contrast, neuronal membranes, and, especially those of the synaptic region, are extremely rich in GM1 and polysialogangliosides (GD_{1a}, GD_{1b}, GT_{1b}, and, at a lesser extent, GQ_{1b}). Minor compounds, including fucosylated-sialylated glycolipids are also present in the brain and up to 30 different sialic acid-containing glycolipids have been isolated from the developing rat cerebellum (Zanetta *et al.*, 1980a) and from the rat brain synaptosomal plasma membranes (in preparation). In some cases, glycolipids show developmental regulations, specially those bearing the HNK-1 epitope (Kuchler *et al.*, 1991). Before synaptogenesis, neuronal membranes are rich in such compounds, whereas glycolipid-bound HNK-1 is absent in adult except in myelin (Kuchler *et al.*, 1991). Subtle structural variations such as O-acylations of sialic acid residues (that may be of fundamental biological importance) have been only marginally studied.

1.1.2. The nervous glycoproteins

With the exception of myelin, the brain tissue shows a considerable heterogeneity of the glycoprotein populations, the majority of them (90%) being associated with plasma membranes.

Although glycoproteins are minor constituents of myelin (about 5% of total protein; Zanetta *et al.*, 1977a), the CNS myelin contains a major glycoprotein, the myelin-associated glycoprotein, MAG (Matthieu, 1981; Quarles *et al.*, 1983). This 95 kDa molecule contains essentially N-glycans (7–9 potential N-glycosylation sites). In fact, two different alternative splicing variants are expressed at different periods (Lai *et al.*, 1987) that also differ in their glycan structures based on their affinity for lectins (Badache *et al.*, 1992). The ''early'' form (expressed during the period of proliferation of myelinating cells and at the beginning of myelination) possesses a longer intracytoplasmic domain with consensus sequence for tyrosine phosphorylations and at least one $Man_6GlcNAc_2$ N-glycan recognized by the endogenous lectin CSL (Badache *et al.*, 1992). The ''late'' form has a shorter intracytoplasmic domain without consensus sequences for phosphorylations (Lai *et al.*, 1987) and is not recognized by lectin CSL (Badache *et al.*, 1992). Both forms have at least one N-glycan bearing the HNK-1 epitope. Besides the two previous membrane-bound MAG, an endogenous proteolytic product of MAG containing the extracellular domain of the molecule (d-MAG) is accumulated in the peri-axonal cytoplasmic collar, the no man's land between axons and the surrounding myelin between the Ranvier's nodes. Minor low M_r glycoproteins are also present in CNS myelin (Badache *et al.*, 1992), as well as high M_r glycoproteins like Ng-CAM (Martini and Schachner, 1986). CNS myelin contains a very minor glycoprotein called the myelin-oligodendrocyte glycoprotein (MOG), a member of the superfamily of immunoglobulins (Amiguet *et al.*, 1992; Phandhim *et al.*, 1993), bearing the HNK-1 epitope. Ultrastructural studies indicated that this molecule was restricted to the CNS myelin and here, presented an exclusive localization on the more external raws of the myelin wrapping. The PNS myelin contains high levels of glycoproteins. It contains the two forms of MAG with developmental characteristics similar to those described in the CNS. But the major glycoprotein is the P0 glycoprotein (Kitamura *et al.*, 1976; Lemke and Axel, 1985; Sakamoto *et al.*, 1987), a 29 kDa transmembrane molecule belonging to the superfamily of immunoglobulins and bearing a single N-glycan. Part of the molecules have N-glycans variant recognized by lectin CSL (Badache *et al.*, 1992), and some have the HNK-1 epitope (Uyemura and Kitamura, 1991). Another low M_r glycoprotein is known as the PAS/PII (Kitamura *et al.*, 1976), or PMP-22 glycoprotein (Hammer *et al.*, 1993). Astrocytic glycoproteins have been only marginally studied although their membranes are rich in glycoproteins (Kuchler *et al.*, 1989a). Interestingly, reactive astrocytes specifically over-express the polysialosyl NeuNAc(α2–8) sequence characteristic of N-CAM, but only intracellularly (Lehmann *et al.*, 1993).

Neuronal plasma membranes, and especially those of the synaptic region, are particularly rich in glycoprotein (about 50% of the total proteins; Zanetta *et al.*, 1975 and 1977b) and show an extremely high heterogeneity that includes enzymes, ion channels, receptors for neurotransmitters, etc. One interesting point is related to developmental studies: some neuronal membrane glycoproteins show developmental regulations. This is the case for the N-CAMs (a single gene and 192 alternative splicing variants), the embryonic form bearing the specific oligosaccharide structure for N-CAM of linear poly-sialic acid sequence Neu5Ac(α2-8) on N-glycans. But low M_r glycoproteins, transiently expressed during development, are extremely abundant. The major molecular entity is a molecule identified as B4-B5 doublet (Zanetta *et al.*, 1978) or P31 (Nédelec *et al.*, 1992) or CD24 or nectadrin or heat stable antigen, a GPI-anchored glycoprotein marker for human B cells. Although the apparent M_r of this compound is 31 kDa by PAGE/SDS, its polypeptide chain contains only about 30 amino acid residues. The mass is essentially due to 3-4 N-glycans, possibly 1-2 O-glycans, the glycan of the GPI anchor, and the phosphatidylinositol moiety. Because of the weakness of the polypeptide chain, this compound has been largely underestimated using classical protein techniques. In fact, at the 13th postnatal rat cerebellum, this compound represents 15–25% of total proteins, whereas its percentage in adult is negligible both in terms of protein and in terms of mRNA expression (Shirasawa *et al.*, 1993). At least one of the N-glycan is of the oligomannoside $Man_6GlcNAc_2$ type since it binds to lectin CSL (Kuchler *et al.*, 1989b). The HNK-1 epitope is expressed on a few numbers of glycan variants during a shorter period. Another major glycoprotein, transiently expressed and also GPI-anchored is the Thy-1 glycoprotein, that, as CD24 is completely binding to concanavalin A and to CSL. These two major glycoproteins, representing between 25 and 35% of the total protein of the rat cerebellum at the 13th postnatal day, are essentially localized, at this period, on the plasma membranes of neurons (cell body, dendrites and axons).

1.1.3. The nervous proteoglycans

Several proteoglycans were isolated from the nervous tissue displaying different localizations. For example, the PGM1 proteoglycan (Normand *et al.*, 1988) is localized inside specific axons and interacts with specific neurofilament structures. The brevican in a chondroitin sulfate proteoglycan of the aggrecan family, synthesized by astrocytes and localized at the surface of the astrocytes surrounding synaptic complexes (like the cerebellar glomeruli; Yamada *et al.*, 1997). Neurocan is also a member of the aggrecan family presenting a hyaluronan binding N-terminal domain and a C-type lectin C-terminal domain (Rauch *et al.*, 1997). Glypicans are heparan sulfate proteoglycans with a glycosyl-phosphatidylinositol anchor suggested to be ligands of heparin-binding growth factors (Carey *et al.*, 1993; Litwak *et al.*, 1994; Saunders *et al.*, 1997). Agrin and cerebroglycans are also heparan sulfate proteoglycans associated with the developing axonal pathways (Halfter *et al.*, 1997; Ivins *et al.*, 1997). However, it is not known if these constituents play adhesive or anti-adhesive roles.

1.2. Nervous Lectins

Several molecules endowed with carbohydrate-binding properties have been identified in neural cells, independently from the E-selectin that localizes on endothelial cells of the nervous tissue vessels. They comprise galactose-binding proteins of the galectin family, mannose-binding lectins of the CSL and R1 families, heparin-binding proteins and a ganglioside binding protein: the MAG. But affino-histochemical methods suggest that lectins with other carbohydrate-binding properties (Fuc, GlcNAc, GalNAc and Neu5Ac) are also present in the nervous tissue (Bardosi *et al.*, 1990; Kuchler *et al.*, 1990 and 1992; Adam *et al.*, 1993).

1.2.1. The nervous galectins

Galectins constitute a family of galactose-binding calcium-independent lectins conserved during evolution (Barondes *et al.*, 1994a and b). Nine different genes have been identified in human tissue, but a few studies were concerned with their function in the nervous tissue. Besides the CBP-35 (Moutsatsos *et al.*, 1986), endowed with a nuclear localization in all cells so far analyzed, the only galectin actually studied in nervous tissue is the galectin 1 (Joubert *et al.*, 1989; Kuchler *et al.*, 1989d; Puche *et al.*, 1997). A function in neurite fasciculation and/or in the guidance of axonal growth with extracellular constituents was documented (Mahanthappa *et al.*, 1994; Puche *et al.*, 1997), the lectin recognizing two different ligands: the first is a glycolipid containing a β-lactosaminic sequence, the second being a glycan of laminin (Puche *et al.*, 1997). But it was also suggested that galectins could be involved in the transport of unidentified ligands to the post-synaptic areas during the formation of synapses in the rat CNS (Kuchler *et al.*, 1989d). This traffic could be associated with neurofilaments, since galectin 1 binds actin (Joubert *et al.*, 1992), a neurofilament constituent that is concentrated in the dendritic spines of post-synaptic neurons.

1.2.2. The lectin MAG

Several studies emphasized the carbohydrate-binding properties of MAG. Although the strict specificity is contested (either polysialo-gangliosides (Yang *et al.*, 1996, Collins *et al.*, 1997a and b), different gangliosides (Yang *et al.*, 1996; Tang *et al.*, 1997; Tropak and Roder, 1997) cr strict specificity for GM_1 (Zanetta *et al.*, submitted), it is clear that MAG binds sialylated glycolipids in a calcium-independent mechanism. The putative involvement of this activity in neuropathology depends greatly on the actual oligosaccharide specificity of MAG. Indeed, the GM_1 specificity suggests a function of MAG in maintaining the contact between axons and myelinating cells in the lateral loops of the Ranvier's nodes. Such a contact may be disturbed in autoimmune diseases in which antibodies against GM1 or against MAG are produced.

1.2.3. The lectins of the CSL family

The CSL lectins (Zanetta *et al.*, 1987a) constitute a family of calcium-independent lectins specific for $Man_6GlcNAc_2Asn$ oligomannosides (Marschal *et al.*, 1989) comprising different oligomers (about 40 identical subunits) made of different monomers presenting a cell type specificity. All neural cells (except the non-myelinating Schwann cells of the PNS) contain CSL, a 43 kDa form in myelinating cells and 31.5 and 33 kDa forms in astrocytes and neurons. These intracytoplasmic molecules are developmentally externalized by astrocytes and myelinating cells and, as polyvalent molecules, make bridges between surface glycoprotein ligands of the CSL producing cells and of neurons. Two different functions have been documented during development: the one is the contact guidance of neuron migration of cerebellar granule cells along the Bergmann astrocytic fibers (Lehmann *et al.*, 1990), and the second is the initial contact between axons and myelinating cells. The major neuronal ligands (on the cell body of neurons, in the first case, and on the axons, in the second) are CD24 and Thy-1. The major myelinating cell ligand is the early form of MAG, but other glycoprotein ligands are present (Badache *et al.*, 1992). In contrast, astrocytes show complex profiles of CSL ligands (Kuchler *et al.*, 1989a). In adult, the CSL is only externalized in three areas: the area of contacts between axons and myelinating cells in the lateral loops of the nodes of Ranvier, in compact myelin where CSL plays an adhesive role, and at the surface of the cilia of ependymal cells and in the junctions between these cells (Perraud *et al.*, 1988). In the first case, a small amount of the ''early'' MAG and of CD24 are the CSL ligands (Kuchler *et al.*, 1989c). In the second case, the ligands in CNS compact myelin are low M_r glycoproteins, whereas the major ligand in the PNS myelin is the P0 glycoprotein (Kuchler *et al.*, 1989c). Evidence was provided that monovalent anti-CSL antibodies are demyelinating both *in vitro* and *in vivo* (Kuchler *et al.*, 1988 and see below).

1.2.4. Lectin R1

Lectin R1 (Zanetta *et al.*, 1985) is a calcium-independent lectin specific for oligomannosides with 5 and 8 mannose residues (Marschal *et al.*, 1989) expressed only in neurons throughout the different neural cells (Dontenwill *et al.*, 1985). In the cerebellum, this intracellular molecule is over-expressed and externalized at the surface of parts of the dendritic tree of large neurons (Purkinje cells) at the period preceding synaptogenesis. It recognizes oligomannosidic N-glycans of a few axonal glycoproteins (including CD24 and Thy-1 as major constituents) and the complex R1/ligands is transported, through double walled coated vesicles (Dontenwill *et al.*, 1983 and 1985), into the lysosomes of the Purkinje cells, where the ligands are degraded, the lectin being probably recycled in large part to other areas of the dendritic tree. In the adult CNS, the lectin is poorly expressed and only intracellularly. But, after lesion of the parallel fibers of the adult rat cerebellum, the induced neo-synaptogenesis is accompanied by an over-expression and re-externalization of R1 in the area of wounding, whereas the granule cells, whose axons were cut, re-expressed the CD24 on their axons (Lehmann *et al.*, 1993). The same processes of internalization occur as during the normal synaptogenesis. Treatments of young animals with inhibitors of N-glycosylation (Zanetta *et al.*, 1987b) and modifications of the levels of the thyroid hormone (Zanetta *et al.*, 1985c) suggest that this mechanism corresponds to the recognition of axons by their target neurons necessary for the formation of correct synapses.

1.2.5. The heparin-binding lectins

A molecule isolated by its affinity to heparin (Rauvala, 1984; Mähönen and Rauvala, 1985; Rauvala and Pihlaskari, 1987; Rauvala *et al.*, 1988; Merrenmies *et al.*, 1991), showing a developmental regulation during brain development, that enhance the neurite outgrowth in cultures of neurons, was identified as amphoterin (Merenmies *et al.*, 1991). This molecule shows a cationic region followed by an anionic region. This intracellular molecule can be externalized *in vitro* and binds to the substrate attached material. A special concentration was found on the *filopodia* of growing processes of neurons. Besides heparin, this protein binds to sulfated glycolipids (Mohan *et al.*, 1992), i.e. sulfatides and glycolipids bearing the HNK-1 epitope. The heparin-binding growth-associated molecule (HB-GAM) is a protein isolated as a factor enhancing neurite outgrowth, presenting 50% homology with molecules of the "midkine" family, a family of differentiation/growth factors conserved during evolution. It is expressed along axonal pathways and in the target areas during the early stages of axonal growth (Nolo *et al.*, 1995, 1996). The hypothesis was proposed that this molecule, associated with the structures of the pathways for axonal growth, accumulates at the growth cone-target area, during the period of synaptogenesis, at least in the peripheral nervous system. It endogenous ligand is the heparan sulfate chain of N-syndecan (Raulo *et al.*, 1994), a proteoglycan particularly rich in 2-O-sulfated iduronic acid residue (Kinnunen *et al.*, 1996).

2. NEUROPATHOLOGIES ASSOCIATED WITH GLYCOPROTEIN MUTATIONS

Several pathologies apparently result from mutations of the genes coding for glycoproteins. Recent studies suggest that mutations into two different PNS glycoproteins are responsible for the Charcot-Marie-Tooth diseases. Charcot-Marie-Tooth diseases comprise a large variety of human genetic disorders provoking PNS dysmyelinations (not demyelination) with different severities. These genetic disorders are characterized by an abnormal myelination, with the formation of the characteristic morphological structure of "onion bulbs", i.e. distended circular protoplasmic processes of Schwann cells, instead of compact myelin.

2.1. Charcot-Marie-Tooth 1A

The most common, the Charcot-Marie-Tooth 1A disease, is mostly associated with the duplication of the gene of the PMP-22 (PAS-PII) glycoprotein (Marrosu *et al.*, 1997; Navon *et al.*, 1996; Warner *et al.*, 1996), although mutations in the gene of connexin 32 were also detected in some patients (Bruzzone *et al.*, 1994). The PMP-22 glycoprotein accumulates in the cell body of Schwann cell and in its processes, forming the onion bulb, instead of accumulating in compact myelin (Hanemann *et al.*, 1994; Haney *et al.*, 1996; Harding, 1995; Kamholtz *et al.*, 1994).

2.2. Charcot-Marie-Tooth 1B

The Charcot-Marie-Tooth 1B diseases constitute a family of diseases associated with different mutations in the gene of the P0 glycoprotein (Gabreelsfesten *et al.*, 1996; Hayasaka *et al.*, 1993; Latour *et al.*, 1995; Su *et al.*, 1993; Tachi *et al.*, 1996). These mutations are concerned with single amino-acid mutations or deletions in the extracellular domain of the P0 molecule. These mutations could affect the conformation of this extracellular domain of P0, possibly affecting the homophilic interactions P0-P0 suggested to be involved in myelin compaction (Filbin *et al.*, 1990). However, several studies reported that this adhesive function of P0 is essentially due to its carbohydrate moiety (Filbin and Tennekoon, 1991; Yazaki *et al.*, 1992). The mutation occurring at the single N-glycosylation site of glycoprotein P0 (Blanquet-Grossard *et al.*, 1996) can explain the absence of myelin compaction due to the absence of its N-glycan. But, in the other cases, it is difficult to understand how a single change in the amino-acid sequence of the extracellular domain of P0 can affect myelination, unless it is assumed that the sequence or the conformation of the polypeptide chain of P0 is a determining parameter for the correct synthesis of its N-glycan or for the incorporation and transport of the molecule.

3. NEUROPATHOLOGIES ASSOCIATED WITH GLYCAN SYNTHESIS OR DEGRADATION DEFICIENCIES

3.1. Carbohydrate-Deficient Glycoprotein Syndromes

Carbohydrate-deficient glycoprotein syndromes (CDG) are extremely rare, but severe, human diseases affecting young children, termed CDG-I and CDG-II, CDG-III, CDG-IV and CDG-V. A significant number of data are only available for CDG-I and CDG-II, which differ in their features and inborn error mechanisms (see for review, Hagberg *et al.*, 1992a; Jaeken *et al.*, 1997).

3.1.1. Carbohydrate-deficient glycoprotein I (CDG-I)

CDG-I is a rare autosomal recessive disorder characterized by major nervous system involvement. Clinical features are severe mental retardation, cerebellar atrophy and ataxia, polyneuropathies, hepatomegaly and dysmorphy. It is biochemically characterized by the absence of glycosylation of some N-glycosylation sites of glycoproteins normally N-glycosylated (Yamashita *et al.*, 1993; Iourin *et al.*, 1996). It is due to a deficiency in phosphomannomutase (Vanschaftingen and Jaeken, 1995), but not to a deficient synthesis of dolichyl phosphate or N-acetylglucosaminyl-pyrophosphoryl-dolichol (Yasugi *et al.*, 1994). According to Panneerselvam and Freeze (1996), mannose corrects the altered N-glycosylation found in carbohydrate-deficient glycoprotein syndrome fibroblasts, especially in the presence of a reduced glucose supply. These authors suggest that a mannose-rich diet could be a way to circumvent major features due to this inherited disease. However, major steps of the nervous system development are occurring prenatally in human and it is not certain that such mannose supply given to the mother could be efficient on the embryo. Nevertheless, considering the major CNS glycoproteins of the neuronal membranes during specific stages of development (especially Thy-1 and CD24 on axons, MAG and P0 on myelinating cells), it is expected that important mechanisms of brain development can be perturbed (contact guidance of neuron migration, synaptogenesis and myelination). The quite specific cerebellar atrophy observed in this disease may be indicative in this respect. Indeed, the hypo-glycosylation of major, transiently expressed glycoproteins (Thy-1 and CD24) at the surface of migrating neurons and of axons in the cerebellum (particularly rich in neurons and axons throughout the different brain regions) could affect massively the contact guidance and synaptogenesis mechanisms (see above, and below the effects of thyroid hormones). The brain stem atrophy may also result from a reduced glycosylation of the same axonal glycoproteins and of the glycoproteins of the myelinating cells. Unfortunately, animal models of this disease are lacking to analyze the effects of this enzymatic defect into details.

3.1.2. Carbohydrate-deficient glycoprotein II

CDG-II is an extremely rare disease resulting from point mutations in the gene of UDP-GlcNAc: α-6-D-mannoside β-1-2-N-acetyl-glucosaminyl-transferase (GnT-II; Jaeken *et al.*, 1994 and 1996; Charuk *et al.*, 1995; Tan *et al.*, 1996). In the absence of this enzyme, complex type N-glycans are not synthesized, the latter leading to hybrid type N-glycans possessing a non substituted Man(α1-6) residue of the core of N-glycans. This indicates that complex N-glycans are dispensable for the normal human brain development. However, it is quite surprising that the completely defective function of a key enzyme in the biosynthesis of complex type N-glycans is not lethal, as observed for the knock-out experiments of the gene of N-acetyl-glucosaminyl-transferase I (GnT-I) in mice (Metzler *et al.*, 1994), where embryo survive only to the mid-gestation period. Assuming that the biosynthetic pathways in mice and human involves the same enzymes, it is suggested that complex type N-glycans (in contrast with hybrid types and oligomannosidic types), are not important for a development of the nervous system compatible with life.

3.2. Altered Levels of Thyroid Hormone

Thyroid deficiency during the *in utero* and neonatal period in man causes mental retardation. In severe cretinism, the intelligence quotient may remain subnormal throughout life, despite continuous treatments with thyroid hormones, whereas anomalies can be overcome by hormone administration in slight cases. This indicated that the thyroid hormone plays a role in critical stages of the brain development brain. Both the prenatal depletion of the thyroid gland by intubation of propyl-thiouracil to the pregnant female in rats or injections of the thyroid hormone after birth, induce a decrease in body weight but a specific effect on the CNS development, with major changes in the cerebellum. Major morphological changes compared to control animals in hypothyroid rats are a later proliferation of granule cells (Nicholson and Altman, 1972a; Clos and Legrand, 1973), a normal migration rate, an increased cell death in the internal granular layer in the second postnatal week and a reduced number of synapses in the molecular layer of 30 days old rats (Nicholson and Altman, 1972b). In the hyperthyroid rats, the pattern is the same with the exception that the granule cell proliferation occurs earlier than in the controls. A small reduction of myelination was observed by several authors (Clos *et al.*, 1973), the expression of MAG and carbonic anhydrase II being retarded in hypothyroid rats and accelerated in hyperthyroid rats, but the adult levels being the same compared to the control animals. Thus, major effects were concerned with the number of synapses in 30 days old animals, due to a decreased density of synaptic profiles in hypothyroid rats and due to a reduced dendritic domain in hyperthyroid rats. Furthermore, the synaptogenesis is retarded in hypothyroid rats and accelerated in hyperthyroid rats, suggesting that the thyroid hormone acts as a timer for neuronal development.

A few studies were concerned with major biochemical changes in the developmental patterns of the cerebella of hypo-and hyper-thyroid rats. Major modifications were concerned with glycoproteins. Indeed, when analyzing the major glycoproteins binding to concanavalin A, strong developmental differences were observed. In normal animals, Thy-1 and CD24 present a peak expression at the 13th postnatal day then drop rapidly at the 18th postnatal day. This peak is absent in hypothyroid rats (Zanetta *et al.*, 1985b) and the expression of these molecules is reduced (to 25–30% that of the control animals). In contrast, the expression of the glycoproteins binding to concanavalin A is normal in the adult. In hyperthyroid rats, the peak is present but earlier (10 th postnatal day and drop to the 15th day) and the expression of the previous glycoproteins is reduced (to 50% that of the control animals). These glycoproteins are ligands of lectin R1 involved in the first contact between neurons before synaptogenesis. The developmental curve of R1 (timing of over-expression and of externalization) was not significantly changed in experimental animals as compared with controls (Zanetta *et al.*, 1985b). Therefore, the reduction in the number of synapses observed in hypothyroid rats (Nicholson and Altman, 1972b) could be due to a very important decrease in the synthesis of these GPI-anchored glycoproteins, especially at the period where R1 is externalized. The reduction in the number of synapses in hyperthyroid rats (Nicholson and Altman, 1972b) could be explained because these glycoproteins are transiently synthesized at an earlier period than that of the externalization of R1. The increased cell death in the internal granular layer could be an "en cascade" degeneration of granule cells (Sotelo and Changeux, 1974) whose axons did not succeed making synapses with their target neurons.

3.3. Inhibitors of N-Glycan Synthesis

Inhibitors of N-glycan synthesis (tunicamycin, deoxynojirimycin and castanospermine) can induce severe impairments of the nervous system development. Although taking of these drugs in food is unlikely for human (not for animals), the incidence of these substances will be examined here, because they were suggested to be efficient drugs against cancer and AIDS (Taylor *et al.*, 1994; Fenouillet and Jones, 1995).

Short-term (3 days) treatments of young animals (postnatal day 10) with inhibitors of N-glycosylation (tunicamycin and castanospermine) provoke an important decrease in the brain weight and specially of the cerebellum. Although a few studies were performed, this decreased weight is accompanied by a significant loss of the number of synapses in the cerebellar molecular layer, followed by the cell death of granule cells not having succeeded to form synapses (Zanetta *et al.*, 1987b). This mechanism is very similar to the "en cascade" degeneration observed in the "*staggerer*" mutant mice (Sotelo and Changeux, 1974). Consistent with this interpretation is the observation (Mallet *et al.*, 1976) that staggerer neurons lack a 31 kDa protein, that could correspond to the CD24 molecule. The effects of these substances are reduced in adult animals. However, tunicamycin (which inhibits the formation of the dolichol-P-P-GlcNAc, the first step of the formation of the lipid intermediate of the synthesis of N-glycans) and deoxynojirimycin and castanospermine (inhibitors of the processing of immature N-glycans favoring the accumulation of $Glc_{2-3}Man_9GlcNAc_2$ N-glycans), when injected into the blood, primarily affect the biosynthesis of circulating cells and of endothelial cells. The effect on immune cells may be extremely damaging for the function of the immune system. Indeed, such glycans are not recognized by the two lectins playing an essential role as activation molecules, interleukin 2 (Zanetta *et al.*, 1996) and CSL (Zanetta *et al.*, 1995). Consequently, prolonged treatments with these substances devoid to inhibit HIV-1 virus replication and/ or cancer cell proliferation, induce a severe immuno-deficiency, an effect contrasting to that expected for therapy (Zanetta *et al.*, 1998a and b).

In contrast, therapeutic improvement could be expected in diseases characterized by an hyper-activation of the immune system and particularly in auto-immune diseases. For example, the experimental allergic encephalomyelitis (EAE) can be completely suppressed by pre-treatments of the EAE-sensitive Lewis rat strains by castanospermine prior to the injection of MBP-specific cells. This pre-treatment inhibits the homing of cells on the brain vessels on which they are binding in animals with EAE (Willenborg *et al.*, 1989). This suggests, in fact, that specific brain vessels of EAE-sensitive Lewis rat strains possess a specific homing system for activated cells of the immune system, involving N-glycans.

3.4. Genetic Deficiencies in Glycosidases

A large variety of diseases are related to deficiencies in the enzymes catabolyzing glycoconjugates, provoking an accumulation of partially degraded products into lysosomes in a category of diseases having the general terminology of ''lysosomal storage disorders''. Besides approximately common features (coarse faces, visceromegaly, skeletal deformities with characteristics specific for each deficiency) and common morphological anomalies (cell vacuolizations, hyperplasic lysosomes), these diseases are affecting the nervous tissue at different degrees. These diseases are characterized, as in other tissues, by the accumulation of undegraded material into the different cells of the nervous tissue, both in the CNS and in the PNS. Therefore, it is expected that, as in other tissues, the accumulated material (in majority oligosaccharides) will affect the normal function of the nervous tissue, primarily as inhibitors of the normal metabolism of the cells. However, specific severities, specific accumulations are observed, that suggest more specific interferences, especially with ontogenetic processes and cell contact mechanisms or signal transduction pathways independent on ontogenetic processes. For example, carbohydrate-lectin interactions involved in cell contact mechanisms may be specifically perturbed by the *in situ* accumulation of specific oligosaccharides. Furthermore, it has been demonstrated that the normal function of neurotransmitter receptors (Sieghart *et al.*, 1993; Fishburn *et al.*, 1995; Kawamoto *et al.*, 1995; Nishizaki and Sumikawa, 1994), ions channels (Zona *et al.*, 1990) are abolished upon abnormal glycosylation. Consequently, it is expected that the accumulation of specific glycans can perturb such functions. Since an extensive recent review appeared recently on this subject, we will refer to that (Michalski, 1996), discussing only points which could be explained by interferences of the accumulated compounds with mechanisms of cell adhesion and recognition and of signal transduction in the nervous tissue.

3.4.1. Mannosidoses

Because of the extremely high amount of transiently expressed glycoproteins endowed with oligomannosidic N-glycans during development, it is expected that deficient mannosidase activities will induce severe perturbations of the nervous system development.

3.4.1.1. β-mannosidosis

β-mannosidosis is an extremely rare human disease. The reason is probably the lethality at the fetal stage. Only patients with the "adult form" may survive. It is due to the defect in the activity of the lysosomal β-mannosidase which cleaves the β-mannose residue of the manno-trioside core of N-glycans. The major oligosaccharides excreted in the human urine are Man(β1-4)GlcNAc and the Man(β1-4)GlcNAc(β1-4)GlcNAc. Additional oligosaccharides NeuAc(α2-3)Man(β1-4)GlcNAc probably results from the abnormal sialylation of the major accumulated oligosaccharide. Significant morphological studies are only accessible for the goat disease (with axonal and myelin lesions in the CNS primarily to a deficit in oligodendrocytes; Jones *et al.*, 1983), but an extrapolation to the human disease cannot be performed since the accumulated oligosaccharides are clearly different from the ones found in human. The reason of the severity of this disease is unknown. A lectin potentially binding to the major accumulated oligosaccharide has been identified as the "core-specific lectin", CSL (Colley and Baenziger, 1987). It should be emphasized that this "core specific lectin CSL" is unrelated to another CSL, the "cerebellar soluble lectin" (Zanetta *et al.*, 1987a). Because of the absence of the core-specific lectin in the nervous tissue and its abundance in the blood, it may be suggested that major interferences of the Man(β1-4)GlcNAc(β1-4)GlcNAc oligosaccharide could take place with its function in the immune system, with specific brain damages as a consequence.

3.4.1.2. α-mannosidosis

Human α-mannosidosis is a rare autosomal recessive disease due to the inheritance of a deficiency in lysosomal A and B α-mannosidases. Two types of diseases are considered termed as type I and II, type I representing a severe infantile phenotype. Features include mental retardation similar to the Hurler's syndrome, facial dysmorphism, psychomotor retardation, deafness and skeletal lesions (Öckerman, 1967; Kjellman *et al.*, 1969). The effects on the brain development are not extremely dramatic considering the enormous amounts of oligomannosides to be degraded in a short period of time (Zanetta *et al.*, 1980b, 1982 and 1983). This can be explained, at least in part, by the presence of a cytosolic α-mannosidase activity, the activity increasing at the time where these glycoprotein glycans are degraded (Reeber *et al.*, 1980). The level of lysosomal activity is relatively poor compared to that of the previous one, in such a way that the catabolism of these glycans can occur. The degradation could also partially avoid the lysosomal compartment since a cytosolic endo-N-acetyl-glucosaminidase is present, that also seems to be induced by the large quantity of substrates to be degraded (Cook *et al.*, 1984). Nevertheless, due to the strict specificity of the cerebellar soluble lectin CSL for Man$_6$GlcNAc$_c$2Asn (Marschal *et al.*, 1989), the oligomannosides accumulated in α-mannosidosis can not interfere with its adhesive functions in contact guidance of neuron migration and in myelination because they are terminated by a single GlcNAc residue at their terminal reducing end. The Man$_8$GlcNAc and overall the Man$_5$GlcNAc accumulated in α-mannosidosis (Yamashita *et al.*, 1979; Daniel *et al.*, 1981) could interfere with the function of the lectin R1 in interneuronal recognition. This does not seem to be the case, although the presence of ectopic dendrites on large neurons (Goodman *et al.*, 1991) may be significant of an anomaly of R1 expression. The observation that these ectopic dendrites are not specific for this lysosomal storage disorder (Jones *et al.*, 1983) rather suggests that they result from an anomaly of the metabolism of these large neurons subsequent to the accumulation of undegraded material.

Throughout the different lysosomal storage diseases, α-mannosidosis is characterized by a severe immunodeficiency, the patients being prone to various bacterial, mycobacterial and viral infections (Öckerman, 1967; Kjellman *et al.*, 1969). This immunological profile can be in large part be explained by the effect of circulating oligomannosides on the lectin activity of interleukin 2 (Zanetta *et al.*, 1996, 1998a and b). Indeed, the concentration of $Man_5GlcNAc$ and $Man_6GlcNAc$ (the high affinity oligosaccharide ligands of IL-2) in patients with α-mannosidosis (Öckerman, 1969; Yamashita *et al.*, 1980; Daniel *et al.*, 1981; Matsuura *et al.*, 1981) is by far higher than the one required (0.75 μM; Zanetta *et al.*, 1996) to inhibit the binding of IL-2 to its ligands of the CD3/TCR complex. As a result, IL-2-dependent signaling will be dramatically abolished and it may be suggested that the immunodeficiency observed in α-mannosidosis affects essentially the IL-2-dependent mechanisms, a view strongly reinforced by the analogy with the immunodeficiency observed in animals after knock-out of the gene for IL-2 (Horak, 1995) and in the human severe combined immunodeficiency disease syndrome (Di Santo *et al.*, 1990). Such a primary effect of the accumulated oligomannosides on the IL-2-dependent mechanisms may explain the absence of relationships between the severity of different α-mannosidoses and the genetic defects observed in these diseases (Cooper, 1990; Bennett, 1995).

3.4.2. α-Fucosidosis

α-Fucosidoses are generally segregated into two forms (1 and 2), the type 1 presenting a higher severity and a general deterioration of the neurological functions at the age of 1 and 2 years followed by death before the age of 5 years. However, a systematic study of 77 patients (Willems *et al.*, 1991) rather suggested a continuous spectrum of clinical profiles, that can not be explained by differences in residual fucosidase activities. Glycans accumulated in the urine are either fucosylated on the first GlcNAc of the core of N-glycans or on branches of complex or hybrid type N-glycans (Strecker *et al.*, 1978). The reason of the severity of this disease might be related to two observations: *i)* fucose-rich N-glycans are particularly abundant on brain membrane glycoproteins, specially on neuronal membranes (Zanetta *et al.*, 1977b); *ii)* lectin activities specific for fucose have been identified by histo-affinity histochemistry in animals (Kuchler *et al.*, 1990), presenting specific concentrations during the early stages of the neuronal development. For example, migrating granule cells in the rat cerebellum (this stage occurs in the embryonic stage in human) accumulate fucose-binding sites both in the nucleus and in the cytosol, with a particular concentration in the growing processes of these neurons. The accumulation of fucose-containing oligosaccharides in developing neurons may affect the formations of neuronal processes, and consequently, the formation of normal neuronal circuitry. Besides developmental interferences, the accumulated oligosaccharides could interfere with the function of fucose containing glycoconjugates in the processes of learning in the adult. Indeed, it has been demonstrated in chicken and rats, that during learning fucose is specifically (compared to the other monosaccharides) incorporated into glycoconjugates of the synaptosomal plasma membranes, this incorporation was independent of protein synthesis (Mileusnic *et al.*, 1995; Sandi and Rose, 1997; Lorenzini *et al.*, 1997). This suggests that fucose-containing glycoconjugates are involved in the formation of new neuronal circuitry, possibly through adhesion processes. It may be suggested that the accumulation of fucose-containing oligosaccharides or glycoasparagine in fucosidosis will perturb the formation of the new circuitry, and, as a consequence, a complete disorganization of the neuronal connections, responsible for the progressive decerebration of the patients.

3.4.3. Aspartyl-glucosaminuria

Aspartylglucosaminuria is characterized by a quite normal development up to 2–4 years followed by a constant and regular regression of motor and intellectual skills. The disease is due to the absence of the aspartyl-glucosaminidase activity, that results in the accumulation in urine of glycans containing GlcNAc(β1-)Asn. The relatively late appearance of the disease suggests that the glycans accumulated in the nervous tissue of these patients do not interfere with the ontogenetic processes, but with the function of the central and peripheral nervous system. One possible explanation is that the major accumulated catabolite GlcNAc(β1-)Asn interferes with endogenous lectins (nuclear and cytosolic; Holt *et al.*, 1987) recognizing GlcNAc residues revealed using GlcNAc-neoglycoproteins (Kuchler *et al.*, 1990). Assuming such hypothesis, it is suggested that this glycan could interfere with the function of O-linked GlcNAc residues (Hart *et al.*, 1988) in signal transduction pathways, and/or neurotubule and neurofilament assembly (Arnold *et al.*, 1996; Dong *et al.*, 1996). On the one hand, an abnormal polymerization of the cytoskeleton elements will impair the normal transport of macromolecules to the presynaptic and postsynaptic part of the synapse, inducing a defect in molecules necessary for a normal release of neurotransmitters and for a normal receptor/re-uptake of the neurotransmitters. Therefore, the defective neuronal activity, both in the CNS and PNS could result from defective neuronal transmission induced by a reduced abundance of molecules involved in this essential mechanism. But, on the other hand the O-GlcNAc-dependent signal transduction pathways (Hart *et al.*, 1996), occurring in the cytosol and in the nucleus (Hart, 1997) or in the traffic between cytosol and nucleus (Heesepeck *et al.*, 1995; Duverger *et al.*, 1996) could be also perturbed, inducing a specific dysfunction of the neural cells (not only neuron).

3.4.4. α-N-acetyl-galactosaminidase deficiency (Schindler and Kanzaki diseases)

Schindler disease is a neurodegenerative disease in which patients develop until 9–12 months then later present a rapid regression of development associated with a brain atrophy. It is due to the defect in α-N-acetyl-galactosaminidase activity, an enzyme involved in the degradation essentially of O-glycans and glycolipids (mostly gangliosides in the nervous tissue). One interpretation based on the structure of the accumulated O-glycopeptides is that the defective enzyme has an endo-N-acetyl-galactosaminidase activity which cleaves the GalNAc residues from Ser/Thr of the polypeptide backbone. There is no interpretation of the postnatal brain atrophy. However, the major glycopeptides accumulated in the urine of these patients have the same composition as the major O-glycans found in neuronal plasma membranes (in preparation). Although the function of these O-glycans is unknown, it may be suggested that the accumulated O-glycopeptides may interfere with cell adhesion mechanisms that are essential for maintaining the normal circuitry.

3.4.5. Gangliosidosis

Although the severity of the neurological symptoms may vary from one lysosomal storage disease to the others, the major morphological events are the accumulation of material in large neurons, which may perturb the neuronal function and may be responsible for the neurological symptoms. For example, although GM_1 is the major ganglioside of myelin, its accumulation in myelinating cells or myelin observed in GM_1 gangliosidosis (a disease resulting from a defect of β-galactosidase) is minimal when compared to neurons, because GM_1 itself, and polysialo-gangliosides giving GM_1 during their catabolism, are by far more concentrated in neurons. The same is true for the GM_2 gangliosidosis, in which the defect concerns the β-N-acetyl-galactosaminidase.

3.4.6. Lysosomal storage disorders affecting myelin

The only lysosomal storage disorders due to a glycosidase defect affecting specifically CNS and PNS myelin are those concerned with the degradation of sulfatides and galactosyl-ceramides, because these constituents are exclusively myelin-specific. These diseases are dysmyelinating, i.e. myelin is not formed properly. However, this dysmyelination rather results from a dysfunction of myelinating cells or of macrophages involved in the turn-over of myelin constituents than from a perturbance of the myelination process itself due to an excess of these myelin constituents. The first, known as metachromatic leucodystrophy, results from a deficiency of arylsulfatase A (sulphatyl-galactosyl-ceramidase) or from a deficiency of the activator of this enzyme. The second, known as Krabbe's leucodystrophy or PAS-positive globoid cells, results from a deficit in galactoceramidase which provokes the accumulation of galactosyl-ceramide.

3.5. Drug-induced Dysfunction of Lysosomes

Substances which elevate the lysosomal pH, and, specially, the anti-paludean substance chloroquine are particularly relevant to the field of neuropathology, since they provoke at certain developmental stages of the CNS, dramatic accumulations of undegraded material into large neurons. This is particularly demonstrative in the young rat cerebellum, in which major glycoproteins (CD24 and Thy-1) of the surface of parallel fibers are degraded in their large integrating target neurons, the Purkinje cells. The cell bodies and dendrites of these cells accumulate very large amounts of undegraded glycoproteins in association with the R1 lectin in structures termed as "myelin like bodies" corresponding to non-functional lysosomes accumulating material (Dontenwill *et al.*, 1983). Due to differences in the time course development of the cerebellum in rat (postnatal) and human (prenatal), it is expected that taking of anti-paludean substances of this kind during pregnancy will affect severely the development of the CNS (and particularly the cerebellum) with irremediable consequences in the adult function because of the death of large, multi-innervated neurons.

4. NEUROPATHOLOGIES INVOKED BY ANTI-GLYCAN ANTIBODIES

Several neurological diseases are associated with the presence (or the increased levels) of antibodies to glycoconjugates. These increased blood or cerebrospinal fluid (CSF) concentrations of antibodies against endogenous constituents and inflammatory aspects suggest that these pathologies belong to the autoimmune type. Most of the identified anti-glycan auto-antibodies are directed against glycolipids (anti-GM_1, anti-Gal-Cer, anti-sulfatides) which appear to be more or less associated with specific diseases. However, cares have to be taken in the interpretation and in the diagnosis values of these determinations, because data differs largely depending on the methodological approaches (ELISA versus overlay techniques, and, especially, the nature of the saturating agent before incubation of human antibodies with the antigen).

4.1. Anti-GM_1 Antibodies

Serum anti- GM_1 antibodies are naturally occurring in the blood of individuals with titers in the range of 1/400 dilution as determined by ELISA or overlay techniques, essentially as IgM. However, GM_1 like substances (possibly GM_1 having lost parts of its ceramide portion) are present in serum and in bovine serum albumin preparations used for the saturation of microwells or overlays after immobilization of the antigen (Zanetta *et al.*, submitted). This problem can be easily circumvented using periodate treated BSA (Zanetta *et al.*, 1993) as a saturating agent. Antibodies to gangliosides, including GM_1 have been identified in multiple sclerosis (MS; Aron *et al.*, 1980). A disease resembling MS was obtained, immunizing rabbits with GM_1 (Cohen *et al.*, 1981), but these data were not confirmed thereafter. In contrast, high titers are found in dys-immune peripheral motor neuropathies with multifocal conduction blocks (MMNCB) and in the acute polyradicu-loneuritis, the Guillain-Barré syndrome (Simone *et al.*, 1993; Jacobs *et al.*, 1996; Bech *et al.*, 1997; Baumann *et al.* 1998). Interestingly, a significant correlation was established between polyradiculoneuritis and previous infections by *Campylobacter jejuni* (Vriesen-dorp *et al.*, 1993;. Jacobs *et al.*, 1996). Since *Campylobacter jejuni* glycans have a surface oligosaccharide possessing the GM_1 structure, it is suggested that the auto-immunity against GM_1 in this disease is induced by chronic infections by this pathogen.

4.2. Anti-Galactosyl-Ceramide and Anti-Sulfatide Antibodies

Anti-galacosyl-ceramides have been found (Aron *et al.*, 1980) or not (Rostami *et al.*, 1983) in the CSF or serum of patients with various neurological diseases suggesting a PNS or CNS demyelination. The presence of high amounts of galactosyl-ceramides in CNS and PNS myelin suggested that the over-production of these auto-antibodies could be responsible for demyelination. In fact, there is only a few evidence that anti-Gal-Cer (generally IgM) are actually demyelinating both *in vivo* and *in vitro* (Saida *et al.*, 1979; Hahn *et al.*, 1993; Rosenbluth *et al.*, 1996), specially because the clear epitope domain is poorly described. Anti-sulfatides antibodies have been only marginally detected in demyelinating diseases including MS (Kirschning *et al.*, 1995) and AIDS, that are associated with PNS diseases (Degasperi *et al.*, 1996) and polyneuropathy syndromes (Pestronk *et al.*, 1991). In fact, the function of these compounds in the process of myelin compaction is questioned by the experiments of the knock-out of the gene for the UDP-Gal:Ceramide galactosyltransferase, the unique enzyme involved in the synthesis of Gal-Cer and, consequently, of its sulfated derivative. Indeed homozygotes –/– developed normal myelin although these two compounds are absent. The concomitant synthesis of relatively high amounts of Glc-Cer in –/– homozygotes, absent in +/+ animals, may be a natural way to circumvent an essential function of the Gal-Cer and sulfatide in the process of myelination.

4.3. Anti HNK-1 Antibodies

Antibodies to the myelin-associated glycoproteins were found to be associated with IgM type paraproteinaemis (McGarry *et al.*, 1983; Steck *et al.*, 1983; Kaku *et al.*, 1994; Valldeoriola *et al.*, 1993), a PNS demyelinating disease. However, the later determination of the IgM characteristic of this disease revealed that the antibody was not directed against the protein part of MAG but against one of its carbohydrate epitope, the glucuronic acid 3-sulfate group, also recognized by a monoclonal antibody against an epitope of the natural killer cells, HNK-1 (Abo and Balch, 1981; Inuzuka *et al.*, 1984; Chou *et al.*, 1985; Baba *et al.*, 1986). This epitope is, in fact, not specific for the MAG but is also present on several nervous glycoproteins (MAG, P0, MOG, N-CAM, Ng-CAM, CD24, etc. (Shashoua *et al.*, 1986; Burger *et al.*, 1989; Kuchler *et al.*, 1991) and on specific glycolipids. In human, the morphological anomalies are concerned with the abnormal compaction of outermost myelin lamellae, contrasting with the exclusive localization of MAG in the peri-axonal cytoplasmic collar and in the areas of junctions between axons and myelinating cells. Since this disease is specific for the PNS myelin, the HNK-1 epitope can not correspond to the MOG (endowed with such a localization but absent from the PNS).

5. NEUROPATHOLOGIES INVOKED BY ANTI-GLYCOPROTEIN OR ANTI-LECTIN ANTIBODIES

A number of demyelinating diseases, including multiple sclerosis (MS), have been suggested to result from autoimmune attacks against myelin protein and glycoproteins. The major protein constituents of the CNS myelin were initially suggested to be autoimmune targets of MS.

5.1. Antibodies to the P0 and PMP-22 Glycoproteins

The experimental allergic neuritis (EAN) is a PNS disease induced by the immunization against the two major glycoproteins of the PNS, i.e. P0 and PMP-22 glycoproteins. Since EAN was considered as an animal model of the Guillain-Barré syndrome, it was expected that anti-P0 and/or anti-PMP-22 glycoprotein antibodies were present in the blood of the patients with the Guillain-Barré syndrome. In fact, anti-P0 and anti-P2 antibodies are found in these patients with a variable but always relatively low proportion depending on the methodology used for the detection of these antibodies (Archelos *et al.*, 1993; Nobile-orazio *et al.*, 1994). However, anti-glycolipid antibodies are also present in variable proportions.

5.2. Antibodies to the Myelin-Oligodendrocyte Glycoprotein (MOG)

The myelin-oligodendrocyte glycoprotein is a member of the superfamily of immunoglo-bulins having a specific CNS localization in oligodendrocytes and in the external areas of myelin. This antigen was demonstrated to induce relapsing-remitting experimental allergic encephalomyelitis in Lewis rat strains sensitive to EAE (Linington and Lassman, 1987; Linington *et al.*, 1984 and 1993; Schluesener *et al.*, 1987; Sun *et al.*, 1991; Adelmann *et al.*, 1995; Derosbo *et al.*, 1990 and 1995; Mendel *et al.*, 1995; Bennum *et al.*, 1996; Bernard *et al.*, 1997). Therefore, it is considered as an important immunological target in the human multiple sclerosis. Unfortunately, antibodies to the MOG were only marginally described in MS (Xiao *et al.*, 1991). The previous manuscript reported the presence of auto-antibodies to a 90 kDa component in the CSF of 10% of the patients with MS (the M_r of MOG is about 25 K). These antibodies might reveal the MAG antigen, as observed in some demyelinating diseases, except MS (Zanetta *et al.*, 1994). Although experiments suggest that MOG could be an important cellular immunological target in the EAE animal model of MS, no evidence was provided that anti-MOG antibodies are actually present in MS and that anti-MOG antibodies are actually able to provoke a demyelination similar to that observed in MS.

5.3. Antibodies to Galectins

Antibodies to human galectin-1 have been found in significantly higher levels in the serum of patients with neurological diseases (including multiple sclerosis) than in control individuals suggesting an impairment of the immune system (Lutomski *et al.*, 1997). However, it should be remembered that antibodies to galectin-3 are naturally occurring in the blood of control human (Mathews *et al.*, 1995). Therefore, the nature of the actual auto-antigen remains unknown. Galectin-1 was proposed to share a common epitope tetrapeptide (sequence WGAE) with the bovine myelin basic protein (Abbot *et al.*, 1889 and 1991). Since this motive is not conserved in the different species, it is difficult to understand how the bovine myelin basic protein used generally to induce EAE could be related to the human galectin-1 in which this sequence is absent.

5.4. Antibodies to the Cerebellar Soluble Lectin (CSL) in Multiple Sclerosis

The CSL lectin is present in compact myelin and anti-CSL Fab fragments are able to dissociate the myelin formed by cultured rat oligodendrocytes (Kuchler *et al.*, 1988). The analysis of 1388 CSF samples from patients for which most biological parameters were accessible (Zanetta *et al.*, 1994) for the presence of anti-CSL antibodies using an immunoblotting technique with a partially purified rat CSL (Zanetta *et al.*, 1990a) showed that anti-CSL antibodies are found in 96.71% of patients with definite MS and 95.09% of patient with probable MS diagnoses. For possible MS patients, the score is 69.47%. The specificity for MS is in all cases higher than 85% for the patients below 50 years. For other diseases than MS and for the age class below 50 years, the presence of anti-CSL antibodies is more associated with demyelinating diseases than the others (myelitis, polyradiculo-neuritis and infectious diseases with CNS demyelination like AIDS (Hagberg *et al.*, 1992b)), although they are rarely present in apparently unrelated symptoms. All patients with Sjögren disease showed anti-CSL antibodies in their CSF. The comparison of the detection of anti-CSL antibodies with other biological parameters indicates that it is not related to a special Ig concentration in the CSF or a particular inflammatory profile.

The majority of the anti-CSL antibodies present in the CSF of patients with MS (and Sjögren disease) inhibited the CSL-induced agglutination of fixed red blood cells covalently coupled with RNAse B glycoasparagines, whereas the majority of those found in the previous false positive CSF did not. When a monoclonal antibody against CSL endowed with the same inhibitory property or the hybridoma cells producing this antibody were injected intraventricularly in adult Wistar rats, they crossed the ependymal barrier, partially destroying it. The antibody penetrated farther into the parenchyma in the area of the *corpus callosum*, defined by the absence of an astrocytic layer below the ependymal layer, and invaded the myelinating tracks for very long distances. In this area, demyelination occurred, accompanied by several morphological events detected by ultrastructural studies in the periventricular areas of multiple sclerosis brains. The presence of inhibitory anti-CSL antibodies in the white matter could explain the poor proliferation of oligodendrocytes and the poor remyelination, since CSL is a mitogen for myelinating cells (Badache *et al.*, 1995; Fressinaud *et al.*, 1988). This could explain the periventricular demyelination observed in all patients with MS but not the formation of the classical demyelination plaques.

In fact, very high levels of anti-CSL antibodies were found in the blood of MS patients (Zanetta *et al.*, 1990b) suggesting that MS is a systemic autoimmune disease. However, anti-CSL antibodies were found in a high number of control individuals. In fact, recent studies indicated that two types of anti-CSL antibodies can occur, the one inhibiting, and the other not, the lectin activity of CSL. Since CSL is a "early expressed amplifier of activation signals" (Zanetta *et al.*, 1995), these two types of antibodies found in the blood of patients with MS have opposite effects on the activation processes and are responsible for the aggregation of activated leukocytes around immune complexes. Recent studies (unpublished data) indicated that the brain of patients with MS, as, constitutively, the brain of the Lewis rat strain sensitive to EAE, had vessels, in majority in the center of demyelination plaques, with endothelial cells possessing N-glycan ligands of CSL. The present hypothesis is that the formation of demyelination plaques results from micro-thromboses due to activated leukocytes in specific vessels having this adhesion system for aggregated activated leukocytes. The same type of mechanism could be invoked for EAE, specially since pretreatment of Lewis rats sensitive to EAE with the glycosylation inhibitors deoxynojirimycin and castanospermine eliminates EAE and the adhesion of cells to the sites of EAE (Willenborg *et al.*, 1989).

6. CONCLUSION AND PERSPECTIVES

An important contingent of neurological diseases is related to the dysfunction of the metabolism (biosynthesis or degradation of glycoconjugates) as well as to auto-immune attacks against the specific glycoconjugates or endogenous lectins. Unfortunately, the majority of the anatomo-pathological studies were not performed in human necropsy, systematically to actually identify the major ultrastructural features on brain areas. Because it becomes evident that major cell adhesion processes involve glycobiological interactions, it appears as necessary to associate glycobiologists and anatomo-pathologists on specific diseases, in order to program cooperative researches. The glycobiology domain, as we tried to do in this review article, provides hypothesis, which can orient the anatomo-pathologists to examine specific anomalies, deduced from the biochemical, histo- and immuno-histo-data on specific areas of the nervous tissue. For example, the research of vacuolization in neural cells in lysosome storage disorders is not a priority, because these accumulations are detected just at the level of light microscopy. It would be more appropriate to look ultrastructurally, at specific part of the CNS (and we suggest the cerebellum) with specific staining, to count the concentration of synapses per a same surface area, in order to see if, as we propose, these anomalies can occur in specific "lysosomal storage disorders".

Such a possibility is based on the accessibility of CNS material perfectly fixed as soon as the patient dye. At the present time, there is no understanding, and consequently, no therapies of these diseases. At least, for α-mannosidosis an IL-2 therapy could be suggested. It is not an elimination of the disease, but a major improvement of the patients, until other ways could be found.

REFERENCES

Abbot, W.M., Mellor, A., Edwards, Y., and Feizi, T. (1989) Soluble bovine galactose-binding lectin. cDNA cloning reveals the complete amino-acid sequence and an antigenic relationship with the major encephalitogenic domain of myelin basic protein. *Biochem. J.*, **259**, 283–290.

Abbott, W.M., and Feizi, T. (1991) Antigenic cross-reaction between beta-galactoside binding lectin and myelin basic protein. *J. Neuroimmunol.*, **31**: 179–179.

Abo, T., and Balch, C.M. (1981). A differentiation antigen of human NK and K cells identified by a monoclonal antibody (HNK-1). *J. Immunol.*, **127**, 1024–1029.

Adam, E., Dziegielewska, K.M., Saunders, N.R., and Schumacher, U. (1993). Neuraminic acid specific lectins as markers of early cortical plate neurons. *Int. J. Dev. Neurosci.*, **11**, 451–460.

Adelmann, M., Wood, J., Benzel, I., Fiori, P., Lassmann, H., Matthieu, J.-M., Gardinier, M.V., and Dornmair, K. (1995) The N-terminal domain of the myelin oligodendrocyte glycoprotein (MOG) induces acute demyelinating experimental autoimmune encephalomyelitis in the Lewis rat. *J. Neuroimmunol.*, **63**, 17–27.

Amiguet, P., Gardinier, M.V., Zanetta, J.-P, and Matthieu, J.-M. (1992) Purification and partial structural and functional characterization of mouse myelin/oligodendrocyte glycoprotein. *J. Neurochem.*, **58**, 1676–1682.

Archelos, J.J., Roggenbuck, K., Schneider-Schaulies, J., Toyka, K.V., and Hartung, H.P. (1993) Detection and quantification of antibodies to the extracellular domain of P0 during experimental allergic neuritis. *J. Neurol. Sci.*, **117**, 197–205.

Arnold, C.S., Johnson, G.V.W., Cole, R.N., Dong, D.L.Y., Lee, M., and Hart, G.W. (1996). The microtubule-associated protein tau is extensively modified with O-linked N-acetylglucosamine. Communication. *J. Biol. Chem.*, **271**, 28741–28744.

Aron, R., Crisp, E., Ellison, G.W., Myers, L.M., and Tourtellotte, W.W. (1980). Anti-ganglioside antibodies in multiple sclerosis. *Neurol. Sci.* **46**, 179–186.

Baba, H., Sato, S., Inuzuka, T., Nishisawa, M., Tanaka, M., and Miyatake, T. (1986). Characterization of the antigenic determinant on HSB-2 cells shared with myelin associated glycoprotein (MAG) using monoclonal antibodies. *J. Neuroimmunol.*, **13**, 89–97.

Badache, A., Burger, D., Villarroya, H., Robert, Y., Kuchler, S., Steck, A.J., and Zanetta, J.-P. (1992) Carbohydrate moieties of MAG, P0 and other myelin glycoproteins potentially involved in cell adhesion. *Dev. Neurosci.*, **14**, 342–350.

Badache, A., Lehmann, S., Kuchler-Bopp, S., Hand, N., and Zanetta, J.-P. (1995) An endogenous lectin and its glycoprotein ligands are triggering basal and axon-induced Schwann cell proliferation. *Glycobiology*, **5**, 371–383.

Bardosi, A., Bardosi, L., Lindenblatt, R., and Gabius, H. -J. (1990). Detection and mapping of endogenous receptors for carrier-immobilized constituents of glycoconjugates (lectins) by labelled (neo)glycoproteins and by affinity chromatography in human adult mesencephalon, pons, medulla oblongata and cerebellum. *Histochemistry*, **94**, 285–291.

Barondes, S.H., Cooper, D.N.W., Gitt, M.A., and Leffler, H. (1994a) Galectins. Structure and function of a large family of animal lectins. *J. Biol. Chem.*, **269**, 20807–20810.

Barondes, S.H., Castronovo, V., Cooper, D.N.W., Cummings, R.D., Drickamer, K., Feizi, T., Gitt, M.A., Hirabayashi, J., Hughes, C., Kasai, K., Leffler, H., Liu, F.T., Lotan, R., Mercurio, A.M., Monsigny, M., Pillai, S., Poirer, F., Raz, A., Rigby, P.W. J., Rini, J.M., and Wang, J.L. (1994b). Galectins. A family of animal beta-galactoside-binding lectins. *Cell*, **76**, 597–598.

Baumann, N., Harpin, M.L., Marie, Y., Lemerle, K., Chassande, B., Bouche, P., Meininger, V., Yu, R.K., and Leger, J.M. (1998) Antiglycolipid antibodies in motor neuvopathies. *Ann. N. Y. Acad. Sci.*, **19**, 322–329.

Bech, E., Orntoft, T.F., Andersen, L.P., Skinhoj, P., and Jakobsen, J. (1997). IgM anti-GM1 antibodies in the Guillain-Barre syndrome: A serological predictor of the clinical course. *J. Neuroimmunol.*, **72**, 59-66.

Bennettt, J.K. (1995) Clinical and biochemical analysis of two families with type I and type II mannosidosis. *Amer. J. Med. Genet.*, **55**, 21–26.

Bennun, A., Mendel, I., Bakimer, R., Fridkishareli, M., Teitelbaum, D., Arnon, R., Sela, M., and Derosbo, N.K. (1996) The autoimmune reactivity to myelin oligodendrocyte glycoprotein (MOG) in multiple sclerosis is potentially pathogenic: Effect of copolymer 1 on MOG-induced disease. *J. Neurol.*, **243**, 14–22.

Bernard, C.C.A., Johns, T.G., Slavin, A., Ichikawa, M., Ewing, C., Liu, J., and Bettadapura, J. (1997) Myelin oligodendrocyte glycoprotein: A novel candidate autoantigen in multiple sclerosis. *J. Molecular Med.-Jmm.*, **75**, 77–88.

Blanquet-Grossard, E., Phamdinh, D., Dautigny, A., Latour, P., Bonnebouche, C., Diraison, P., Chapon, F., Chazot, G., and Vandenberghe, A. (1996). Charcot-Marie-Tooth type 1B neuropathy: A mutation at the single glycosylation site in the major peripheral myelin glycoprotein P0. *Hum. Mutat.*, **8**, 185–186.

Bruzzone, R., White, T.W., Scherer, S.S., Fischbeck, K.H., and Paul, D.L. (1994). Null mutations of connexin 32 in patients with X-linked Charcot-Marie-Tooth disease. *Neuron*, **13**, 1253–1260.

Burger, D., Simon, M., Perruisseau, G., and Steck, A.J. (1989). The epitope(s) recognized by HNK-1 antibody and IgM paraprotein in neuropathy is present on several N-linked oligosaccharide structures on human P0 and myelin-associated glycoprotein. *J. Neurochem.*, **54**, 1569–1575.

Cammer, W., and Zhang, H. (1996 a). A reexamination of ganglioside GD3-immunopositive glial cells in the forebrains of rats and mice less than 1 week of age. *J. Histochem. Cytochem.*, **44**, 143–149.

Cammer, W., and Zhang, H. (1996 b). Ganglioside GD3 in radial glia and astrocytes in situ in brains of young and adult mice. *J. Neurosci. Res.*, **46**, 18–23.

Carey D.J., Stahl R.C., Asundi V.K., and Tucker B. (1993) Processing and subcellular distribution of the Schwann cell lipid-anchored heparan sulfate proteoglycan and identification as glypican. *Exp. Cell Res.*, **208**, 10–18.

Charuk, J.H.M., Tan, J., Bernardini, M., Haddad, S., Reithmeier, R.A.F., Jaeken, J., and Schachter, H. (1995). Carbohydrate-deficient glycoprotein syndrome type II. An autosomal recessive N-acetylglucosaminyltransferase II deficiency different from typical hereditary erythroblastic multinuclearity, with a positive acidified-serum lysis test (HEMPAS). *Eur. J. Biochem.*, **230**, 797–805.

Chou, K.H., Ilyas, A.A., Evans, J.E., Quarles, R.H., and Jungalwala, F.B. (1985). Structure of a glycolipid reacting with monoclonal IgM in neuropathy and with HNK-1. *Biochem. Biophys. Res. Commun.*, **128**, 383–388.

Clos, J., and Legrand, J. (1973). Effects of thyroid deficiency on the different cell populations of the cerebellum in the young rat. *Brain Res.*, **63**, 450–455.

Clos, J., Rebière, A., and Legrand, J. (1973). Differential effects of hypothyroidism and undernutrition on the development of glia in the rat cerebellum. *Brain Res.*, **63**, 445–449.

Cohen, O., Sela, B., Schwartz, M., Eshaar, N., and Cohen, I.R. (1981). Multiple sclerosis-like disease induced in rabbits by immunisation with brain gangliosides. *Israel J. Med.*, **17**, 711–714.

Colley, K.J., and Baenziger, J.U. (1987) Identification of the post-translational modifications of the core-specific lectin. The core-specific lectin contains hydroxyproline, hydroxylysine, and glucosylgalactosylhydroxylysine residues. *J. Biol. Chem.*, **262**, 10290–10295

Collins, B.E., Yang, L.J.S., Mukhopadhyay, G., Filbin, M.T., Kiso, M., Hasegawa, A., and Schnaar, R.L. (1997a) Sialic acid specificity of myelin-associated glycoprotein binding. *J. Biol. Chem.*, **272**, 1248–1255.

Collins, B.E., Kiso, M., Hasegawa, A., Tropak, M.B., Roder, J.C., Crocker, P.R., and Schnaar, R.L. (1997b) Binding specificities of the sialoadhesin family of I-type lectins. Sialic acid linkage and substructure requirements for binding of myelin-associated glycoprotein, Schwann cell myelin protein, and sialoadhesin. *J. Biol. Chem.*, **272**, 16889–16895.

Cook, N.J., Dontenwill, M., Meyer, A., Vincendon, G., and Zanetta, J.-P. (1984). Postnatal variations of endo-β-D-N-acetyl-glucosaminidase in the developing rat cerebellum. *Dev. Brain Res.* **15**, 298–301.

Cooper, M. (1990) Alpha and beta mannosidoses. *J. Inherit. Metab. Dis.*, **13**, 523–527.

Daniel, P.F., Defeudis, D.F., and Lott, I.T. (1981) Mannosidosis: Isolation and comparison of mannose-containing oligosaccharides from gingiva and urine. *Eur. J. Biochem.*, **114**, 235–237.

Degasperi, R., Sosa, M.A.G., Patarca, R., Batistini, S., Lamoreux, M.R., Raghavan, S., Kowall, N.W., Smith, K.H., Fletcher, M.A., and Kolodny, E.H. (1996). Intrathecal synthesis of anti-sulfatide IgG is associated with peripheral nerve disease in acquired immunodeficiency syndrome. *AIDS Res. Hum. Retroviruses*, **12**, 205–211.

Derosbo, N.K., Honegger, P., Lassmann, H., and Matthieu, J.-M. (1990) Demyelination induced in aggregating brain cell cultures by a monoclonal antibody against myelin oligodendrocyte glycoprotein. *J. Neurochem.*, **55**, 583–587.

Derosbo, N.K., Mendel, I., and Bennun, A. (1995) Chronic relapsing experimental autoimmune encephalomyelitis with a delayed onset and an atypical clinical course, induced in PL/J mice by myelin oligodendrocyte glycoprotein (MOG)-derived peptide: Preliminary analysis of MOG T cell epitopes. *Eur. J. Immunol.*, **25**, 985–993.

Di Santo, J.P., Keever, C.A., Small, T.N., Nichols, G.L., O'Relly, R.J., and Flomenberg N. (1990) Absence of interleukin 2 production in a severe combined immunodeficiency disease syndrome with T cells. *J. Exp. Med.*, **171**, 1697–1704.

Dong, D.L.Y., Xu, Z.S., Hart, G.W., and Cleveland, D.W. (1996). Cytoplasmic O-GlcNAc modification of the head domain and the KSP repeat motif of the neurofilament protein neurofilament-H. *J. Biol. Chem.*, **271**, 20845–20852.

Dontenwill, M., Devilliers, G., Langley, O.K., Roussel, G., Hubert, P., Reeber, A., Vincendon, G., and Zanetta, J.-P. (1983). Arguments in favour of endocytosis of glycoprotein components of the membrane of parallel fibers by Purkinje cells during the development of the rat cerebellum. *Dev. Brain Res.*, **10**, 287–299.

Dontenwill, M., Roussel, G., and Zanetta, J.-P. (1985). Immunohistochemical localization of a lectin like molecule R1 during the postnatal development of the rat cerebellum. *Dev. Brain Res.* **17**, 245–252.

Duverger, E., Roche, A.-C., and Monsigny, M. (1996). N-acetylglucosamine-dependent nuclear import of neoglycoproteins. *Glycobiology*, **6**, 381–386.

Fenouillet, E., and Jones, I.M. (1995) The glycosylation of human immunodeficiency virus type 1 transmembrane glycoprotein (gp41) is important for the efficient intracellular transport of the envelope precursor gp160. *J. Gen. Virol.*, **76**, 1509–1514.

Filbin, M.T., and Tennekoon, G.I. (1991) The role of complex carbohydrates in adhesion of the myelin protein P0. *Neuron*, **7**, 845–855.

Filbin, M.T., Walsh, F.S., Trapp, B.., Pizzey, J.A., and Tennekoon, G.I. (1990) Role of myelin P0 protein as a homophilic adhesion molecule. *Nature*, **344**, 871–872.

Fishburn, C.S., Elazar, Z., and Fuchs, S. Differential glycosylation and intracellular trafficking for the long and short isoforms of the D-2 dopamine receptor. *J. Biol. Chem.*, **270** (1995) 29819–29824.

Fressinaud, C., Kuchler, S., Sarlieve, L. L., Vincendon, G., and Zanetta, J.-P. (1988). Adhesion on a matrix made of the endogenous lectin CSL induces increased proliferation of oligodendrocytes. *C. R. Acad. Sci. Paris*, **307**, 863–868.

Gabreelsfesten, A.A.W.M., Hoogendijk, J.E., Meijerink, P.H.S., Gabreels, F.J.M., Bolhuis, P.A., Vanbeersum, S., Kulkens, T., Nelis, E., Jennekens, F.G.I., Devisser, M., Vanengelen, B.G.M., Vanbroeckhoven, C., and Mariman, E.C.M. (1996). Two divergent types of nerve pathology in patients with different P0 mutations in Charcot-Marie-Tooth disease. *Neurology*, **47**, 761–765.

Goodman, L.A., Livingston, P.O., and Walkley, S.U. (1991) Ectopic dendrites occur only on cortical pyramidal cells containing elevated GM2 ganglioside in alpha-mannosidosis. *Proc. Natl. Acad. Sci. USA*, **88**, 11330–11334.

Hagberg, B., Blennow, G., and Stibler, H. (1992a). The birth and infancy of a new disease. The carbohydrate-deficient glycoprotein syndrome. *Fet. Perinat. Neurol.*, **1**, 314–319.

Hagberg, L., Norkrans, G., Zanetta, J.-P., Lehmann, S., and Bergström, M. (1992b) Cerebrospinal fluid anti-cerebellar soluble lectin antibodies in human defficiency virus type 1 infection. *J. Neuroimmunol.*, **36**, 245–249.

Hahn, A.F., Feasby, T.E., Wilkie, L., and Lovgren, D. (1993). Antigalactocerebroside antibody increases demyelination in adoptive transfer experimental allergic neuritis. *Muscle Nerve*, **16**, 1174–1180.

Halfter, W., Schurer, B., Yip, J., Yip, L., Tsen, G., Lee, J.A., and Cole, G.J. (1997) Distribution and substrate properties of agrin, a heparan sulfate proteoglycan of developing axonal pathways. *J. Comp. Neurol.*, **383**, 1–17.

Hammer, J.A., Oshannessy, D.J., Deleon, M., Gould, R., Zand, D., Daune, G., and Quarles R.H. (1993) Immunoreactivity of PMP-22, P0, and other 19 to 28 kDa glycoproteins in peripheral nerve myelin of mammals and fish with HNK1 and related antibodies. *J. Neurosci. Res.*, **35**, 546–558.

Hanemann, C.O., Stoll, G., Durso, D., Fricke, W., Martin, J.J., Vanbroeckhoven, C., Mancardi, G.L., Bartke, I., and Muller, H.W. (1994). Peripheral myelin protein-22 expression in Charcot-Marie-Tooth disease type 1A sural nerve biopsies. *J. Neurosci. Res.*, **37**, 654–659.

Haney, C., Snipes, G.J., Shooter, E.M., Suter, U., Garcia, C., Griffin, J.W., and Trapp, B.D. (1996). Ultrastructural distribution of PMP22 in Charcot-Marie-Tooth disease type 1A. *J. Neuropathol. Exp. Neurol.*, **55**, 290–299.

Harding, A.E. (1995). From the syndrome of Charcot-Marie and Tooth to disorders of peripheral myelin proteins. *Brain*, **118**, 809–818.

Hart, G.W., Holt, G.D., and Haltiwanger, R.S. (1988). Nuclear and cytoplasmic glycosylation: novel saccharide linkages in unexpected places. *Trends Biochem. Sci.*, **13**, 380–384.

Hart, G.W. (1997). Dynamic O-linked glycosylation of nuclear and cytoskeletal proteins. *Annu. Rev. Biochem.*, **66**, 315–335.

Hart, G.W., Kreppel, L.K., Comer, F.I., Arnold, C.S., Snow, D.M., Ye, Z.Y., Cheng, X.G., Dellamanna, D., Caine, D.S., Earles, B.J., Akimoto, Y., Cole, R.N., and Hayes, B.K. (1996). O-GlcNAcylation of key nuclear and cytoskeletal proteins: Reciprocity with O-phosphorylation and putative roles in protein multimerization. *Glycobiology*, **6**, 711–716.

Hayasaka, K., Ohnishi, A., Takada, G., Fukushima, Y., and Murai, Y. (1993). Mutation of the myelin Po gene in Charcot-Marie-Tooth neuropathy type-1. *Biochem. Biophys. Res. Commun.*, **194**, 1317–1322.

Heesepeck, A., Cole, R.N., Borkhsenious, O.N., Hart, G.W., and Raikhel, N.V. (1995). Plant nuclear pore complex proteins are modified by novel oligosaccharides with terminal N-acetylglucosamine. *Plant Cell*, **7**, 1459–1471.

Holt, G.D., Snow, C.M., Senior, A., Haltiwanger, R.S., Gerace, L., and Hart, G.W. (1987). Nuclear pore complex glycoproteins contain cytoplasmically disposed O-linked N-Acetylglucosamine. *J. Cell Biol.*, **104**, 1157–1164.

Horak, I. (1995) Immunodeficiency in IL-2-knockout mice. *Clin. Immunol. Immunopathol.*, **76**, 172–173.

Inuzuka, T., Quarles, R.H., Noronha, A.B., Dobersen, M.J., and Brady, R.O. (1984). A human lymphocyte antigen is shared with agroup of glycoproteins in peripheral nerve. *Neurosci. Lett.*, **51**, 105–111.

Iourin, O., Mattu, T.S., Mian, N., Keir, G., Winchester, B., Dwek, R.A., and Rudd, P.M. (1996). The identification of abnormal glycoforms of serum transferrin in carbohydrate deficient glycoprotein syndrome type I by capillary zone electrophoresis. *Glycoconjugate J.*, **13**, 1031–1042.

Ivins, J.K., Litwack, E.D., Kumbasar, A., Stipp, C.S., and Lander, A.D. (1997) Cerebroglycan, a developmentally regulated cell-surface heparan sulfate proteoglycan, is expressed on developing axons and growth cones. *Dev. Biol.*, **184**, 320–332.

Jacobs, B.C., Vandoorn, P.A., Schmitz, P.I.M., Tiogillen, A.P., Herbrink, P., Visser, L.H., Hooijkaas, H., and Vandermeche, F.G.A. (1996). Campylobacter jejuni infections and anti-GM1 antibodies in Guillain-Barre syndrome. *Ann. Neurol.*, **40**, 181–187.

Jaeken, J., Schachter, H., Carchon, H., Decock, P., Coddeville, B., and Spik, G. (1994). Carbohydrate deficient: Glycoprotein syndrome type II: A deficiency in Golgi localised N-acetyl-glucosaminyltransferase II. *Arch. Dis. Child*, **71**, 123–127.

Jaeken, J., Spik, G., and Schachter, H. (1996) Carbohydrate-deficient glycoprotein syndrome type II. in J. Montreuil, J.F.G. Vliengenhart, and H. Schachter eds. *"Glycoproteins and disease"* Elsevier, Amsterdam, pp. 457–467.

Jaeken, J., Matthijs, G., Barone, R., and Carchon, H. (1997). Carbohydrate deficient glycoprotein (CDG) syndrome type I. *J. Med. Genet.*, **34**, 73–76.

Joubert, R., Kuchler, S., Zanetta, J.-P., Bladier, D., Avellana-Adalid, V., Caron, M., Doinel, C., and Vincendon, G. (1989) Immunohistochemical localization of a β-galactoside-binding lectin in rat central nervous system. I. Light and electron-microscopical studies on developing cerebral cortex and corpus callosum. *Dev. Neurosci.*, **11**, 397–413.

Joubert, R., Caron, M., Avellana-Adalid, V., Mornet, D., and Bladier, D. (1992). Human brain lectin. A soluble lectin that binds actin. *J. Neurochem.* **58**, 200–203.

Kaku, D.A., England, J.D., and Sumner, A.J. (1994). Distal accentuation of conduction slowing in polyneuropathy associated with antibodies to myelin-associated glycoprotein and sulphated glucuronyl paragloboside. *Brain*, **117**, 941–947.

Kamholz, J., Shy, M., and Scherer S. (1994) Elevated expression of messenger RNA for peripheral myelin protein 22 in biopsied peripheral nerves of patients with Charcot Marie Tooth disease type 1A. *Ann. Neurol.*, **36**, 451–452.

Kawamoto, S., Hattori, S., Sakimura, K., Mishina, M., and Okuda, K. (1995) N-linked glycosylation of the alpha-amino-3-hydroxy-5-methylisoxazole-4-propionate(AMPA)-selec tive glutamate receptor channel alpha 2 subunit is essential for the acquisition of ligand-binding activity. *J. Neurochem.*, **64**, 1258–1266.

Kinnunen, T., Raulo, E., Nolo, R., Maccarana, M., Lindahl, U., and Rauvala, H. (1996). Neurite outgrowth in brain neurons induced by heparin. binding growth-associated molecule (HB-GAM) depends on the specific interaction of HB-GAM with heparan sulfate at the cell surface. *J. Biol. Chem.*, **271**, 2243–2248.

Kirschning, E., Rutter, G., Uhlig, H., and Dernick, R. (1995). A sulfatide-reactive human monoclonal antibody obtained from a multiple sclerosis patient selectively binds to the surface of oligodendrocytes. *J. Neuroimmunol.*, **56**, 191–200.

Kitamura, K., Suzuki, M., and Uyemura, K. (1976). Purification and partial characterization of the two glycoproteins in bovine peripheral nerve myelin membrane. *Biochim. Biophys. Acta*, **455**, 806–816.

Kjellman, B., Gamstorp, I., Brun, A., Öckerman, P.A., and Palmgren, B. (1969) Mannosidosis: A clinical and histopathologic study. *J. Pediat.*, **75**, 366–373.

Kuchler, S., Fressinaud, C., Sarliève, L.L., Vincendon, G., and Zanetta, J.-P. (1988). Cerebellar soluble lectin is responsible for cell adhesion and participates in myelin compaction in cultured rat oligodendrocytes. *Dev. Neurosci.*, **10**, 199–212.

Kuchler, S., Perraud, F., Sensenbrenner, M., Vincendon, G., and Zanetta, J.-P. (1989a). An endogenous lectin found in rat astrocyte cultures has a role in cell adhesion but not in cell proliferation. *Glia*, **2**, 437–445.

Kuchler, S., Rougon, G., Marschal, P., Lehmann, S., Reeber, A., Vincendon, G., and Zanetta, J.-P. (1989b) Localization of a transiently expressed glycoprotein in developing cerebellum delineating its possible ontogenetic roles. *Neuroscience*, **56**, 22–32.

Kuchler, S., Herbein, G., Sarliève, L.L., Vincendon, G., and Zanetta, J.-P. (1989c) An endogenous lectin CSL interacts with major glycoprotein components in peripheral nervous system myelin. *Cell. Molec. Biol.*, **45**, 33–43.

Kuchler, S., Joubert, R., Avellana-Adalid, V., Caron, M., Bladier, D., Vincendon, G., and Zanetta, J.-P. (1989d) Immunohistochemical localization of a β-galactoside-binding lectin in rat central nervous system. II. Light- and electron-microscopical studies in developing cerebellum. *Dev. Neurosci.*, **11**, 414–427.

Kuchler, S., Zanetta, J.-P., Vincendon, G., and Gabius, H.-J. (1990) Detection of binding sites for biotinylated neoglycoproteins and heparin (endogenous lectins) during cerebellar ontogenesis in the rat. *Eur. J. Cell Biol.*, **52**, 87–97.

Kuchler, S., Zanetta, J.-P., Bon, S., Zaepfel, M., Massoulié, J., and Vincendon, G. (1991). Expression and localization in the developing cerebellum of the carbohydrate epitopes revealed by Elec-39 an IgM monoclonal antibody related to HNK-1. *Neuroscience*, **41**, 551–562.

Kuchler, S., Zanetta, J.-P., Vincendon, G., and Gabius, H.-J. (1992) Detection of binding sites for biotinylated neoglycoproteins and heparin (endogenous lectins) during cerebellar ontogenesis in the rat: an ultrastructural study. *Eur. J. Cell Biol.*, **59**, 373–381.

Lai, C., Brow, M.A., Nave, K.-A., Noronha, A.B., Quarles, R.H., Bloom, F.E., Milner, R.J., and Sutcliffe, J.G. (1987) Two forms of 1B236/myelin-associated glycoprotein, a cell adhesion molecule for postnatal development, are produced by alternative splicing. *Proc. Natl. Acad. Sci. USA*, **84**, 4337–4341.

Latour, P., Blanquet, F., Nelis, E., Bonnebouche, C., Chapon, F., Diraison, P., Ollagnon, E., Dautigny, A., Phamdinh, D., Chazot, G., Boucherat, M., Vanbroeckhoven, C., and Vandenberghe, A. (1995). Mutations in the myelin protein zero gene associated with Charcot-Marie-Tooth disease type 1B. *Hum. Mutat.*, **6**, 50–54.

Lehmann, S., Kuchler, S., Théveniau, M., Vincendon, G., and Zanetta, J.-P. (1990). An endogenous lectin and one of its neuronal glycoprotein ligands are involved in contact guidance of neuron migration. *Proc. Natl. Acad. Sci. USA*, **87**, 6455–6459.

Lehmann, S., Kuchler, S., Gobaille, S., Marschal, P., Badache, A., Vincendon, G., and Zanetta, J.-P. (1993). Lesion-induced re-expression of neonatal recognition molecules in adult rat cerebellum. *Brain Res. Bull.*, **30**, 515–521.

Lemke, G., and Axel, R. (1985) Isolation and sequence of a cDNA encoding the major structural protein of peripheral myelin. *Cell*, **40**, 501–508.

Linington, C., and Lassmann, H. (1987) Antibody responses in chronic relapsing experimental allergic encephalomyelitis: correlation of serum demyelinating activity with antibody titre to the myelin-oligodendrocyte glycoprotein (MOG). *J. Neuroimmunol.*, **17**, 61–69.

Linington, C., Webb, M., and Woodhams, P.L. (1984) A novel myelin associated glycoprotein defined by a mouse monoclonal antibody. *J. Neuroimmunol.*, **6**, 387–396.

Linington, C., Berger, T., Perry, L., Weerth, S., Hinzeselch, D., Zhang, Y.P., Lu, H.C., Lassmann, H., and Wekerle H. (1993) T-cells specific for the myelin oligodendrocyte glycoprotein mediate an unusual autoimmune inflammatory response in the central nervous system. *Eur. J. Immunol.*, **23**, 1364–1372.

Litwack, E.D., Stipp, C.S., Kumbasar, A., and Lander, A.D. (1994) Neuronal expression of glypican, a cell-surface glycosylphosphatidylinositol-anchored heparan sulfate proteoglycan, in the adult rat nervous system. *J. Neurosci.*, **14**, 3713–3724.

Lorenzini, C.G., Baldi, E., Bucherelli, C., Sacchetti, B., and Tassoni, G. (1997) 2-deoxy-D-galactose effects on passive avoidance memorization in the rat. *Neurobiol. Learn. Mem.*, **68**, 317–324.

Jones, M.Z, Cunningham, J.G., Dade, A.W., Alessi, D.M., Mostosky, U.V., Vorro, J.R., Benitez, J.T., and Lovell, K.L. (1983) Caprine b-mannosidosis: clinical and pathological features. *J. Neuropathol. Exptl. Neurol.*, **42**, 268–285.

Lutomski, D., R. Joubert-Caron, C. Lefebure, J. Salama, C. Belin, D. Bladier, and M. Caron (1997) Anti-galectin-1 autoantibodies in serum of patients with neurological diseases. *Clin. Chim. Acta*, **262**, 131–138.

Mahanthappa, N.K., Cooper, D.N.W., Barondes, S.H., and Schwarting, G.A. (1994) Rat olfactory neurons can utilize the endogenous lectin, L-14, in a novel adhesion mechanism. *Development*, **120**, 1373–1384.

Mähönen, Y., and Rauvala, H. (1985). Adhesive membrane protein of the rat brain enhances neurite outgrowth of neuroblastoma cells. *Eur. J. Cell Biol.*, **36**, 91–97.

Mallet, J., Huchet, M., Shelanski, M., and Changeux, J.-P. (1974). Protein differences associated with the absence of granule cells in the cerebellum of the mutant weaver mouse and from X-irradiated rat. *FEBS Lett.*, **46**, 243–246.

Marrosu, M.G., Vaccargiu, S., Marrosu, G., Vannelli, A., Cianchetti, C., and Muntoni, F. (1997). A novel point mutation in the peripheral myelin protein 22 (PMP22) gene associated with Charcot-Marie-Tooth disease type 1A. *Neurology*, **48**, 489–493.

Marschal, P., Reeber, A., Neeser, J.-R., Vincendon, G., and Zanetta, J.-P. (1989) Carbohydrate and glycoprotein specificity of two endogenous cerebellar lectins. *Biochimie*, **71**, 645–653.

Martini, R., and Schachner, M. (1986) Immunoelectron microscopic localization of neural cell adhesion molecules (L1, N-CAM, and MAG) and their shared carbohydrate epitope and myelin basic protein in developing sciatic nerve. *J. Cell Biol.*, **103**, 2439–2448.

Mathews, K.P., Konstantinov, K.N., Kuwabara, I., Hill, P.N., Hsu, D.K., Zuraw, B.L., and Liu, F.T. (1995) Evidence for IgG autoantibodies to galectin-3, a beta-galactoside-binding lectin (Mac-2, epsilon binding protein, or carbohydrate binding protein 35) in human serum. *J. Clin. Immunol.*, **15**, 329–337.

Matsuura, F., Nunez, H.A., Grabowski, G.A., and Sweeley, C.C. (1981) Structural studies of urinary oligosaccharides from patients with mannosidosis. *Arch. Biochem. Biophys.*, **207**, 337–352.

Matthieu, J.-M. (1981). Glycoproteins associated with myelin in the central nervous system. *Neurochem. Int.*, **3**, 355–363.

McGarry, R.C., Helfand, S.L., Quarles, R.H., and Roder, J.C. (1983). Recognition of myelin-associated glycoprotein by monoclonal antibody HNK-1. *Nature*, **306**, 376–378.

Mendel, I., Derosbo, N.K., and Bennun, A. (1995) A myelin oligodendrocyte glycoprotein peptide induces typical chronic experimental autoimmune encephalomyelitis in H-2(b) mice: Fine specificity and T cell receptor V beta expression of encephalitogenic T cells. *Eur. J. Immunol.*, **25**, 1951–1959.

Merenmies, J., Pihlaskari, R., Laitinen, J., Wartiovaara, J., and Rauvala, H. (1991). 30-kDa heparin-binding protein of brain (amphoterin) involved in neurite outgrowth. Amino acid sequence and localization in the filopodia of the advancing plasma membrane. *J. Biol. Chem.* **266**, 16722–16729.

Metzler, M., Gertz, A., Sarkar, M., Schachter, H., Schrader, J.W., and Marth, J.D. (1994). Complex asparagine-linked oligosaccharides are required for morphogenic events during post-implantation development. *EMBO J.*, **13**, 2056–2065.

Michalski, J.-C. (1996) Normal and pathological catabolism of glycoproteins. in J. Montreuil, J.F.G. Vliegenthart and H. Schachter eds. *"Glycoproteins and disease"*, Elsevier, Amsterdam, pp. 55–97.

Miller, R.H., and Raff, M.C. (1984) Fibrous and protoplasmic astrocytes are biochemically and developmentally distinct. *J. Neurosci.*, **4**, 585–592.

Mileusnic, R., Rose, S.P., Lancashire, C., and Bullock, S. (1995) Characterisation of antibodies specific for chick brain neural cell adhesion molecules which cause amnesia for a passive avoidance task. *J. Neurochem.*, **64**, 2598–2606.

Mohan, P.S., Laitinen, J., Merenmies, J., Rauvala, H., and Jungalwala, F.B. (1992). Sulfoglycolipids bind to adhesive protein amphoterin (P30) in the nervous system. *Biochem. Biophys. Res. Commun.*, **182**, 689–696.

Molander, M., Fredman, P., Persson, H., and Berthold, C.-H. (1996) Immunoreactivity of gangliosides GD1b and GM1 in CNS and PNS sections investigated with monoclonal antibodies. *J. Neurochem.*, **66**, suppl. 2, S100.

Moutsatsos, I.K., Davis, J.M., and Wang, J.L. (1986). Endogenous lectins from cultured cells: subcellular localization of carbohydrate binding protein 35 in 3T3 fibroblasts. *J. Cell Biol.*, **102**, 477–483.

Mullin, B.R., Decandis, F.X., Montanaro, A.J., and Reid, J.D. (1981). Myelin basic protein interacts with the myelin specific ganglioside GM4. *Brain Res.*, **222**, 218–221.

Navon, R., Seifried, B., Galon, N.S., and Sadeh, M. (1996). A new point mutation affecting the fourth transmembrane domain of PMP22 results in severe, de novo Charcot-Marie-Tooth disease. *Hum. Genet.*, **97**, 685–687.

Nédelec, J., Pierres, M., Moreau, H., Barbet, J., Naquet, P., Faivre-Sarrailh, C., and Rougon, G. (1992) Isolation and characterization of a novel glycosyl-phosphatidylinositol-anchored glycoconjugate expressed by developing neurons. *Eur. J. Biochem.*, **203**, 433–442.

Nicholson, J.L., and Altman, J. (1972 a). The effects of early hypo and hyperthyroidism on the development of rat cerebellar cortex. I. Cell proliferation and differentiation. *Brain Res.*, **44**, 13–23.

Nicholson, J.L., and Altman, J. (1972 b). The effects of early hypo and hyperthyroidism on the development of the cerebellar cortex. II. Synaptogenesis in the moleular layer. *Brain Res.*, **44**, 25–36.

Nishizaki, T., and Sumikawa, K. (1994) Tunicamycin alters channel gating characteristics of junctional and extrajunctional acetylcholine receptors expressed in Xenopus oocytes. *Neurosci. Lett.*, **170** 273–276.

Nobile-Orazio, E., Manfredini, E., Sgarzi, M., Spagnol, G., Allaria, S., Quadroni, M., and Scarlato G. (1994) Serum IgG antibodies to a 35-kDa P0-related glycoprotein in motor neuron disease. *J. Neuroimmunol.*, **53**, 143–151.

Nolo, R., Kaksonen, M., and Rauvala, H. (1996). Developmentally regulated neurite outgrowth response from dorsal root ganglion neurons to heparin-binding growth- associated molecule (HB-GAM) and the expression of HB-GAM in the targets of the developing dorsal root ganglion neurites. *Eur. J. Neurosci.*, **8**, 1658–1665.

Nolo, R., Kaksonen, M., Raulo, E., and Rauvala, H. (1995). Co-expression of heparin-binding growth-associated molecule (HB-GAM) and N-syndecan (syndecan-3) in developing rat brain. *Neurosci Lett.*, **191**, 39–42.

Normand, G., Kuchler, S., Meyer, A., Vincendon, G., and Zanetta, J.-P. (1988) Isolation and immunohisto-chemical localization of a chondroitin sulfate proteoglycan from adult rat brain. *J. Neurochem.*, **51**, 665–676.

Öckerman, P.A. (1967) A generalized storage disorder resembling Hurler's syndrome. *Lancet*, **2**, 239–241.

Öckerman, P.A. (1969) Mannosidosis: Isolation of oligosaccharide storage material from brain. *J. Pediat.*, **75**, 360–365.

Ozawa, H., Kotani, M., Kawashima, I., Numata, M., Ogawa, T., Terashima, T., and Tai, T. (1993). Generation of a monoclonal antibody specific for ganglioside GM4. Evidence for GM4 expression on astrocytes in chicken cerebellum. *J. Biochem.*, **114**, 5–8.

Panneerselvam, K., and Freeze, H.H. (1996). Mannose corrects altered N-glycosylation in carbohydrate-deficient glycoprotein syndrome fibroblasts. *J. Clin. Invest.*, **97**, 1478–1487.

Perraud, F., Kuchler, S., Gobaille, S., Labourdette, G., Vincendon, G., and Zanetta, J.-P. (1988). Endogenous lectin CSL is present on the membrane of cilia of rat brain ependymal cells. *J. Neurocytol.*, **17**, 745–751.

Pestronk, A., Li, F., Griffin, J., Feldman, E.L., Cornblath, D., Trotter, J., Zhu, S., Yee, W.C., Phillips, D., Peeples, D.M., and Winslow, B. (1991). Polyneuropathy syndromes associated with serum antibodies to sulfatide and myelin-associated glycoprotein. *Neurology*, **41**, 357–362.

Phamdinh, D., Dellagaspera, B., Derosbo, N.K., and Dautigny A. (1995) Structure of the human Myelin/Oligodendrocyte glycoprotein gene and multiple alternative spliced isoforms. *Genomics*, **29**, 345–352.

Puche, A.C., Poirier, F., Hair, M., Bartlett, P.F., and Key, B. (1996). Role of galectin-1 in the developing mouse olfactory system. *Dev. Biol.*, **179**, 274–287.

Quarles, R.H., Barbarash, G.R., Figlewitz, D.A., and McIntyre, L.J. (1983). Purification and partial characterization of the myelin-associated glycoprotein from adult rat brain. *Biochim. Biophys. Acta*, **757**, 140–143.

Rauch, U., Clement, A., Retzler, C., Frohlich, L., Gohring, W., and Faissner, A. (1997) Mapping of a defined neurocan binding site to different domains of tenascin-C. *J. Biol. Chem.*, **272**, 26905–26912.

Raulo, E., Chernousov, M.A., Carey, D.J., Nolo, R., and Rauvala, H. (1994). Isolation of a neuronal cell surface receptor of heparin binding growth-associated molecule (HB-GAM). Identification as N-syndecan (syndecan-3). *J. Biol. Chem.*, **269**, 12999–13004.

Rauvala, H. (1984). Neurite outgrowth of neuroblastoma cells: dependence on adhesion surface cell surface interactions. *J. Cell Biol.*, **98**, 1010–1016.

Rauvala, H., and Pihlaskari, R. (1987). Isolation and some characteristics of an adhesive factor of brain that enhances neurite outgrowth in central nervous system. *J. Biol. Chem.*, **262**, 16625–16635.

Rauvala, H., Merenmies, J., Pihlaskari, R., Korkolainen, M., Huhtala, M.L., and Panula, P. (1988). The adhesive and neurite-promoting molecule P30: analysis of the amino terminal sequence and production of antipeptide antibodies that detect P30 at the surface of neuroblastoma cells and of brain neurons. *J. Cell Biol.*, **107**, 2293–2305.

Reeber, A., Vincendon, G., and Zanetta, J.-P. (1980). Transient Concanavalin A-binding glycoproteins of the parallel fibers of the developing rat cerebellum: evidence for the destruction of their glycans. *J. Neurochem.*, **35**, 1273–1277.

Rosenbluth, J., Liang, W.L., Liu, Z.G., Guo, D.Z., and Schiff, R. (1996). Expanded CNS myelin sheaths formed in situ in the presence of an IgM antigalactocerebroside-producing hybridoma. *J. Neurosci.*, **16**, 2635–2641.

Rostami, A., Eccleston, P.A., Silberberg, D.H., Manning, M.C., Burns, J.B., and Lisak, R.P. (1983). Absence of antibodies to galactocerebroside in the sera and cerebrospinal fluid of human demyelinating disorders. *Neurology*, **33**, 130–134.

Saida, K., Saida, T., Brown, M.J., and Silverberg, D.H. (1979). In vivo demyelination induced by intraneural injection of antigalactocerebroside serum—A morphological study. *Am. J. Pathol.*, **95**, 99–116.

Sakamoto, Y., Kitamura, K., Yoshimura, K., Nishijima, T., and Uyemura, K. (1987). Complete amino acid sequence of P0 protein in bovine peripheral nerve myelin. *J. Biol. Chem.*, **262**, 4208–4214.

Sandi, C., and Rose, S.P. (1997) protein synthesis- and fucosylation-dependent mechanisms in corticosterone facilitation of long-term memory in the chick. *Behav. Neurosci.*, **111**, 1098–1104.

Saunders, S., Paine-Saunders, S., and Lander, A.A. (1997) Expression of the cell surface proteoglycan glypican-5 is developmentally regulated in kidney, limb, and brain. *Dev. Biol.*, **190**, 78–93.

Schluesener, H.J., Sobel, R.A., Linington, C., and Weiner, H.L. (1987) A monoclonal antibody against the myelin-oligodendrocyte glycoprotein induces relapses and demyelination in central nervous system autoimmune disease. *J. Immunol.*, **139**, 4016–4021.

Shashoua, V.E., Daniel, P.F., Moore, M.E., and Jungalwala, F.B. (1986). Demonstration of glucuronic acid on brain glycoproteins which react with HNK-1 antibody. *Biochem. Biophys. Res. Commun.*, **138**, 902–909.

Shirasawa, T., Akashi T., Sakamoto K., Takahashi H., Maruyama N., and Hirokawa K. (1993) Gene expression of CD24 core peptide molecule in developing brain and developing non-neural tissues. *Develop. Dynam.*, **198**, 1–13.

Sieghart, W., Item, C., Buchstaller, A., Fuchs, K., Hoger, H., and Adamiker, D. (1993) Evidence for the existence of differential O-glycosylated alpha5-subunits of the gamma-aminobutyric Acid(A) receptor in the rat brain. *J. Neurochem.*, **60** 93–98.

Simone, I.L., Annunziata, P., Maimone, D., Liguori, M., Leante, R., and Livrea, P. (1993). Serum and CSF anti-GM1 antibodies in patients with Guillain-Barre syndrome and chronic inflammatory demyelinating polyneuropathy. *J. Neurol. Sci.*, **114**, 49–55.

Sotelo, C., and Changeux, J.-P. (1974). Trans-synaptic degeneration "en cascade" in the cerebellar cortex of staggerer mutant mouse. *Brain Res.*, **67**, 519–526.

Steck, A.J., Murray, N., Vandevelde, M., and Zurbriggen, A. (1983). Human monoclonal antibodies to myelin-associated glycoprotein. *J. Neuroimmunol.*, **5**, 145–156.

Strecker, G., Fournet, B., and Montreuil, J. (1978) Structure of the three major fucosyl-glycoasparagines accumulating in the urine of a patient with fucosidosis. *Biochimie*, **60**, 725–734.

Su, Y., Brooks, D.G., Li, L.Y., Lepercq, J., Trofatter, J.A., Ravetch, J.V., and Lebo, R.V. (1993). Myelin protein zero gene mutated in Charcot-Marie-Tooth type-1B patients. *Proc. Natl. Acad. Sci. USA*, **90**, 10856–10860.

Sun, J.B., Link, H., Olsson, T., Xiao, B.G., Andersson, G., Ekre, H.P., Linington, C., and Diener, P. (1991) T-cell and B-cell responses to myelin-oligodendrocyte glycoprotein in multiple sclerosis. *J. Immunol.*, **146**, 1490–1495.

Tachi, N., Kozuka, N., Ohya, K., Chiba, S., Sasaki, K., Uyemura, K., and Hayasaka, K. (1996). A new mutation of the Po gene in patients with Charcot-Marie-Tooth disease type 1B: Screening of the Po gene by heteroduplex analysis. *Neurosci. Lett.*, **204**, 173–176.

Tan, J., Dunn, J., Jaecken, J., and Schachter, H. (1996) Mutations in the MGAT2 gene controlling complex N-glycan synthesis cause carbohydrate-deficient glycoprotein syndrome type II, an autosomal recessive disease with defective brain development. *Am. J. Hum. Genet.*, **59**, 810–817.

Tang, S., Shen, Y.J., Debellard, M.E., Mukhopadhyay, G., Salzer, J.L., Crocker, P.R., and Filbin, M.T. (1997) Myelin-associated glycoprotein interacts with neurons via a sialic acid binding site at ARG118 and a distinct neurite inhibition site. *J. Cell Biol.*, **138**, 1355–1366.

Taylor, D.L., Kang, M.S., Brennan, T.M., Bridges, C.G., Sunkara, P.S., and Tyms, A.S. (1994) Inhibition of alpha-glucosidase I of the glycoprotein-processing enzymes by 6-O-butanoyl castanospermine (MDL 28, 574) and its consequences in human immunodeficiency virus-infected T cells. *Antimicrob. Agents Chemother.*, **38**, 1780–1787.

Tropak, M.B., and Roder, J.C. (1997) Regulation of myelin-associated glycoprotein binding by sialylated cis-ligands. *J. Neurochem.*, **68**, 1753–1763.

Uyemura, K., and Kitamura, K. (1991). Comparative studies on myelin proteins in mammalian peripheral nerve. *Comp. Biochem. Physiol.*, **98**, 63–72.

Valldeoriola, F., Graus, F., Steck, A.J., Munoz, E., Delafuente, M., Gallart, T., Ribalta, T., Bombi, J.A., and Tolosa, E. (1993). Delayed appearance of anti-myelin-associated glycoprotein antibodies in a patient with chronic demyelinating polyneuropathy. *Ann. Neurol.*, **34**, 394–396.

Vanschaftingen, E., and Jaeken, J. (1995). Phosphomannomutase deficiency is a cause of carbohydrate-deficient glycoprotein syndrome type I. *FEBS Lett.*, **377**, 318–320.

Vriesendorp, F.J., Mishu, B., Blaser, M.J., and Koski, C.L. (1993). Serum antibodies to GM1, GD1b, peripheral nerve myelin, and Campylobacter jejuni in patients with Guillain-Barre syndrome and controls-Correlation and prognosis. *Ann. Neurol.*, **34**, 130–135.

Warner, L.E., Roa, B.B., and Lupski, J.R. (1996). Absence of PMP22 coding region mutations in CMT1A duplication patients: Further evidence supporting gene dosage as a mechanism for Charcot-Marie-Tooth disease type 1A. *Hum. Mutat.*, **8**, 362–365.

Willems, P.J., Gatti, R., Darby, J.K., Romeo, G., Durand, P., Dumon, J.E., and O'Brien, J.S. (1991) Fucosidosis revisted: a review of 77 patients. *Am. J. Med. Genet.*, **38**, 111–131.

Willenborg, D.O., Parish, C.R., and Cowden, W.B. (1989) Inhibition of experimental allergic encephalomyelitis by the α-glucosidase inhibitor castanospermine. *J. Neurol. Sci.*, **90**, 77–85.

Xiao, B.G., Linington, C., and Link, H. (1991) Antibodies to myelin-oligodendrocyte glycoprotein in cerebrospinal fluid from patients with multiple sclerosis and controls. *J. Neuroimmunol.*, **31**, 91–96.

Yamada, H., Fredette, B., Shitara, K., Hagihara, K., Miura, R., Ranscht, B., Stallcup, W.B., and Yamaguchi, Y. (1997) The brain chondroitin sulfate proteoglycan brevican associates with astrocytes ensheathing cerebellar glomeruli and inhibits neurite outgrowth from granule neurons. *J. Neurosci.*, **17**, 7784–7795.

Yamashita, K., Tachibana, Y., Mihara, K., Okada, S., Yabuuchi, H. and Kobata, A. (1980) Urinary oligosaccharides of,mannosidosis. *J. Biol. Chem.*, **255**: 5126–5133.

Yamashita, K., Ideo, H., Ohkura, T., Fukushima, K., Yuasa, I., Ohno, K., and Takeshita, K. (1993). Sugar chains of serum transferrin from patients with carbohydrate deficient glycoprotein syndrome. Evidence of asparagine-N-linked oligosaccharide transfer deficiency. *J. Biol. Chem.*, **268**, 5783–5789.

Yang, L.J.S., Zeller, C.B., Shaper, N.L., Kiso, M., Hasegawa, A., Shapiro, R.E., and Schnaar, R.L. (1996) Gangliosides are neuronal ligands for myelin-associated glycoprotein. *Proc. Natl. Acad. Sci. USA*, **93**, 814–818.

Yasugi, E., Nakasuji, M., Dohi, T., and Oshima, M. (1994). Major defect of carbohydrate-deficient-glycoprotein syndrome is not found in the synthesis of dolichyl phosphate or N-acetylglucosaminyl-pyrophosphoryl-dolichol. *Biochem. Biophys. Res. Commun.*, **200**, 816–820.

Yazaki, T., Miura, M., Asou, H., Kitamura, K., Toya, S., and Uyemura, K. (1992) Glycopeptide of P0 protein inhibits homophilic cell adhesion. *FEBS Lett.* **307**, 361–366.

Zanetta, J.-P., Morgan, I.-G., and Gombos, G. (1975) Synaptosomal plasma membrane glycoprotein: fractionation by affinity chromatography on Concanavalin A. *Brain Res.*, **83**, 337–348.

Zanetta, J.-P., Sarlième, L.L., Mandel, P., Vincendon, G., and Gombos, G. (1977a) Fractionation of glycoproteins associated to adult rat brain myelin fractions. *J. Neurochem.*, **29**, 827–828.

Zanetta, J.-P., Reeber, A., Vincendon, G., and Gombos, G. (1977b) Synaptosomal plasma membrane glycoproteins. II. Isolation of fucosyl glycoproteins by affinity chromatography on the Ulex europeus lectin specific for L-fucose. *Brain Res.*, **138**, 317–328.

Zanetta, J.-P., Roussel, G., Ghandour, M.S., Vincendon, G., and Gombos, G. (1978). Postnatal development of rat cerebellum: massive and transient accumulation of Concanavalin A binding glycoproteins in parallel fiber axolemma. *Brain Res.*, **142**, 301–319.

Zanetta, J.-P., Vitiello, F., and Vincendon, G. (1980a) Gangliosides from rat cerebellum: demonstration of a considerable heterogeneity using a new solvent for thin-layer chromatography. *Lipids*, **15**, 1055–1061.

Zanetta, J.-P., Federico, A., and Vincendon, G. (1980b). Glycosidases and cerebellar ontogenesis in the rat. *J. Neurochem.*, **34**, 831–834.

Zanetta, J.-P., Meyer, A., Dontenwill, M., Basset, P., and Vincendon, G. (1982). Purification and properties of α-D- mannosidase from adult rat brain and interaction with its antibodies. *J. Neurochem.*, **39**, 1601–1606.

Zanetta, J.-.P., Roussel, G., Dontenwill, M., and Vincendon, G. (1983). Immunohistochemical localization of α-D- mannosidase during the cerebellar development of the rat. *J. Neurochem.*, **40**, 202–208.

Zanetta, J.-P., Dontenwill, M., Meyer, A., and Roussel, G. (1985a). Isolation and immunohistochemical losalization of a lectin like molecule from the rat cerebellum. *Develop. Brain Res.*, **17**, 233–243.
Zanetta, J.-P., Dontenwill, M., Reeber, A., Vincendon, G., Lagrand, C., Clos, J., and Legrand, J. (1985b). Con A-binding glycoproteins in the developping cerebellum of control and hypothyroid rats. *Dev. Brain Res.*, **21**, 1–6.
Zanetta, J.-P., Meyer, A., Kuchler, S., and Vincendon, G. (1987a) Isolation and immunochemical study of a soluble cerebellar lectin delineating its structure and function. *J. Neurochem.*, **49**, 1250–1257.
Zanetta, J.-P., Dontenwill, M., Reeber, A., and Vincendon, G. (1987b) Expression of recognition molecules in the cerebellum of young and adult rats. *NATO ASI Series H*, **2**, 92–104.
Zanetta, J.-P., Warter, J.-M., Kuchler, S., Marschal, P., Rumbach, L., Lehmann, S., Tranchant, C., Reeber, A., and Vincendon, G. (1990a) Antibodies to cerebellar soluble lectin CSL in multiple sclerosis. *Lancet*, **335**, 1482–1484.
Zanetta, J.-P., Warter, J.-M., Lehmann, S., Tranchant, C., Kuchler, S., and Vincendon G. (1990b) Presence of antibodies to lectin CSL in the blood of multiple sclerosis patients. *C. R. Acad. Sci. Paris*, **311**, 327–331.
Zanetta, J.-P., Tranchant, C., Kuchler-Bopp, S., Lehmann, S., and Warter, J.-M. (1994). Presence of anti-CSL antibodies in the cerebrospinal fluid of patients. A sensitive and specific test in the diagnosis of multiple sclerosis. *J. Neuroimmunol.*, **52**, 175–182.
Zanetta, J.-P., Wantyghem, J., Kuchler-Bopp, S., Badache, A., and Aubery, M. (1995) Human lymphocyte activation is associated with the early and high level expression of the endogenous lectin CSL at the cell surface. *Biochem. J.*, **311**, 629–636.
Zanetta, J.-P., Alonso, C., and Michalski, J.-C. (1996). Interleukin 2 is a lectin that associates its receptor with the T cell receptor complex. *Biochem. J.*, **318**, 49–53.
Zanetta, J.-P., Bonaly, R., Maschke, S., Strecker, G., and Michalski, J.-C. (1998a). Differential binding of lectins IL-2 and CSL to Candida albicans and cancer cells. *Glycobiology*, **8**, 221–225.
Zanetta, J.-P., Bonaly, R., Maschke, S., Strecker, G., and Michalski, J.-C. (1998b). Hypothesis: Immunodeficiencies in α-mannosidosis, mycosis, aids and cancer: a common mechanism of inhibition of the function of interleukin 2 by oligomannosides. *Glycobiology*, **8**, V-XI.
Zona, C., Eusebi, F., and Miledi, R. (1990) Glycosylation is required for maintenance of functional voltage-activated channels in growing neocortical neurons of the rat. *Proc. Roy. Soc. B*, **239**, 119–127.

B6. Low-Tech Mannose Therapy for Protein Glycosylation Deficiencies

Hudson H. Freeze, Hinrich K. Harms and Thorsten Marquardt

1. INTRODUCTION

Gene and enzyme replacement therapies are the most sophisticated, cutting-edge technologies available to treat genetic diseases (Rader, 1997; Yogalingam *et al.*, 1997; Kozarsky *et al.*, 1996; Vomdahl *et al.*, 1998; O'Connor *et al.*, 1998; Gornati *et al.*, 1998; Kikuchi *et al.*, 1998). These ''high-tech'' approaches are challenging, multi-faceted, specific, and innovative, but they are also research-intensive, costly, and restrictive. It is unlikely that most genetic diseases will be treated this way in the near future, especially orphan diseases with only a few hundred or a few thousand cases. There is a ''low-tech'' therapy that relies on dietary manipulation.

This time-honored approach is based on a simple tenet: if a substance is toxic, avoid it; if a substance is deficient, provide it as a supplement. Inherited disorders such as galactosemia, fructosemia and phenylketonuria have long relied on avoiding the toxic substances (Petry and Reichardt, 1998; Chung, 1997; Holton, 1996). This approach is also applicable to induced conditions. For instance, celiac disease patients avoid gluten and pregnant women are told to avoid alcohol to prevent fetal alcohol syndrome (Mattson and Riley, 1998). Dietary supplements can also prevent the development of serious problems in the developing fetus, e.g., folic acid lowers the risk of spina-bifida (Butterworth and Bendich, 1996).

In this chapter we will discuss the basis and development of a recent proposal to use the simple sugar mannose as a low-tech dietary supplement for patients with a set of rare genetic disorders, collectively called Congenital Disorders of Glycosylation (CDG). These were previously called Carbohydrate Deficient Glycoprotein Syndromes, but the nomenclature was changed to accommodate the rapidly expanding spectrum. Although the results of short-term clinical studies on the most common type of CDG patients have not been positive, it has fostered new insights into glycobiology and medicine and gives a new appreciation of mannose metabolism in cells and mannose absorption in experimental animals and humans. Serendipitously, an unusual CDG patient on mannose therapy was essentially cured of a potentially fatal genetic disorder. This led to the discovery of a new glycosylation disorder as well as its primary defect. It also validated the use of mannose for ''low-tech'' therapy.

2. CONGENITAL DISORDERS OF GLYCOSYLATION

2.1. Clinical and Biochemical Features of Congenital Disorders of Glycosylation (CDG)

CDG is the collective name given to a set of rare autosomal genetic disorders that show different defects in N-glycosylation (Schachter *et al.*, 1998; McDowell and Gahl, 1997; Krasnewich and Gahl, 1997; Jaeken *et al.*, 1993a; Jaeken *et al.*, 1991; Hagberg *et al.*, 1993). Most patients share an overlapping set of symptoms such as severe mental and psychomotor retardation, dysmorphic features, gastrointestinal tract problems leading to a failure to thrive and to liver pathology. Coagulation problems may arise due to deficiencies in serum glycoproteins, including ATIII and protein C. Some of the patients are tube-fed, and have limited mobility. Mortality is about 20% within the first two years.

CDG can be biochemically confirmed in patients who show these symptoms by a simple and reliable serum transferrin isoelectric focusing (IEF) test (Stibler and Jaeken, 1990; Stibler, 1991; Jaeken and Stibler, 1989; Lof *et al.*, 1993; De Jong *et al.*, 1995). Transferrin has two N-glycosylation sites that usually have biantennary chains with a total of four sialic acids i.e., tetrasialo transferrin. Small amounts of hexasialo, trisialo and disialo species also exist. Misglycosylation can result from a failure to add entire N-linked chains or from faulty processing of normal precursor chains (Yamashita *et al.*, 1993b; Yamashita *et al.*, 1993a; Coddeville *et al.*, 1998) (Figure 1). In either case, the number of sialic acids changes, and so does the IEF pattern. Altered forms are called carbohydrate-deficient transferrins (CDT). Such IEF alterations can also occur temporarily as a result of recent alcohol consumption or in uncontrolled fructosemia and galactosemia (Stibler *et al.*, 1997; Jaeken *et al.*, 1996; Adamowicz and Pronicka, 1996; Charlwood *et al.*, 1998; Landberg *et al.*, 1995), but are reversible by simple dietary control. Various types of CDG can be distinguished by the specific transferrin IEF pattern. Clearly, there are many ways to create a similar transferrin IEF pattern, and thus, more than one primary defect can produce the same CDT pattern. Other serum proteins besides transferrin, such as acid 1-glycoprotein, α1-antitrypsin, and protein C, also show aberrant IEF forms (Harrison *et al.*, 1992; Henry *et al.*, 1997; Stibler *et al.*, 1998; Krasnewich *et al.*, 1995). Misglycosylation may also reduce the circulating levels or activity of glycoproteins such as ATIII, Factor XI, and thyroxine binding globulin (Stibler *et al.*, 1998). β-Trace protein from brain is misglycosylated in CDG patients (Pohl *et al.*, 1997).

2.2. CDG and Transferrin IEF

Four different types of transferrin IEF patterns were traditionally used to define types of CDG (see Figure 2). However, finding that different primary defects produce the same transferrin pattern limits the utility of transferrin-based typing as a method of clinical classification. Transferrin IEF remains the best way of biochemically detecting a CDG, regardless of the specific defect (Stibler *et al.*, 1998; Vreken *et al.*, 1998).

CDG Type Ia is by far the most frequent, best understood and most studied type (McDowell and Gahl, 1997; Krasnewich and Gahl, 1997; Schachter *et al.*, 1998). The total amount of transferrin remains the same, but there is about a 2-fold reduction of the normal (tetrasialo) form and a corresponding increase in disialo and asialo forms. This pattern results from an absence of entire N-linked chains at some of the normal glycosylation sites (Yamashita *et al.*, 1993b; Yamashita *et al.*, 1993a; Wada *et al.*, 1992). Various defects

Comparison of Glycosylation Abnormalities and IEF Patterns

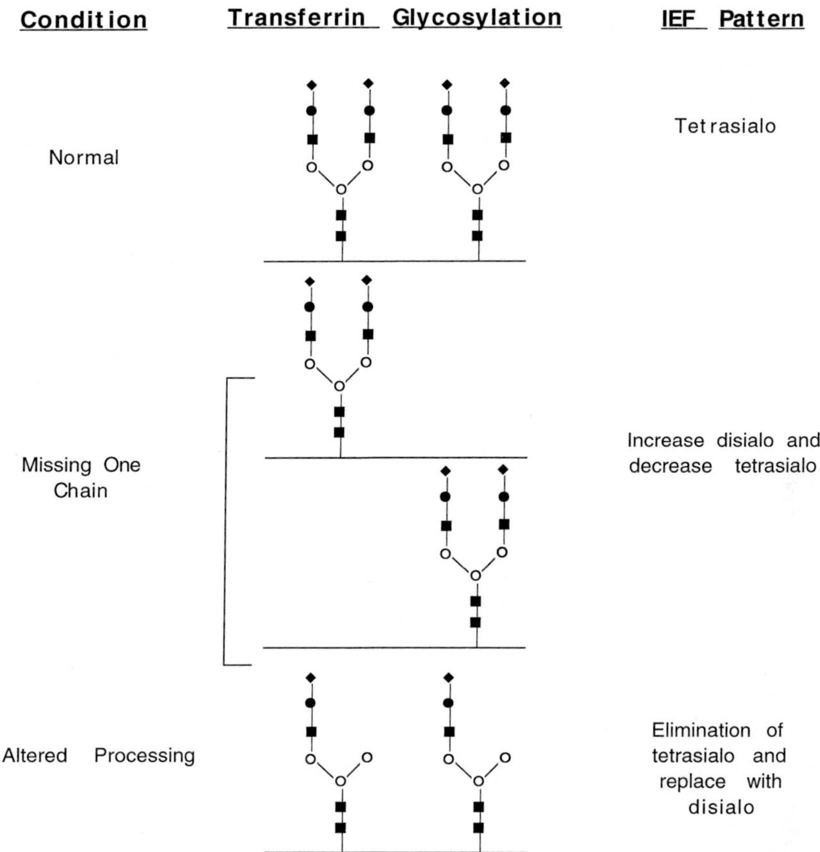

Figure 1: Examples of transferrin misglycosylation and the effects on sialylation and isoelectric focusing patterns.
Normal transferrin has primarily two disialylated, biantennary N-linked sugar chains with a total of 4 sialic acids i.e., the tetrasialo form at the top of the figure. If one sugar chain is entirely missing, the tetrasialo form is replaced by molecules with only a single chain containing 2 sialic acids; the disialo form shown in the middle. Absence of both chains produces an asialo form (not shown). A different disialo form can arise when both chains are added normally, but defects in oligosaccharide processing prevent the full addition of sialic acids, as in the example shown at the bottom. Symbols used: (◯) Man, (■) GlcNAc, (●) Gal, and (◆) Sia.

cause this pattern, which are shown in Figure 2, patients 1 and 2, and are discussed below. In another pattern (patient 3), the tetrasialo form is nearly totally replaced by the disialo form. This pattern is caused by the absence of GlcNAc transferase II activity which results in the production of two monoantennary chains (Jaeken *et al.*, 1993b; Tan *et al.*, 1996). The disorder is very severe and only two patients have been identified with this pattern. A third pattern in patient 4 shows small increases in tri-, di-, mono-, and asialo forms that together account for only about 10% of the total transferrin (Stibler *et al.*, 1993). In contrast to all other CDG, no abnormally glycosylated transferrins can be detected with SDS-PAGE, a complementary procedure for CDG diagnosis (Figure 2). The slight IEF changes observed here are also observed in newborns (Marquardt, manuscript in preparation). Taken together, the very small fraction of hyposialylated proteins may mean

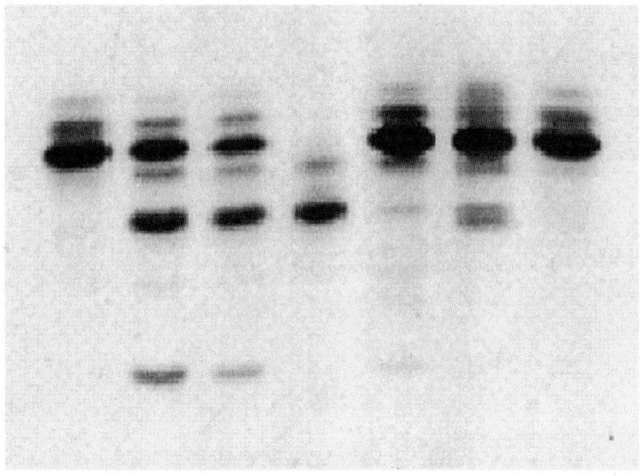

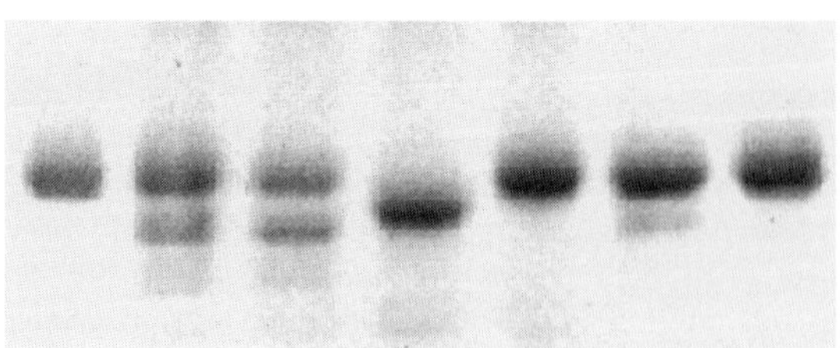

Figure 2: Isoelectric focusing pattern and SDS-PAGE analysis of transferrin abnormalities in CDG patients. Top panel shows serum transferrin isoelectric focusing (IEF) patterns from controls (c), and five different CDG patients, 1-5. Distinct patterns were traditionally used to define CDG Types, but this is no longer used, since this important method cannot indicate the primary defect. Different mutations in the glycosylation pathway can yield the same IEF pattern, such as that seen in patients 1 and 2 who have completely different clinical presentations. The bottom panel shows the same samples analyzed by SDS-PAGE. All except the pattern for patient 4 show differences in migration compared to the controls.

that this might not be a CDG syndrome. The pattern shown in patient 5 shows an increase in the disialo form, but almost no asialo form (Stibler *et al.*, 1995).

2.3. Primary Defects in CDG

The majority of known CDG Type 1 cases (80%) show a deficiency of phosphomanno-mutase activity (PMM, Man-6-P<–>Man-1-P) (Matthijs *et al.*, 1998; Jaeken *et al.*, 1997; Van Schaftingen and Jaeken, 1995; Matthijs *et al.*, 1997a) and are called CDG-Ia (see Figure 3). There is variation in the clinical severity among diagnosed patients, but no obvious correlation with the amount of residual enzymatic activity. Since Man-1-P is the immediate precursor to GDP-Man, which is needed for addition of mannose to all glycoconjugates, this defect is predicted to reduce the amount of all mannose containing macromolecules including N-linked oligosaccharides, glycophospholipid anchors, O-mannosylated proteins (Yuen *et al.*, 1997; Chiba *et al.*, 1997), and proteins with the newly identified C-mannosylation (Doucey *et al.*, 1998; Krieg *et al.*, 1998; Krieg *et al.*, 1997). Deficiency in PMM leads to unoccupied N-glycosylation sites in many proteins. Fucosylated glycoconjugates may also be affected by PMM deficiency since GDP-Fuc can be derived from GDP-Man (Figure 3). Since the relative amounts of fucose derived from GDP-Man and directly from the diet are unknown, the importance of reduced fucosylation to CDG pathology is also unknown.

Two genes, *PMM1* on chromosome 22 (Matthijs *et al.*, 1997b) and *PMM2* on chromosome 16 encode active enzymes in humans, but only defects in *PMM2* are known to cause CDG Type 1 and are compatible with the linkage data showing the defect is on chromosome 16 (Matthijs *et al.*, 1997a). Active *PMM1* and *PMM2* have been expressed in bacteria (Matthijs *et al.*, 1997b; Kjaergaard *et al.*, 1999; Pirard *et al.*, 1999). The very substantial amount of residual glycosylation occurring in *PMM2*-deficient children is probably due to residual *PMM2* activity. Several prominent mutations have been mapped to the *PMM2* gene in more than 50 patients (Matthijs *et al.*, 1998).

About 20% of patients with a transferrin pattern typical of CDG-Ia have normal PMM activity, and therefore, must have other defects. Considering the large number of steps involved in N-glycosylation, there are many possible lesions. Some patients make a lipid-linked oligosaccharide (LLO) precursor containing no glucose residues instead of one with the normal 3 glucose residues (Burda *et al.*, 1998). The underglucosylated precursor is transferred to protein at a lower rate than the normal precursor, thus generating underglycosylated proteins (Jakob *et al.*, 1998). The defect in these patients is the first α-1,3 glucosyl transferase needed for LLO synthesis (Burda *et al.*, 1998; Imbach *et al.*, 1999). These mutations are specific for the N-linked pathway and would not affect other mannose containing glycans. This may explain why these patients are reported to be less severely affected than those deficient in PMM. Other defects also produce a Type 1 transferrin pattern as discussed below. Two more defects have been identified in CDG. One (CDG-Ie) is a deficiency in Dol-P-Man synthesis resulting from mutations in the *DPM1* gene, which encodes the catalytic subunit of Dol-P-Man synthase (Kim *et al.*, 2000; Imbach *et al.*, 2000). The other closely-related defect (CDG-Id) is a deficiency in the first mannosyl transferase in LLO biosynthesis, which uses Dol-P-Man as a donor and converts $Man_5GlcNAc_2$-P-P-Dol to $Man_6GlcNAc_2$-P-P-Dol on the luminal face of the ER (Korner *et al.*, 1999).

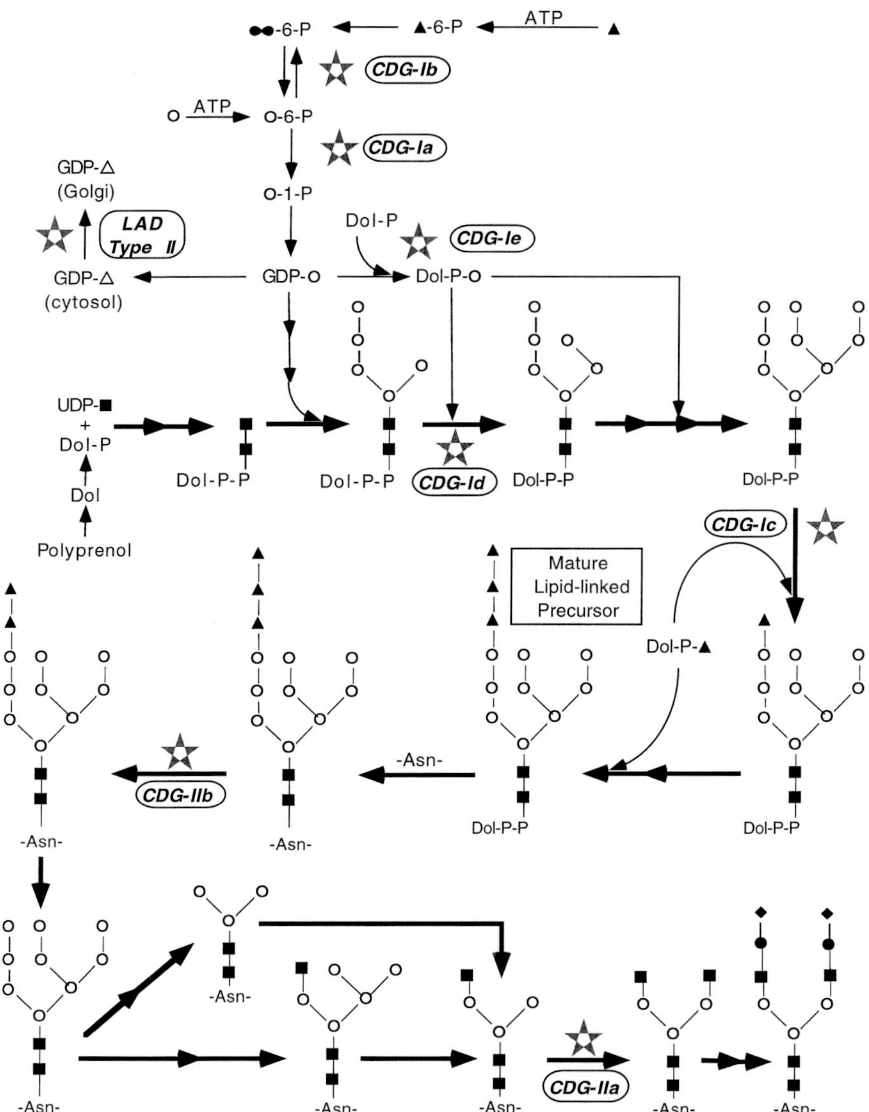

Figure 3: Simplified N-linked Oligosaccharide Precursor Biosynthesis and Processing Pathways.

This abbreviated pathway sketches an outline for the formation and utilization of sugar precursors for N-linked oligosaccharide biosynthesis. Steps indicated by the ⭐ are known to be deficient in various forms of CDG. The symbols used to denote monosaccharides are the same as in Figure 1. Additional symbols: (▲) glucose and (●●) fructose.

Mannose (○) delivered through the mannose transporter can be used directly for glycoprotein synthesis by conversion to Man-6-P (○-6-P) using hexokinase and ATP. Alternatively, fructose-6-P (●●-6-P) from glucose (▲) is converted to Man-6-P using phosphomannose isomerase (PMI, the defective enzyme in CDG-Ib). Phosphomannomutase (PMM) converts Man-6-P into Man-1-P (○-1-P). One of the two genes coding for PMM, *PMM2*, is defective in CDG-Ia. Man-1-P and UTP

Figure 3 (*continued*)

form GDP-Man (GDP-○) via GDP-Man pyrophosphorylase. GDP-Man has several fates. It can be incorporated into glycoproteins directly or converted into dolichol-P-Man (Dol-P-○) or into GDP-Fuc (GDP-△). Fuc is found in glycolipids and both N- and O-linked oligosaccharides. The conversion of GDP-Man to GDP-Fuc occurs in the cytosol, and its import into the Golgi is deficient in the rare disorder, Leukocyte Adhesion Deficiency Type II. Dol-P-Man can be incorporated into glycoproteins or glycophospholipid anchors and into C-mannosylated proteins (Doucey *et al.*, 1998; Krieg *et al.*, 1998; Krieg et al., 1997).

Dolichol, the lipid carrier for N-linked oligosaccharides, is formed from long chain unsaturated polyisoprenoids. The proximate double bond is reduced by polyprenol reductase to form dolichol which is phosphorylated prior to reaction with UDP-GlcNAc (UDP-■), to form Dol-P-P-GlcNAc, the first step in synthesis of the precursor for N-linked chains. A second GlcNAc (■) residue is added to form the chitobiose GlcNAc-GlcNAc core. Five Man are added from GDP-Man, each one using a separate mannosyl transferase. An additional 4 mannosyl transferases are required to add the final 4 Man residues derived from Dol-P-Man. Mutations in the first of these four α-mannosyl transferases causes CDG-Id. Defective production of Dol-P-Man causes CDG-Ie. Three separate α-glucosyl transferases use Dol-P-Glc (Dol-P-▲) to add three tandem Glc residues to the lipid-bound sugar chain. Addition of the first Glc residue is deficient in CDG-Ic. The mature lipid linked oligosaccharide (LLO) chain composed of 2 GlcNAc, 9 Man and 3 Glc is transferred to Asn-X-Thr/Ser sequons on proteins in the ER. Processing of the high mannose-type N-linked oligosaccharide begins with the removal of all Glc by a set of two α-glucosidases. Defects in the first α-glucosidase cause CDG-IIb. This is followed by removal of a portion of the Man residues using two different α-mannosidases. A GlcNAc residue is transferred to the chain followed by removal of two more Man units. Another option for the processing of the high mannose-type chain is to use a different combination of α-mannosidases to trim the sugar chain (Chui *et al.* 1997) followed by the addition of a GlcNAc using GlcNAc transferase I. A second GlcNAc is added using GlcNAc transferase II (MGAT2, which is deficient in CDG-IIa). Further build up of the chain continues by the addition of Gal (●) and finally Sia (◆) residues to generate a disialo-biantennary chain. Different branching patterns and other chain extensions can occur on other proteins, but the final structure shown is typical of serum glycoproteins proteins synthesized by the liver. Defective synthesis of the LLO itself, any of its precursors, or the oligosaccharide processing steps may alter protein glycosylation, suggesting that we are only seeing the "tip of the iceberg" (Freeze, 1998).

3. CELLULAR LABELING STUDIES ON CDG FIBROBLASTS

3.1. Labeling Studies on CDG

Early on, several laboratories showed that cultured fibroblasts from CDG patients incorporated less tracer [2-^{3}H]mannose into glycoproteins than controls (Powell *et al.*, 1994; Panneerselvam and Freeze, 1996b; Krasnewich *et al.*, 1995; Korner *et al.*, 1998). The medium contained reduced glucose, but complete absence of glucose is avoided since many previous studies showed that it induces truncated lipid-linked oligosaccharide formation and protein underglycosylation (Baumann and Jahreis, 1983; Turco and Pickard, 1982; Turco, 1980; Spiro *et al.*, 1983; Gershman and Robbins, 1981; Stark and Heath, 1979). Incorporation of [2-^{3}H]mannose into the lipid-linked precursor, phosphorylated intermediates, and newly synthesized glycoproteins were variably reduced 3- to 10-fold. In addition, the size of the LLO was reduced from 9 Man residues to ones having only 4 or 5 Man residues. These oligosaccharides were resistant to endoglycosidase H digestion when first transferred to protein; however, subsequent processing was normal.

3.2. Effects of Mannose on N-linked Glycosylation of PMM-Deficient Cells

Simply adding 300–500 μM mannose to the culture medium of CDG-Ia fibroblasts normalized the incorporation of [2-^{3}H]mannose into glycoproteins and various precursors, and also produced normal-sized LLO (Panneerselvam and Freeze, 1996b). When these chains were transferred to newly synthesized proteins, they were all sensitive to Endo H digestion. Thus, adding mannose caused both qualitative and quantitative changes in glycosylation. The size of the GDP-Man pool is very small in CDG fibroblasts, but it becomes nearly normal when mannose is added (Rush *et al.*, 2000). Since the only known pathway for making GDP-Man is via Man-1-P, it is possible that residual activity due to *PMM2* was able to convert the increased amount of Man-6-P into Man-1-P. Alternatively, a Man-1-kinase may exist, but such an enzyme has not been identified in mammals. The corrective effect was specific for mannose. Addition of ten-fold more glucose or galactose to the culture medium could not substitute for mannose in correcting the incorporation, size of the LLO, EndoH sensitivity, or size of the GDP-Man pool (Panneerselvam and Freeze, 1996b; Panneerselvam *et al.*, 1997b; Freeze, unpublished observations).

The failure of glucose to correct the defects was surprising since mannose is assumed to be produced from glucose through phosphomannose isomerase (PMI, Fru-6-P<–>Man-6-P), although this assumption had not been tested. Clearly, two decades of [2-^{3}H]mannose labeling studies show it is highly efficient as a direct precursor (Figure 3). These results raise the question of whether mannose itself or mannose derived from glucose is preferred for glycosylation when both glucose and mannose are available. Direct use of mannose would require a sufficient concentration of mannose in the blood that could be delivered in the presence of physiological concentrations of glucose.

4. CELLULAR MANNOSE METABOLISM

Mannose Transporter

The first practical issue in mannose utilization is how mannose enters cells. Previous reports showed that humans have about 50 μM mannose in the blood (Akazawa *et al.*, 1986). Although mannose can be transported by the well-known hexose transporters that normally carry glucose and fructose (GLUTs), it was unlikely that 50 μM mannose could enter cells very efficiently in the presence of 5 mM glucose (Gould and Holman, 1993; Gould and Bell, 1990). However, it turns out that mammalian cells have a facilitated diffusion mannose-specific transporter. The transporter is found in many cell types including hepatoma, glial-like cells, macrophage, mast cells, MDCK, and CaCo2 cells (Panneerselvam and Freeze, 1996a). It is highly mannose specific and relatively insensitive to inhibition by glucose even at 5 mM. The K_{uptake} of mannose is about 30–70 μM which suggests that it drives uptake at physiological concentrations. The calculated V_{max} of 5–33 nmole/hr/mg protein is in excess of the amount of mannose that would be required for glycoprotein synthesis in cells with a doubling time of 24 hr. Polarized epithelial CaCo2 cells have two mannose transporters (Ogier-Denis *et al.*, 1994; Ogier-Denis *et al.*, 1994; Ogier-Denis *et al.*, 1988; Ogier-Denis *et al.*, 1990). One is an active transporter on the apical surface, and the other is a facilitated diffusion transporter on the basolateral surface, suggesting that mannose is actively delivered to the blood. Mammalian sera contain 35–110 μM mannose (Alton *et al.*, 1998).

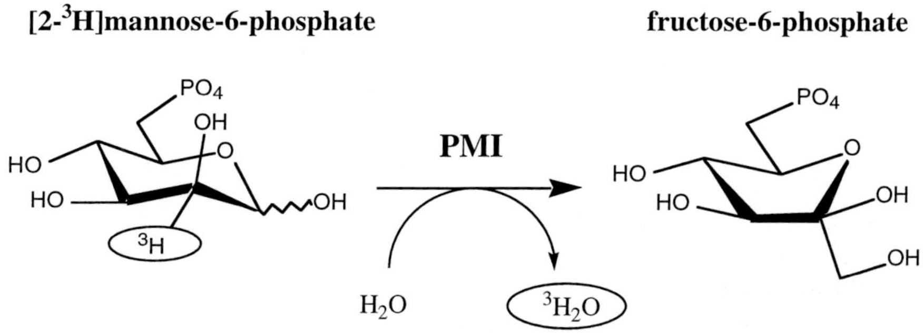

[2-³H]mannose-6-phosphate **fructose-6-phosphate**

Figure 4: Phosphomannose isomerase converts [2-³H] Mannose-6-P into ³HOH.
PMI catalyzes the reversible conversion of Man-6-P ↔ Fru-6-P, but a single round of PMI-catalyzed conversion of [2-³H]Man-6-P to Fru-6-P releases ³HOH which rapidly equilibrates with the endogenous water in the cell. The label is permanently lost and cannot reform [2-³H]Man-6-P.

Once mannose enters the cell, hexokinase converts it to Man-6-P which can be used for glycosylation. Alternatively, Man-6-P can be converted to Fru-6-P by phosphomannose isomerase (Figure 3). The relative proportion of mannose metabolized by these alternative routes can be determined by labeling cells with [2-³H]Man. If [2-³H]Man-6-P is converted to Fru-6-P, ³HOH is lost from the 2-position and rapidly equilibrates with non-labeled water whose concentration is 55 M. This prevents reformation of [2-³H]Man-6-P (Figure 4). Some cells have been shown to incorporate a small amount of label into glycogen as glucose, but the label is not at the 2-position (Rodriguez and Fliesler, 1990). The majority (83-90%) of [2-³H]mannose taken up through the transporter of fibroblasts and hepatoma cells is converted to ³HOH (Panneerselvam *et al.*, 1997a; Alton *et al.*, 1998). Calculations based on the kinetic parameters indicated that the transporter could probably supply sufficient mannose for glycoprotein synthesis. This hypothesis was directly confirmed in cultured cells. Short and long-term labeling of human fibroblasts using either [2-³H]mannose or [³H]glucose at their physiological concentrations of 50 μM and 5 mM, respectively, showed that mannose accounted for 65–75% of the label in PNGaseF-released oligosaccharides. Isotope dilution experiments with [2-³H]mannose in human fibroblasts and human hepatoma cells incubated under similar conditions showed that the specific activity of mannose in glycoproteins was only about 20% lower than that of [2-³H]mannose added to the medium for labeling. This means that only about 20% of the total mannose could have come from other sources. Therefore, direct mannose utilization accounts for up to 80%. Labeling with [³H]glucose under physiological conditions was extremely inefficient because most of the glucose is simply catabolized to lactic acid and water. Moreover, the large amount of [³H]glucose required (1-2 mCi/ml) to see significant conversion into [³H]mannose in oligosaccharides caused radiation-induced cell death. Glucose conversion was estimated to contribute 20-23%. The toxicity and extremely low efficiency of glucose labeling explained why a direct comparison of mannose and glucose utilization had not been tested directly in cultured cells. These results demonstrate that even though glucose is 100-fold more abundant than mannose in the blood, tissue culture cells prefer direct incorporation of mannose for glycoprotein synthesis over conversion from glucose (Panneerselvam *et al.*, 1997a).

5. MANNOSE UTILIZATION BY HIGHER ORGANISMS

In yeast and probably other lower organisms, nearly all mannose for glycosylation is derived from glucose. Mutation of PMI in yeast is lethal, but addition of mannose, together with glucose, rescues the mutant (Smith *et al.*, 1992). Yeast has a high demand for mannose since about 14% of the dry weight is found as cell wall mannans. The heavy synthetic demand is accommodated by a PMI activity nearly 100-fold higher than that found in mammalian cells. Among the 16 different sugar transporters identified in yeast by sequence homology analysis, none has yet been identified as a mannose-specific transporter (Bisson *et al.*, 1993). This is not surprising since the primary purpose of these transporters is to supply the major carbon and energy needs of the cell as well as the raw materials for glycoprotein synthesis.

In higher organisms, where the amount of mannose required for glycoprotein synthesis (~0.1–0.2% of protein mass) is 100-fold lower than yeast, it is reasonable that specialized mechanisms may have developed. Finding both energy-dependent and energy-independent mannose transporters in mammals suggest additional specialization beyond that seen in lower organisms. Mannose can be absorbed from the intestine by a transporter distinct from the active transporters used for glucose (Ogier-Denis *et al.*, 1994). A similar mannose transporter is found in kidney tubules (Silverman *et al.*, 1970).

A survey of about 35 different species of mammals shows that mannose concentration in serum ranges from about 35–110 μM with a mean of 75 μM, suggesting that mannose is generally available for direct utilization (Alton *et al.*, 1998). Human amniotic fluid (Akazawa *et al.*, 1986) and bovine milk contain mannose. Rats efficiently absorb mannose from the gut and rapidly deliver it into the blood stream (Alton *et al.*, 1998). Very little (<1%) is lost in the feces. Large amounts of mannose (0.2 gm/kg body weight) can be readily absorbed and raise the blood mannose levels from 70 μM to 250–300 μM. [2-^{3}H] Mannose half-life in blood is about 30 minutes, and after a 15–30 min. lag, label begins to reappear in the circulation as either ^{3}HOH or ^{3}H-labeled glycoproteins. Very little is metabolized in the blood itself. Overall, about 2% of the radioactivity is quickly incorporated into glycoproteins in all organs, but primarily those made by the liver and intestine. During the first hour, about 95% of the label in glycoproteins is found in the liver, plasma and intestine. Labeling for 4–8 hr leads to secretion of liver-derived glycoproteins into the circulation. The label in glycoproteins plateaus and gradually declines over the next several days as the serum glycoproteins turn over. [2-^{3}H] Mannose is also incorporated into the placenta and fetus, including the fetal brain, showing that mannose crosses the placental barrier.

One problem using [2-^{3}H]mannose for labeling studies in the whole organism is that the rapid diffusion of ^{3}HOH makes it impossible to determine which cells actually metabolized the [2-^{3}H]mannose via the PMI-pathway. A comparison of the specific activities of PMI in the different organs shows that liver is the lowest, consistent with the results of preference for mannose in culture and its major role in the uptake of mannose from the blood. Calculating the rates of mannose clearance from the blood and hepatic glycoprotein biosynthetic rate suggests that direct use of mannose could provide substantial mannose needed for plasma glycoprotein synthesis (Alton *et al.*, 1998).

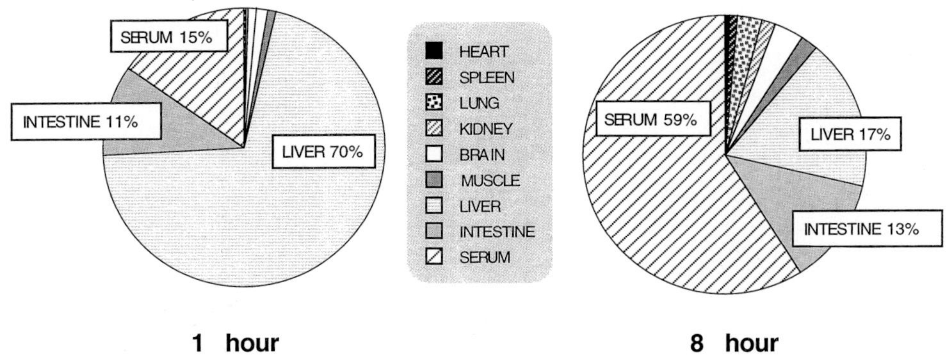

Figure 5: Proportion of radiolabeled glycoproteins in various rat tissues following a gavage of [2-^{3}H]mannose.
Rats were given [2-^{3}H]mannose and the total amount of radiolabeled incorporated into all glycoproteins in the blood and in each organ were defined as 100% at 1 and 8 h. The liver and intestine are clearly the major producers of glycoproteins during the first hour after dosing, while other organs make a relatively small contribution. During the next seven hours, liver-derived glycoproteins appear in the blood (Alton *et al.*, 1998).

5.1. Source of Mannose in the Blood

The source of mannose in the blood is not known, but it must come from the diet, oligosaccharide processing, or from glucose. There are essentially no dietary studies on the mannose content of food or its bioavailability. α-Mannosidases I and II that are normally used for oligosaccharide processing are found on brush border membranes of the intestine (Velasco *et al.*, 1993), suggesting they may help degrade glycoproteins in food. Calculations of estimated mannose content in food and demands of the liver suggest that dietary mannose could provide a portion, but probably not all, of the mannose needed for glycosylation. No studies have been done on mannose salvage from glycoproteins, but galactose, amino sugars, and sialic acids are very efficiently reutilized following glycoprotein digestion (Rome and Hill, 1986; Kingsley *et al.*, 1986; Krieger *et al.*, 1989; Schauer, 1988; Martin *et al.*, 1998). Endocytosis of sugar-rich plasma glycoproteins and salvage could provide an alternate to intracellular generation of each monosaccharide for glycoprotein biosynthesis. Undoubtedly, glucose provides mannose for glycoprotein synthesis, but it may not do so uniformly within each cell. Rather, exogenous mannose, whether ultimately derived from glucose or the diet may be used more often. Higher organisms and cells in tissue culture are well designed to utilize exogenous mannose for glycoprotein synthesis.

5.2. Mannose Studies in Humans and Results on PMM-Deficient CDG-Ia Patients

The finding that mannose was able to correct the glycosylation in PMM-deficient fibroblasts suggested that mannose might be of therapeutic value for CDG patients (Panneerselvam and Freeze, 1996b). In addition, the discovery of mannose transporters and mannose utilization in rats suggested that mannose could be absorbed and utilized. Previous human absorption studies in 1963 indicated that a single oral dose of 0.5 gm/kg body weight given to one subject led to diarrhea within a few hours (Wood and Cahill, 1963). More extensive human studies in 1997 showed that mannose could be well tolerated up to 0.15–0.2 gm/kg body weight and that multiple doses would maintain blood mannose levels up to about 200 μM (Alton *et al.*, 1997). There were no side-effects of mannose. CDG patients tend to have low mannose levels of 10–30 μM instead of 55 μM seen in normal individuals (Panneerselvam *et al.*, 1997b); however, CDG patients also absorbed mannose.

6. MANNOSE THERAPY FOR PMM-DEFICIENT CDG PATIENTS

Several mannose supplementation studies were carried out on different groups of confirmed CDG-Ia patients (Mayatepek *et al.*, 1997; Marquardt *et al.*, 1997; Kjaergaard *et al.*, 1998; Mayatepek and Kohlmuller, 1998). Various serum glycoproteins, including transferrin IEF pattern, were monitored. Assuming a normal turnover rate of serum glycoproteins in CDG patients, underglycosylated forms should be gradually replaced with the fully glycosylated forms having normal IEF patterns. Since some misglycosylated serum glycoproteins have reduced biological activity, or increased clearance, mannose supplementation might improve these features as well.

Despite the encouraging work in cells and animals, none of the clinical studies to date have provided positive results. In one study (Mayatepek *et al.*, 1997), a single patient was continuously infused with mannose to raise the serum level to 2 mM for several weeks. There was no improvement in the underglycosylated forms of transferrin seen in IEF. Oral mannose supplements raised the serum mannose to 200 μM and over six months, aberrant bands appeared inconsistently, but there was no obvious clinical or biochemical improvement (Mayatepek and Kohlmuller, 1998). In another short-term, but more extensive study, five Danish patients were given oral mannose doses of 0.1 gm/kg every 3 hr for 9 days (Kjaergaard *et al.*, 1998). Blood mannose increased from an average of 32 mM (22–42 μM) to only about 75 μM, but there was no improvement in mean serum concentration of transferrin, α1-antitrypsin, ATIII and thyroxin-binding globulin or in CDT.

The largest study so far was a 2 month study of 15 type 1 patients (12 PMM-deficient, 1 glucosyltransferase-deficient, 1 with an unknown defect) given oral mannose from 0.1 to 0.15 gm/kg bw five times per day. Mannose was well tolerated with no side effects and all patients showed increases in serum mannose concentrations. However, a high variability in serum mannose levels was found. Whereas some children reached mannose levels that were just over 50 μM, others had peak levels of more than 300 μM. None of the children showed any change in the transferrin IEF pattern nor in blood coagulation parameters. One of the children with PMM-deficiency was kept on mannose therapy with an even higher dose of 0.2 gm/kg bw five times per day for 1 year. No changes in the clinical condition or the transferrin glycosylation pattern were observed.

Oral mannose therapy studies involving both PMM-deficient and PMM-normal CDG patients are nearing completion in the United States.

7. A TREATABLE TYPE OF CDG

7.1. An Unusual CDG Patient

As mentioned above, CDG is primarily a neurological disorder with mental retardation, characteristic dysmorphic and skeletal features, accompanied by coagulation, hepatic, and gastrointestinal problems. Although the transferrin IEF test confirms aberrant glycosylation, many alterations could produce the same underglycosylation pattern. Several studies have appeared recently describing a new form of CDG (Niehues *et al.*, 1998; de Koning *et al.*, 1998; Jaeken *et al.*, 1998). One is particularly relevant to low-tech therapy (Niehues *et al.*, 1998).

Immediately after the human tolerance studies were completed, a six-year old boy who had been diagnosed with a CDG type 1 based on his transferrin IEF was in grave condition from intractable massive intestinal bleeding. Compassionate use of oral mannose was begun. However this patient's clinical profile was incompatible with typical CDG Type 1 diagnosis. He had no neurological, dysmorphic or abnormal skeletal features typical of CDG children. Instead, he had a failure to thrive, recurrent and severe hypoglycemia, evidence of thrombotic problems and hepatic and gastrointestinal problems such as severe vomiting and diarrhea. A persistently low ATIII level led to the transferrin IEF test that showed a type 1 pattern indistinguishable from PMM-deficient patients. PMM activity was normal, and serum mannose level was considerably below normal.

The patient's clinical history was also quite different from typical CDG-Ia children who are recognizable at or soon after birth. In contrast, it was only a year after a normal birth that he began showing severe gastrointestinal problems including vomiting, diarrhea, and protein losing enteropathy. An intestinal biopsy performed during surgery indicated villus atrophy and microthrombic lesions in the intestine. Massive precipitates of unidentified proteins were found in the highly distended ER of his enterocytes. Total blood protein was very low and ATIII was about 10–30% of normal. Originally he was suspected of having celiac disease, an immuno-reactive condition that is precipitated in susceptible individuals by consuming gluten-containing products. Dietary management appeared to offer temporary improvement, but the chronic symptoms reappeared including several serious bouts of intestinal bleeding. These may have been caused by anti-coagulant therapy designed to reduce thromboses.

7.2. Successful Mannose Therapy

The patient has now been on continuous oral mannose therapy (0.1–0.15 gm/kg) 3–5 times daily for nearly four years. Within 2 months of starting mannose, protein-losing enteropathy, hypoalbuminemia, ATIII deficiency, hypoglycemia, and chronic symptoms permanently reversed (Figure 6). By 11 months his underglycosylated transferrin and α1-antitrypsin IEF patterns had essentially returned to normal (Figure 7) and have remained there. Based on all previous symptoms, mannose appears to have reversed the metabolic defect.

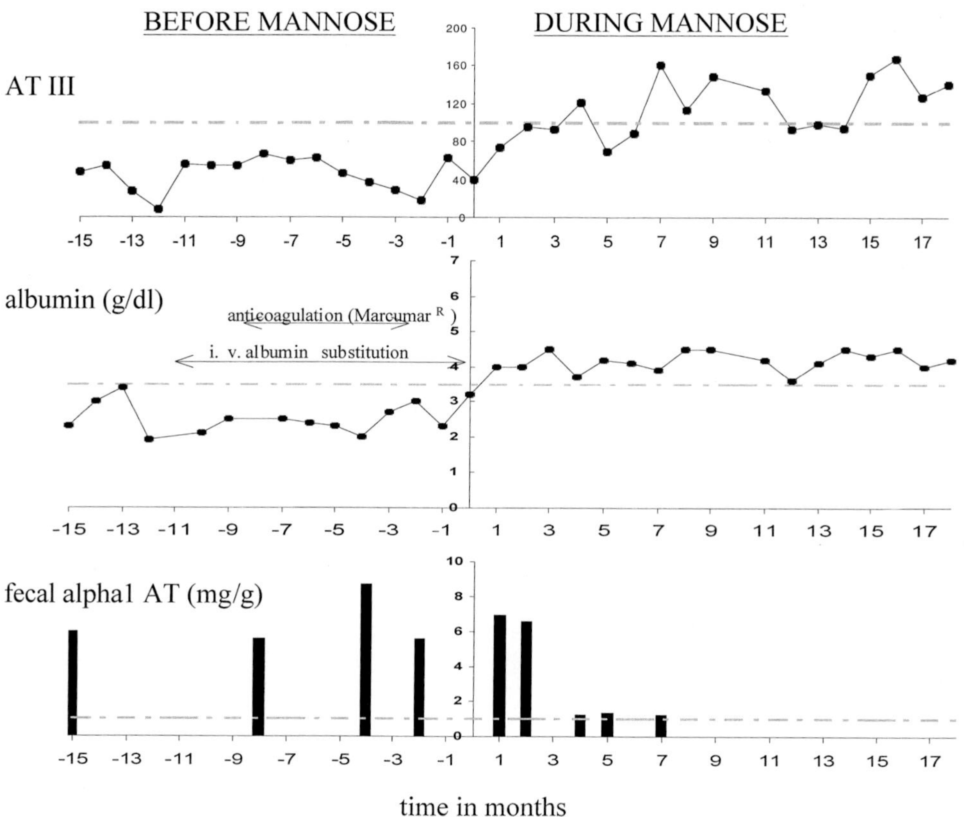

Figure 6: Protein levels in a CDG-Ib patient before and after mannose therapy.
Prior to starting mannose therapy, the patient suffered from reduced levels of ATIII and albumin resulting from protein-losing enteropathy (fecal α1-antitrypsin), and received periodic albumin and immunoglobulin transfusions. Anticoagulant therapy (Marcumar) was included to reduce the risk of thomboses. At time "0" (vertical line) mannose was given orally 5 times a day at 100–150 mg/kg. Soon after beginning mannose therapy, the ATIII and albumin concentrations returned to normal levels (horizontal dotted lines) and protein-losing enteropathy was also reversed.

7.3. Phosphomannose Isomerase Deficiency is the Primary Defect

Direct assay of fibroblast and lymphocyte lysates (Niehues *et al.*, 1998) showed that this patient has an 85–95% deficiency in phosphomannose isomerase (PMI) activity. The disorder is now called CDG-Ib, to distinguish it from PMM-deficient CDG-Ia. Labeling intact cells with [2-^{3}H]mannose, resulted in a 6-fold higher efficiency of [2-^{3}H]mannose incorporation into mannose-containing precursors and glycoproteins, and a corresponding 6-fold decrease in ^{3}HOH production. The structure of metabolically labeled and non-labeled N-linked oligosaccharides from the cells appeared normal. Leukocytes from the mother, father and an unaffected sib had heterozygous levels (34–65%) of PMI activity.

The gene for human PMI has been cloned (Proudfoot *et al.*, 1994) and RT-PCR was used to identify the mutation in the patient. The paternal allele carried a mutation at amino

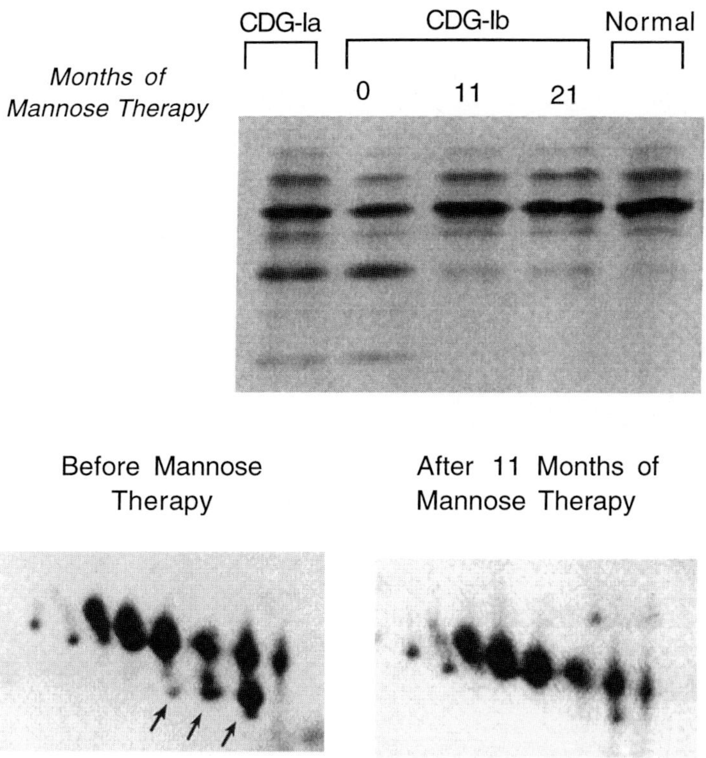

Figure 7: Effects of mannose therapy on the glycosylation of transferrin and a1-antitrypsin in PMI-Deficient patient.
Top panel shows IEF analysis of serum transferrin from a PMM-deficient CDG-Ia patient (lane 1) and the CDG-Ib patient before mannose therapy (lane 2) and then 11 (lane 3) and 21 (lane 4) months after starting mannose therapy. Normal control (lane 5). In the two bottom panels, α1-antitrypsin was analyzed by two-dimensional gel electrophoresis before mannose therapy and after 11 months of mannose therapy. The undersialylated forms (arrows) disappeared or were considerably reduced after mannose therapy. Several other glycoproteins showed similar changes of their isoform distribution following therapy.

acid 219 that converted an R–>Q. This R is a conserved residue in *C. albicans, C. elegans* and mouse PMI. Expression of this allele in COS cells showed 0–10% the activity of a normal control. The maternal allele showed no mutation in the coding region, but this allele was found in only 2 of 24 clones analyzed. This may mean reduced transcription or reduced stability of the maternal allele, which may in effect, make it a silent allele. This explanation would account for the 34% of normal PMI activity seen in the mother's leukocytes.

The results of mannose therapy and the identification of PMI as the defect in this patient show that PMI makes a significant contribution of mannose for glycosylation. It also shows that exogenous mannose corrects the defects in the liver and intestine, the two most active organs in the rat studies on mannose absorption (Alton *et al.*, 1998).

7.4. Other Patients with PMI Deficiencies

In another study (Jaeken *et al.*, 1998), Jaeken *et al.* found two PMI deficient patients with a similar clinical presentation. The first patient had two independent mutations at conserved locations in *S. cerevisiae* and *C. albicans* PMI, one is S102L and the other M138T. The Glu 137 is thought to be involved in the binding of the required Zn^{2+} ion at the active site. Unfortunately, this patient died at age 4 prior to diagnosis and did not have an opportunity for mannose therapy. Another patient had very low PMI activity and abnormal transferrin IEF pattern during the first several months of life. He had vomiting and a spectrum of symptoms including generalized hypotonia, megaloblastic anemia, liver dysfunction, low ATIII, and diffuse hypodensity of white matter. No mutations were reported for his PMI.

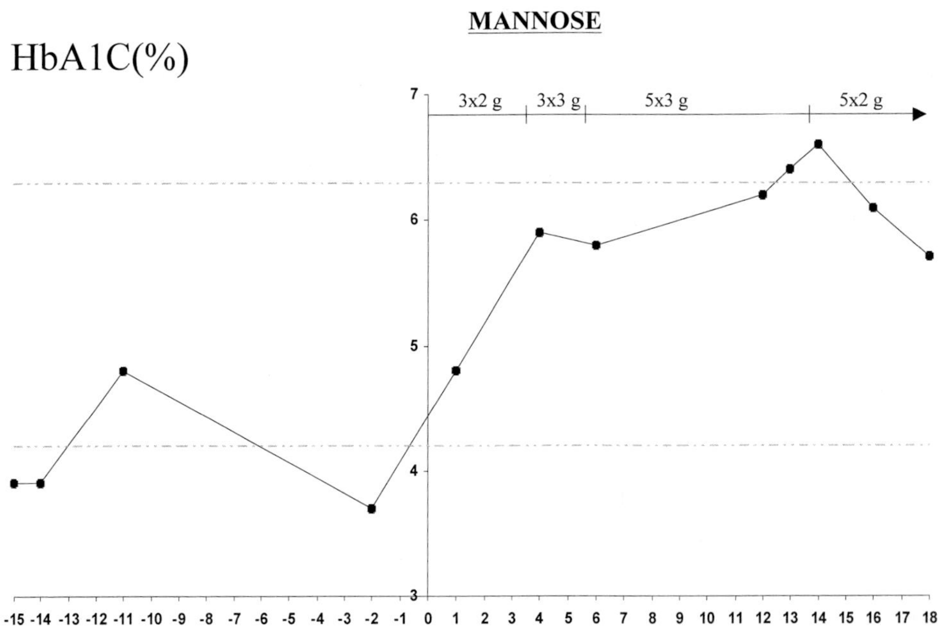

Figure 8: Effects of mannose therapy on the level of hemoglobin A1c.
Glycation is the non-enzymatic reaction of a reducing monosaccharide with the free amino groups of proteins to form Shiff bases. The rate of glycation depends on the reactivity of the monosaccharide and its concentration. Since mannose is about 5-fold more active in glycation than glucose, a potential side-effect of mannose therapy could be increased glycation of proteins such as hemoglobin. Prior to mannose therapy, and consistent with the hypoglycemia seen in the CDG-Ib patient, the level of glycated hemoglobin (HbA1c) was below normal. After initiating mannose therapy (3 × 2 gm/day), glucose levels quickly normalized and HbA1c also approached normal level as indicated by the upper horizontal dotted line. Increasing mannose dosage (5 × 3gm/day) eventually caused a slight elevation of HbA1c above normal, which decreased again when mannose dosage was lowered to 5 × 2gm/day.

At age 4 months he began a diet containing vegetables, fruit and meat. He subsequently stopped vomiting, showed improved weight gain and muscle tone. All biochemical parameters gradually improved and became normal by 10 months of age. Transferrin IEF completely normalized. He is now healthy with normal psychomotor development. This dramatic reversal was tentatively attributed to changing a diet to one containing increased amounts of mannose.

In another study, three teenage sibs of a consanguineous family had only ~10–20% of normal PMI activity (de Koning *et al.*, 1998). They had milder symptoms without demonstrated protein-losing enteropathy, but persistent vomiting and diarrhea resulted in dehydration and hospitalizations. The onset began at 10 months in two of the patients and at 2 months of age in the other. Albumin was consistently low. A liver biopsy was consistent with congenital hepatic fibrosis, and hepatic symptoms ranged from mildly enlarged liver to severe liver failure. Mental and motor development were completely normal. The asymptomatic parents had heterozygous PMI levels. Although the defect was not localized to the *PMI1* gene, it is clear that the same PMI deficiency can generate intestinal and hepatic symptoms of variable severity.

We are now aware of at least four additional patients who are PMI-deficient and exhibit similar symptoms. One patient in the United States, who has severe hypoglycemia, responded well to mannose within a few days of beginning treatment (Babovic-Vuksanovic *et al.*, 1999). Other patients have also been successfully treated with mannose therapy (de Lonlay *et al.*, 1999). Since mannose supplementation may need to be a life-long therapy, possible side effects have to be carefully observed. Mannose is known to be about five-fold more active in non-enzymatic glycation of hemoglobin (HbA1c) than an equivalent concentration of glucose (Bunn and Higgins, 1981). In fact, an increase of HbA1c was observed in our first patient described above during long-term mannose therapy, as shown in Figure 8 (Harms *et al.*, manuscript in preparation). The level of HbA1c decreased with reduced mannose dosage. Long term therapy needs to be monitored carefully for glycation reactions.

How can we explain the phenotypic differences between PMM and PMI-deficient patients? Although they share some symptoms, the major difference between the two disorders is that loss of PMM causes severe mental and psychomotor retardation, while loss of PMI does not. There is considerable evidence that glycosylation is critical for normal fetal brain development (Schachner and Martini, 1995; Kadmon *et al.*, 1990). The residual PMI activity in the CDG-Ib patients may be sufficient to maintain a minimal level of glycosylation *in utero* which is higher than that for the PMM-deficient patients. Mannose in the maternal circulation may be sufficient to maintain glycosylation *in utero* (Akazawa *et al.*, 1986). Currently, there is no information on the transcriptional or developmental regulation of either PMI or PMM, so it is difficult to know whether residual activity or differential regulation of these enzymes may be important.

8. IMPLICATIONS OF THESE FINDINGS

The PMI deficiency broadens our perspective on misglycosylation disorders from an emphasis on neurologic abnormalities toward an expanded appreciation of intestinal abnormalities. Misglycosylation resulting from any number of defects could compromise enterocyte adhesion and disturb the crypt-villus axis. This could lead to protein-losing enteropathies and poor nutrient absorption that generates a host of secondary problems. Such defects could result from a combination of genetic predisposition and environmental factors. For instance, the small intestine is an open and dynamic ecosystem where a variety of resident and transient bacteria vie for restricted niches (Bry *et al.*, 1996). Lectin binding studies show that the intestine has highly localized areas of fucosylated glycan expression. Full expression of the necessary fucosyl transferases depends on colonization by resident bacteria that can utilize L-fucose as a carbon source. Even temporary alterations in the glycosylation of rapidly dividing intestinal epithelial cells could act as a trigger under the right environmental stresses or situations to alter this delicate and dynamic relationship, especially if the glycosylation machinery is already genetically compromised. Bacteria also rely on oligosaccharides to adhere to the gastric epithelial lining, e.g., *H. pylori* binding to Lewis b histo-blood group antigen (Ilver *et al.* 1998).

PMM-deficient patients are clinically the best characterized of the currently known CDG syndromes. However, the clinical variation among the non-PMM deficient patients is even broader than within the PMM-deficient cases (VanOmmen *et al.*, 2000). Some non-PMM deficient patients currently under investigation were detected by keen clinicians who suspected glycosylation abnormalities even though the patients did not present with the typical symptoms of PMM deficiency. Since it has been estimated that 1% of the human genome is devoted to the production or recognition of glycosylated molecules (Varki and Marth, 1995), this may mean that glycosylation-based disorders are considerably more common than is currently appreciated, as suggested in recent editorials (Gahl, 1997; Kornfeld, 1998). Not all such defects can be detected by the serum transferrin IEF test, since transferrin has only a portion of known types of N-linked chains. Other tests for altered glycosylation need to be developed. This is illustrated by the most recently discovered defect, CDG-IIb. It is caused by a deficiency in an α1,2-glucosidase required for the removal of the terminal glucose from the newly transferred N-linked chains on proteins. The transferrin pattern was normal. The deficiency was cleverly detected by detailed analysis of urinary oligosaccharides that showed the accumulation of a tetrasaccharide, Glcα1,2Glcα1,3Glcα1,3Man, resulting from α-endomannosidase cleavage of unprocessed chains (DePraeter *et al.*, 2000).

Low-tech therapy also has been successfully applied to treating a Leukocyte Adhesion Deficiency Type II patient who cannot efficiently import GDP-Fuc into the Golgi (Lubke *et al.*, 1999; Marquardt *et al.*, 1999). Recent studies by Murch suggest that providing GlcNAc to patients with inflammatory bowel disease may have beneficial effects (Murch *et al.*, 1998). The basis of these positive effects is unknown. Nevertheless, the message that altered glycosylation causes a wide variety of clinical (and perhaps sub-clinical) problems should foster excitement and new developments in glycobiology and medicine.

ACKNOWLEDGMENTS

The authors wish to thank the members of their laboratories whose tireless work contributed to these studies and to Geetha Srikrishna for her helpful suggestions and Susan Greaney and Joseph P. Henig for preparation of the manuscript. This work was supported by grants RO1 GM55695 and DK55615 to HF.

REFERENCES

Adamowicz, M. and Pronicka, E. (1996) Carbohydrate deficient glycoprotein syndrome–like transferrin isoelectric focusing pattern in untreated fructosaemia. *Eur. J. Pediatr.*, **155**, 347–348.

Akazawa, S., Metzger, B.E. and Freinkel, N. (1986) Relationships between glucose and mannose during late gestation in normal pregnancy and pregnancy complicated by diabetes mellitus: concurrent concentrations in maternal plasma and amniotic fluid. *J. Clin. Endocrinol. Metab.*, **62**, 984–989.

Alton, G., Kjaergaard, S., Etchison, J.R., Skovby, F. and Freeze, H.H. (1997) Oral ingestion of mannose elevates blood mannose levels: a first step toward a potential therapy for carbohydrate-deficient glycoprotein syndrome type I. *Biochem. Mol. Med.*, **60**, 127–133.

Alton, G., Hasilik, M., Niehues, R., Panneerselvam, K., Etchison, J.R., Fana, F. and Freeze, H.H. (1998) Direct utilzation of mannose for mammalian glycoprotein biosynthesis. *Glycobiology*, **8**, 285–295.

Babovic-Vuksanovic, D., Patterson, M.C., Schwek, W.F., O'Brien, J.F., Vockley, J., Freeze, H.H., Mehta, D.P., Michels, V.V. (1999) Severe hypoglycemia as a presenting symptom of carbohydrate-deficient glycoprotein syndrome. *J Pediatr.*, **135**, 775–781.

Baumann, H. and Jahreis, G.P. (1983) Glucose starvation leads in rat hepatoma cells to partially N-glycosylated glycoproteins including alpha 1-acid glycoproteins. Identification by endoglycolytic digestions in polyacrylamide gels. *J. Biol. Chem.*, **258**, 3942–3949.

Bisson, L.F., Coons, D.M., Kruckeberg, A.L. and Lewis, D.A. (1993) Yeast sugar transporters. *Crit. Rev. Biochem. Mol. Biol.*, **28**, 259–308.

Bry, L., Falk, P.G., Midtvedt, T. and Gordon, J.I. (1996) A model of host-microbial interactions in an open mammalian ecosystem. *Science*, **273**, 1380–1383.

Bunn, H.F. and Higgins, P.J. (1981) Reaction of monosaccharides with proteins: possible evolutionary significance. *Science*, **213**, 222–224.

Burda, P., Borsig, L., de Rijk-van Andel, J., Wevers, R., Jaeken, J., Carchon, H., Berger, E.G. and Aebi, M. (1998) A novel carbohydrate-deficient glycoprotein syndrome characterized by a deficiency in glucosylation of the dolichol-linked oligosaccharide. *J. Clin. Invest.*, **102**, 647–652.

Butterworth, C.E.,Jr. and Bendich, A. (1996) Folic acid and the prevention of birth defects. *Annu. Rev. Nutr.*, **16**, 73–97.

Charlwood, J., Clayton, P., Keir, G., Mian, N. and Winchester, B. (1998) Defective galactosylation of serum transferrin in galactosemia. *Glycobiology*, **8**, 351–357.

Chiba, A., Matsumura, K., Yamada, H., Inazu, T., Shimizu, T., Kusunoki, S., Kanazawa, I., Kobata, A. and Endo, T. (1997) Structures of sialylated O-linked oligosaccharides of bovine peripheral nerve alpha-dystroglycan. The role of a novel O-mannosyl-type oligosaccharide in the binding of alpha-dystroglycan with laminin. *J. Biol. Chem.*, **272**, 2156–2162.

Chui, D., Oh-Eda, M., Liao, Y.F., Panneerselvam, K., Lal, A., Marek, K.W., Freeze, H.H., Moremen, K.W., Fukuda, M.N. and Marth, J.D. (1997) Alpha-mannosidase-II deficiency results in dyserythropoiesis and unveils an alternate pathway in oligosaccharide biosynthesis. *Cell*, **90**, 157–167.

Chung, M.A. (1997) Galactosemia in infancy: diagnosis, management, and prognosis. *Pediatric. Nursing.*, **23**, 563–569.

Coddeville, B., Carchon, H., Jaeken, J., Briand, G. and Spik, G. (1998) Determination of glycan structures and molecular masses of the glycovariants of serum transferrin from a patient with carbohydrate deficient glycoprotein syndrom type II. *Glycoconj. J.*, **15**, 265–273.

De Jong, G., Feelders, R., van Noort, W.L. and Van Eijk, H.G. (1995) Transferrin microheterogeneity as a probe in normal and disease states. *Glycoconj. J.*, **12**, 219–226.

de Koning, T.J., Dorland, L., van Diggelen, O.P., Boonman, A.M.C., de Jong, G.J., van Noort, W.L., De Schryver, J., Duran, M., van den Berg, I.E.T., Gerwig, G.J., Berger, R. and Poll-The, B.T. (1998) A novel disorder of N-glycosylation due to phosphomannose isomerase deficiency. *Biochem. Biophy. Res. Commun.*, **245**, 38–42.

de Lonlay, P., Cuer, M., Vuillaumier-Barrot, S., Beaune ,G., Castelnau, P., Kretz, M., Durand, G., Saudubray, J.M., Seta, N. (1999) Hyperinsulinemic hypogycemia as a presenting sign in phosphomannose isomerase deficiency: A new manifestiation of carbohydrate-deficient glycoprotein syndrome treatable with mannose. *J. Pediatr..* **135**, 379–383.

De Praeter, C.M., Gerwig, G.J., Bause, E., Nuytinck, L.K., Vliegenthart, J.F., Breuer, W., Kamerling, J.P., Espeel, M.F., Martin, J.J., De Paepe, A.M., Chan, N.W., Dacremont, G.A., Van Coster, R.N. (2000) A novel disorder caused by defective biosynthesis of N-linked oligosaccharides due to glucosidase I deficiency. *Am. J. Hum. Genet.*, **66**, 1744–1756.

Doucey, M.A., Hess, D., Cacan, R. and Hofsteenge, J. (1998) Protein C-mannosylation is enzyme-catalysed and uses dolichyl-phosphate-mannose as a precursor. *Mol. Biol. Cell*, **9**, 291–300.

Freeze, H.H. (1998) Disorders in protein glycosylation and potential therapy: tip of an iceberg? *J. Pediatr.*, **133**, 593–600.

Gahl, W.A. (1997) Carbohydrate-deficient glycoprotein syndrome: hidden treasures. *J. Lab. Clin. Med.*, **129**, 394–395.

Gershman, H. and Robbins, P.W. (1981) Transitory Effects of Glucose Starvation onthe Synthesis of Dolichol-linked Oligosaccharides in Mammalian Cells. *J. Biol. Chem.*, **256**, 7774–7780.

Gornati, R., Bembi, B., Tong, X., Boscolo, R. and Berra, B. (1998) Total glycolipid and glucosylceramide content in serum and urine of patients with Gaucher's disease type 3 before and after enzyme replacement therapy. *Clin. Chim. Acta*, **271**, 151–161.

Gould, G.W. and Bell, G.I. (1990) Facilitative glucose transporters: an expanding family. *Trends Biochem. Sci.*, **15**, 18–23.

Gould, G.W. and Holman, G.D. (1993) The glucose transporter family: structure, function and tissue-specific expression. *Biochem. J.*, **295**, 329–341.

Hagberg, B.A., Blennow, G., Kristiansson, B. and Stibler, H. (1993) Carbohydrate-deficient glycoprotein syndromes: peculiar group of new disorders. *Pediatr. Neurol.*, **9**, 255–262.

Harrison, H.H., Miller, K.L., Harbison, M.D. and Slonim, A.E. (1992) Multiple serum protein abnormalities in carbohydrate-deficient glycoprotein syndrome: pathognomonic finding of two-dimensional electrophoresis? *Clin. Chem.*, **38**, 1390–1392.

Henry, H., Tissot, J.D., Messerli, B., Markert, M., Muntau, A., Skladal, D., Sperl, W., Jaeken, J., Weidinger, S., Heyne, K. and Bachmann, C. (1997) Microheterogeneity of serum glycoproteins and their liver precursors in patients with carbohydrate-deficient glycoprotein syndrome type I: apparent deficiencies in clusterin and serum amyloid P. *J. Lab. Clin. Med.*, **129**, 412–421.

Holton, J.B. (1996) Galactosaemia: pathogenesis and treatment. *J. Inherit. Metab. Dis.*, **19**, 3–7.

Ilver, D., Arnqvist, A., Ogren, J., Frick, I.M., Kersulyte, D., Incecik, E.T., Berg, D.E., Covacci, A., Engstrand, L. and Boren, T. (1998) Helicobacter pylori adhesin binding fucosylated histo-blood group antigens revealed by retagging. *Science*, **279**, 373–377.

Imbach, T., Burda P., Kuhnert P., Wevers, R.A., Aebi, M., Berger, E.G., Hennet, T. (1999) A mutation in the human ortholog of the Saccharomyces cerevisiae ALG6 gene causes carbohydrate-deficient glycoprotein syndrome type-Ic. *Proc. Natl. Acad. Sci. U. S. A.*, **96**, 6982–6987.

Imbach, T., Schenk, B., Schollen, E., Burda, P., Stutz, A., Grunewald, S., Bailie, N.M., King, M.D., Jaeken, J., Matthijs, G., Berger, E.G., Aebi, M., Hennet, T. (2000) Deficiency of dolichol-phosphate-mannose synthase-1 causes congenital disorder of glycosylation type Ie. *J. Clin. Invest.*, **105**, 233–239.

Jaeken, J. and Stibler, H. (1989) *A newly recognized inherited neurological disease with carbohydrate deficent secretory glycoproteins.* In L. Wetterberg (ed.),Genetics of Neuropsychiatric Diseases, Wenner-Green International Symposium Series, pp. 69–80.

Jaeken, J., Stibler, H. and Hagberg, B. (1991) The carbohydrate-deficient glycoprotein syndrome. A new inherited multisystemic disease with severe nervous system involvement. *Acta Paediatr. Scand. Suppl.*, **375**, 1–71.

Jaeken, J., Carchon, H. and Stibler, H. (1993a) The carbohydrate-deficient glycoprotein syndromes: pre-Golgi and Golgi disorders? *Glycobiology*, **3**, 423–428.

Jaeken, J., De Cock, P., Stibler, H., Van Geet, C., Kint, J., Ramaekers, V. and Carchon, H. (1993b) Carbohydrate-deficient glycoprotein syndrome type II. *J. Inherit. Metab. Dis.*, **16**, 1041.

Jaeken, J., Pirard, M., Adamowicz, M., Pronicka, E. and Van Schaftingen, E. (1996) Inhibition of phosphomannose isomerase by fructose 1-phosphate: an explanation for defective N-glycosylation in hereditary fructose intolerance. *Ped. Res.*, **40**, 764–766.

Jaeken, J., Artigas, J., Barone, R., Fiumara, A., de Koning, T.J., Poll-The, B.T., de Rijk-van Andel, J.F., Hoffmann, G.F., Assmann, B., Mayatepek, E., Pineda, M., Vilaseca, M.A., Saudubray, J.M., Schluter, B., Wevers, R. and Van Schaftingen, E. (1997) Phosphomannomutase deficiency is the main cause of carbohydrate-deficient glycoprotein syndrome with type I isoelectrofocusing pattern of serum sialotransferrins. *J. Inherit. Metab. Dis.*, **20**, 447–449.

Jaeken, J., Matthijs, G., Saudubray, J.M., Dionisivici, C., Bertini, E., Delonlay, P., Henri, H., Carchon, H., Schollen, E. and Vanschaftingen, E. (1998) Phosphomannose isomerase deficiency—A carbohydrate-deficient glycoprotein syndrome with hepatic-intestinal presentation. *Am. J. Hum. Genet.*, **62**, 1535–1539.

Jakob, C.A., Burda, P., Teheesen, S., Aebi, M. and Roth, J. (1998) Genetic tailoring of n-linked oligosaccharides—the role of glucose residues in glycoprotein processing of saccharomyces cerevisiae in vivo. *Glycobiology*, **8**, 155–164.

Kadmon, G., Kowitz, A., Altevogt, P. and Schachner, M. (1990) Functional cooperation between the neural adhesion molecules L1 and N-CAM is carbohydrate dependent. *J. Cell. Biol.*, **110**, 209–218.

Karsan, A., Cornejo, C.J., Winn, R.K., Schwartz, B.R., Way, W., Lannir, N., Gershoni-Baruch, R., Etzioni, A., Ochs, H.D. and Harlan, J.M. (1998) Leukocyte Adhesion Deficiency Type II is a generalized defect of de novo GDP-fucose biosynthesis. Endothelial cell fucosylation is not required for neutrophil rolling on human nonlymphoid endothelium. *J. Clin. Invest.*, **101**, 2438–2445.

Kikuchi, T., Yang, H.W., Pennybacker, M., Ichihara, N., Mizutani, M., Van Hove, J.L.K. and Chen, Y.T. (1998) Clinical and metabolic correction of pompe disease by enzyme therapy in acid maltase-deficient quail. *J. Clin. Invest.*, **101**, 827–833.

Kim, S., Westphal, V., Srikrishna, G., Mehta, D.P., Peterson, S., Filiano, J., Karnes, P.S., Patterson, M.C., Freeze, H.H. (2000) Dolichol phosphate mannose synthase (DPM1) mutations define congenital disorder of glycosylation Ie (CDG-Ie). *J. Clin. Invest.*, **105**, 191–198.

Kingsley, D.M., Kozarsky, K.F., Hobbie, L. and Krieger, M. (1986) Reversible defects in O-linked glycosylation and LDL receptor expression in a UDP-Gal/UDP-GalNAc 4-epimerase deficient mutant. *Cell*, **44**, 749–759.

Kjaergaard, S., Kristiansson, B., Stibler, H., Freeze, H.H., Schwartz, M., Martinsson, T. and Skovby, F. (1998) Failure of short-term mannose therapy of patients with carbohydrate-deficient glycoprotein syndrome type 1. *Acta Paediatr.*, (In Press).

Kjaergaard, S., Skovby, F., Schwartz, M. (1999) Carbohydrate-deficient glycoprotein syndrome type IA: expression and characterisation of wild type and mutant PMM2 in E. coli. *Eur. J. Hum. Genet.*, **7**, 884–888.

Korner, C., Lehle, L. and Von Figura, K. (1998) Abnormal synthesis of mannose 1-phosphate derived carbohydrates in carbohydrate-deficient glycoprotein syndrome type I fibroblasts with phosphomannomutase deficiency. *Glycobiology*, **8**, 165–171.

Korner, C., Knauer, R., Stephani, U., Marquardt, T., Lehle L., von Figura, K. (1999) Carbohydrate deficient glycoprotein syndrome type IV: deficiency of dolichoyl-P-Man:Man(5)GlcNAc(2)-PP-dolichyl mannosyltransferase. *EMBO J.*, **18**, 6816–6822.

Kornfeld, S. (1998) Diseases of abnormal protein glycosylation—an emerging area. *J. Clin. Invest.*, **101**, 1293–1295.

Kozarsky, K.F., Jooss, K., Donahee, M., Strauss, J.F. and Wilson, J.M. (1996) Effective treatment of familial hypercholesterolaemia in the mouse model using adenovirus-mediated transfer of the VLDL receptor gene. *Nat. Genet.*, **13**, 54–62.

Krasnewich, D. and Gahl, W.A. (1997) Carbohydrate-deficient glycoprotein syndrome. *Adv. Pediatr.*, **44**, 109–140.

Krasnewich, D.M., Holt, G.D., Brantly, M., Skovby, F., Redwine, J. and Gahl, W.A. (1995) Abnormal synthesis of dolichol-linked oligosaccharides in carbohydrate-deficient glycoprotein syndrome. *Glycobiology*, **5**, 503–510.

Krieg, J., Glasner, W., Vicentini, A., Doucey, M.A., Loffler, A., Hess, D. and Hofsteenge, J. (1997) C-Mannosylation of human RNase 2 is an intracellular process performed by a variety of cultured cells. *J. Biol. Chem.*, **272**, 26687–26692.

Krieg, J., Hartmann, S., Vicentini, A., Glasner, W., Hess, D. and Hofsteenge, J. (1998) Recognition signal for C-mannosylation of TRP-7 in RNAse 2 consists of sequence TRP-X-X-TRP. *Mol. Biol. Cell*, **9**, 301–309.

Krieger, M., Reddy, P., Kozarsky, K., Kingsley, D., Hobbie, L. and Penman, M. (1989) Analysis of the synthesis, intracellular sorting, and function of glycoproteins using a mammalian cell mutant with reversible glycosylation defects. *Methods Cell. Biol.*, **32**, 57–84.

Landberg, E., Pahlsson, P., Lundblad, A., Arnetorp, A. and Jeppsson, J.O. (1995) Carbohydrate composition of serum transferrin isoforms from patients with high alcohol consumption. *Biochem. Biophys. Res. Commun.*, **210**, 267–274.

Löf, K., Koivula, T., Seppa, K., Fukunaga, T. and Sillanaukee, P. (1993) Semi-automatic method for determination of different isoforms of carbohydrate-deficient transferrin.*Clin. Chim. Acta.*, **217**, 175–186.

Lubke, T., Marquardt, T., von Figura, K., Korner, C. (1999) A new type of carbohydrate-deficient glycoprotein syndrome due to a decreased import of GDP-fucose into the golgi. *J. Biol. Chem.*, **274**, 25986–25989.

Marquardt, T., Brune, T., Luhn, K., Zimmer, K.P., Korner, C., Fabritz, L., van der Werft, N., Vormoor, J., Freeze, H.H., Louwen, F., Biermann, B., Harms, E., von Figura, K., Vestweber, D., Koch, H.G. (1999) Leukocyte adhesion deficiency II syndrome, a generalized defect in fucose metabolism. *J. Pediatr.*, **134**, 681–688.

Marquardt, T., Hasilik, M., Niehues, R., Herting, M., Muntau, A., Holzbach, U. and Hanefeld, F. (1997) Mannose therapy in carbohydrate-deficient glycoprotein syndrome type 1—first results from the German multicenter study. *Amino Acids*, **12**, 389.

Marquardt ,T., Lühn, K., Srikrishna, G., Freeze, H.H., Harms, E., Vestweber, D. (1999) Correction of leukocyte adhesion deficiency type II (LAD II) with oral fucose. *Blood*, **94**, 3976–3985.

Martin, A., Rambal, C., Berger, V., Perier, S. and Louisot, P. (1998) Availability of specific sugars for glyconjugate biosynthesis–a need for further investigations in man. *Biochimie.*, **80**, 75–86.

Matthijs, G., Schollen, E., Pardon, E., Veiga-Da-Cunha, M., Jaeken, J., Cassiman, J.-J. and Van Schaftingen, E. (1997a) Mutations in PMM2, a phosphomannomutase gene on chromosome 16p13, in carbohydrate-deficient glycoprotein type I syndrome (Jaeken syndrome). *Nat. Genet.*, **16**, 88–92.

Matthijs, G., Schollen, E., Pirard, M., Budarf, M.L., Van Schaftingen, E. and Cassiman, J.J. (1997b) PMM (PMM1), the human homologue of SEC53 or yeast phosphomannomutase, is localized on chromosome 22q13. *Genomics*, **40**, 41–47.

Matthijs, G., Schollen, E., Van Schaftingen, E., Cassiman, J.J. and Jaeken, J. (1998) Lack of homozygotes for the most frequent disease allele in carbohydrate-deficient glycoprotein syndrome type 1A. *Am. J. Hum. Genet.*, **62**, 542–550.

Mattson, S.N. and Riley, E.P. (1998) A review of the neurobehavioral deficits in children with fetal alcohol syndrome or prenatal exposure to alcohol. *Alcohol Clin. Exp.*, **22**, 279–294.

Mayatepek, E., Schroder, M., Kohlmuller, D., Bieger, W.P. and Nutzenadel, W. (1997) Continuous mannose infusion in carbohydrate-deficient glycoprotein syndrome type I. *Acta Paediatr.*, **86**, 1138–1140.

Mayatepek, E. and Kohlmuller, D. (1998) Mannose supplementation in carbohydrate-deficient glycoprotein syndrome Type 1 and phosphomannomutase deficiency. *Eur. J. Pediatr.*, **157**, 605–606.

McDowell, G. and Gahl, W.A. (1997) Inherited disorders of glycoprotein synthesis: cell biological insights. *Proc. Soc. Exp. Biol. Med.*, **215**, 145–157.

Murch, S., Salvatore, S., Heushkel, R., Thomson, M., Davies, S., French, I. and Walker-Smith, J. (1998) Initial trial of N-acetyl glucosamine in children with chronic inflammatory bowel disease. *British Society of Gastroenterology*, (Abstract).

Niehues, R., Hasilik, M., Alton, G., Körner, C., Schiebe-Sukumar, M., Koch, H.G., Zimmer, K.P., Wu, R., Harms, E., Reiter, K., Von Figura, K., Freeze, H.H., Harms, H.K. and Marquardt, T. (1998) Carbohydrate-deficient glycoprotein syndrome type Ib. Phosphomannose isomerase deficiency and mannose therapy. *J. Clin. Invest.*, **101**, 1414–1420.

O'Connor, L.H., Erway, L.C., Vogler, C.A., Sly, W.S., Nicholes, A., Grubb, J., Holmberg, S.W., Levy, B. and Sands, M.S. (1998) Enzyme replacement therapy for murine mucopolysaccharidosis type VII leads to improvements in behavior and auditory function. *J. Clin. Invest.*, **101**, 1394–1400.

Ogier-Denis, E., Codogno, P., Chantret, I. and Trugnan, G. (1988) The processing of asparagine-linked oligosaccharides in HT-29 cells is a function of their state of enterocytic differentiation. An accumulation of Man9,8-GlcNAc2-Asn species is indicative of an impaired N-glycan trimming in undifferentiated cells. *J. Biol. Chem.*, **263**, 6031–6037.

Ogier-Denis, E., Trugnan, G., Sapin, C., Aubery, M. and Codogno, P. (1990) Dual effect of 1-deoxymannojirimycin on the mannose uptake and on the N-glycan processing of the human colon cancer cell line HT-29. *J. Biol. Chem.*, **265**, 5366–5369.

Ogier-Denis, E., Blais, A., Houri, J.J., Voisin, T., Trugnan, G. and Codogno, P. (1994) The emergence of a basolateral 1-deoxymannojirimycin-sensitive mannose carrier is a function of intestinal epithelial cell differentiation. Evidence for a new inhibitory effect of 1-deoxymannojirimycin on facilitative mannose transport. *J. Biol. Chem.*, **269**, 4285–4290.

Panneerselvam, K. and Freeze, H.H. (1996a) Mannose corrects altered N-glycosylation in carbohydrate-deficient glycoprotein syndrome fibroblasts. *J. Clin. Invest.*, **97**, 1478–1487.

Panneerselvam, K. and Freeze, H.H. (1996b) Mannose enters mammalian cells using a specific transporter that is insensitive to glucose. *J. Biol. Chem.*, **19;271**, 9417–9421.

Panneerselvam, K., Etchison, J.R. and Freeze, H.H. (1997a) Human fibroblasts prefer mannose over glucose as a source of mannose for N-glycosylation. Evidence for the functional importance of transported mannose. *J. Biol. Chem.*, **272**, 23123–23129.

Panneerselvam, K., Etchison, J.R., Skovby, F. and Freeze, H.H. (1997b) Abnormal metabolism of mannose in families with carbohydrate-deficient glycoprotein syndrome type 1. *Biochem. Mol. Med.*, **61**, 161–167.

Petry, K.G. and Reichardt, J.K. (1998) The fundamental importance of human galactose metabolism: lessons from genetics and biochemistry. *Trends Genet.*, **14**, 98–102.

Pirard, M., Achouri, Y., Collet, J.F., Schollen, E., Matthijs, G., Van Schaftingen, E. (1999) Kinetic properties and tissular distribution ofmammalian phosphomannomutase isozymes. *Biochem. J.*, **339**, 201–207.

Pohl, S., Hoffmann, A., Rudiger, A., Nimtz, M., Jaeken, J. and Conradt, H.S. (1997) Hypoglycosylation of a brain glycoprotein (beta-trace protein) in CDG syndromes due to phosphomannomutase deficiency and N-acetylglucosaminyl-transferase II deficiency. *Glycobiology*, **7**, 1077–1084.

Powell, L.D., Paneerselvam, K., Vij, R., Diaz, S., Manzi, A., Buist, N., Freeze, H. and Varki, A. (1994) Carbohydrate-deficient glycoprotein syndrome: not an N-linked oligosaccharide processing defect, but an abnormality in lipid-linked oligosaccharide biosynthesis? *J. Clin. Invest.*, **94**, 1901–1909.

Proudfoot, A.E., Turcatti, G., Wells, T.N., Payton, M.A. and Smith, D.J. (1994) Purification, cDNA cloning and heterologous expression of human phosphomannose isomerase. *Eur. J. Biochem.*, **219**, 415–423.

Rader, D.J. (1997) Gene therapy for atherosclerosis.*Int. J. Clin. Lab. Res.*, **27**, 35–43.

Ripka, J., Adamany, A. and Stanley, P. (1986) Two Chinese hamster ovary glycosylation mutants affected in the conversion of GDP-mannose to GDP-fucose.*Arch. Biochem. Biophys.*, **249**, 533–545.

Rodriguez, I.R. and Fliesler, S.J. (1990) Glycogenesis in the amphibian retina: in vitro conversion of [2–3H]mannose to [^{3}H]glucose and subsequent incorporation into glycogen. *Exp. Eye. Res.*, **51**, 71–77.

Rome, L.H. and Hill, D.F. (1986) Lysosomal degradation of glycoproteins and glycosaminoglycans. Efflux and recycling of sulphate and N-acetylhexosamines. *Biochem. J.*, **235**, 707–713.

Rush, J.S., Panneerselvam, K., Waechter, C.J., Freeze, H.H. (2000) Mannose supplementation corrects GDP-mannose deficiency in cultured fibroblasts from some patients with Congenital Disorders of Glycosylation (CDG). *Glycobiology*, In press.

Schachner, M. and Martini, R. (1995) Glycans and the modulation of neural-recognition molecule function. *Trends in Neurosciences*, **18**, 183–191.

Schachter, H., Tan, J., Sarkar, M., Yip, B., Chen, S., Dunn, J. and Jaeken, J. (1998) Defective glycosyltransferases are not good for your health. *Adv. Exp. Med. Biol.*, **435**, 9–27.

Schauer, R. (1988) Sialic acids as antigenic determinants of complex carbohydrates. *Exp. Med. Biol.*, **228**, 47–72.

Silverman, M., Aganon, M.A. and Chinard, F.P. (1970) Specificity of monosaccharide transport in dog kidney. *Am. J. Physiol.*, **218**, 743–750.

Smith, D.J., Proudfoot, A., Friedli, L., Klig, L.S., Paravicini, G. and Payton, M.A. (1992) PMI40, an intron-containing gene required for early steps in yeast mannosylation. *Mol. Cell Biol.*, **12**, 2924–2930.

Spiro, R.G., Spiro, M.J. and Bhoyroo, V.D. (1983) Studies on the Regulation of the Biolsynthesis of Glucose-containing Oligosaccharide-Lipids. *J. Biol. Chem.*, **258**, 9469–9476.

Stark, N.J. and Heath, E.C. (1979) Glucose-dependent glycosylation of secretory glycoprotein in mouse myeloma cells. *Arch. Biochem. Biophys.*, **192**, 599–609.

Stibler, H. and Jaeken, J. (1990) Carbohydrate deficient serum transferrin in a new systemic hereditary syndrome. *Arch. Dis. Child.*, **65**, 107–111.

Stibler, H. (1991) Carbohydrate-deficient transferrin in serum: a new marker of potentially harmful alcohol consumption reviewed. *Clin. Chem.*, **37**, 2029–2037.

Stibler, H., Westerberg, B., Hanefeld, F. and Hagberg, B. (1993) Carbohydrate-deficient glycoprotein (CDG) syndrome—a new variant, type III. *Neuropediatrics*, **24**, 51–52.

Stibler, H., Stephani, U. and Kutsch, U. (1995) Carbohydrate-deficient glycoprotein syndrome—a fourth subtype. *Neuropediatrics*, **26**, 235–237.

Stibler, H., von Dobeln, U., Kristiansson, B. and Guthenberg, C. (1997) Carbohydrate-deficient transferrin in galactosaemia. *Acta Paediatr.*, **86**, 1377–1378.

Stibler, H., Holzbach, U. and Kristiansson, B. (1998) Isoforms and levels of transferrin, antithrombin, alpha(1)-antitrypsin and thyroxine-binding globulin in 48 patients with carbohydrate-deficient glycoprotein syndrome type I.*Scand. J. Clin. Lab. Invest.*, **58**, 55–61.

Sturla, L., Etzioni, A., Bisso, A., Zanardi, D., Deflora, G., Silengo, L., Deflora, A. and Tonetti, M. (1998) Defective intracellular activity of GDP-D-Mannose-4,6-Dehydratase in leukocyte adhesion deficiency type II syndrome. *FEBS Lett.*, **429**, 274–278.

Tan, J., Dunn, J., Jaeken, J. and Schachter, H. (1996) Mutations in the MGAT2 gene controlling complex N-glycan synthesis cause carbohydrate-deficient glycoprotein syndrome type II, an autosomal recessive disease with defective brain development. *Am. J. Hum. Genet.*, **59**, 810–817.

Turco, S.J. (1980) Modification of Oligosaccharide-Lipid Synthesis and Protein Glycosylation in Glucose-Derived Cells. *Arch. Biochem. Biophys.*, **205**, 330–339.

Turco, S.J. and Pickard, J.L. (1982) Altered G-protein glycosylation in vesicular stomatitis virus-infected glucose-deprived baby hamster kidney cells. *J. Biol. Chem.*, **257**, 8674–8679.

Van Ommen, C.H., Peters, M., Barth, P.G., Vreken, P., Wanders, R.J., Jaeken, J. (2000) Carbohydrate-deficient glycoprotein syndrome type 1a: a variant phenotype with borderline cognitive dysfunction, cerebellar hypoplasia, and coagulation disturbances. *J. Pediatr.*, **136**, 400–403.

Van Schaftingen, E. and Jaeken, J. (1995) Phosphomannomutase deficiency is a cause of carbohydrate-deficient glycoprotein syndrome type I. *FEBS Lett.*, **377**, 318–320.

Varki, A. and Marth, J. (1995) Oligosaccharides in verterbrate development. *Semin. Dev. Biol.*, **6**, 127–138.

Velasco, A., Hendricks, L., Moremen, K.W., Tulsiani, D.R., Touster, O. and Farquhar, M.G. (1993) Cell type-dependent variations in the subcellular distribution of alpha-mannosidase I and II. *J. Cell Biol.*, **122**, 39–51.

Vomdahl, S., Niederau, C. and Haussinger, D. (1998) Loss of vision in gauchers-disease and its reversal by enzyme-replacement therapy. *New Eng. J. Med.*, **338**, 1471–1472.

Vreken, P., Rusch, H., Huijben, K., Wevers, R.A., Mayatepek, E., Kohlmuller, D., Kristiansson, B., Borulf, S., Conradi, N., Erlansonalbertsson, C., Ryd, W. and Stibler, H. (1998) Anion-exchange chromatography versus isoelectric focusing of transferrin in diagnosing the carbohydrate-deficient glycoprotein syndrome. *J. Inherit. Metabol. Dis.*, **21**, 447–448.

Wada, Y., Nishikawa, A., Okamoto, N., Inui, K., Tsukamoto, H., Okada, S. and Taniguchi, N. (1992) Structure of serum transferrin in carbohydrate-deficient glycoprotein syndrome. *Biochem. Biophys. Res. Commun.*, **189**, 832–836.

Wood, F.C. and Cahill, G.F. (1963) Mannose Utilization in Man. *J. Clin. Invest.*, **42**, 1300–1312.

Yamashita, K., Ideo, H., Ohkura, T., Fukushima, K., Yuasa, I., Ohno, K. and Takeshita, K. (1993a) Sugar chains of serum transferrin from patients with carbohydrate deficient glycoprotein syndrome. Evidence of asparagine-N-linked oligosaccharide transfer deficiency. *J. Biol. Chem.*, **268**, 5783–5789.

Yamashita, K., Ohkura, T., Ideo, H., Ohno, K. and Kanai, M. (1993b) Electrospray ionization-mass spectrometric analysis of serum transferrin isoforms in patients with carbohydrate-deficient glycoprotein syndrome. *J. Biochem. (Tokyo).*, **114**, 766–769.

Yogalingam, G., Bielicki, J., Hopwood, J.J. and Anson, D.S. (1997) Feline mucopolysaccharidosis type VI: correction of glycosaminoglycan storage in myoblasts by retrovirus-mediated transfer of the feline N-acetylgalactosamine 4-sulfatase gene. *DNA Cell. Biol.*, **16**, 1189–1194.
Yuen, C.T., Chai, W., Loveless, R.W., Lawson, A.M., Margolis, R.U. and Feizi, T. (1997) Brain contains HNK-1 immunoreactive O-glycans of the sulfoglucuronyl lactosamine series that terminate in 2-linked or 2,6-linked hexose (mannose). *J. Biol. Chem.*, **272**, 8924–8931.

B7. Glycoconjugates as Vectors for Gene and Oligonucleotide Delivery

Michel Monsigny, Christophe Quétard, Eric Duverger, Chantal Pichon, Valérie Altemayer, Sylvain Bourgerie, Violaine Carriere, Patrick Midoux, Roger Mayer and Annie-Claude Roche

Glycoconjugates are recognized by cell surface as well as intracellular lectins in a sugar dependent manner. Synthetic glycoconjugates such as neoglycoproteins, glycosylated polymers, glycopeptides or glyco-oligonucleotides are useful tools to characterize endogenous sugar-binding proteins and also to target drugs, oligonucleotides and genes. In this chapter, we present a review of different ways of preparing sugar derivatives suitable to synthesize glycosylated fluorescent probes and glycosylated carriers by using either simple sugars or complex oligosaccharides isolated from natural compounds. We describe the preparation of the glycoconjugates themselves and finally we summarize several applications of them, including their use in characterizing endogenous lectins, and in efficiently introducing oligonucleotides and genes into defined types of cells.

1. INTRODUCTION

The use of glycoconjugates in order i) to characterize endogenous lectins, ii) to study their intracellular localization and motion and iii) to help drugs, oligonucleotides and genes to be selectively taken up by animal cells, is based on the discovery of G. Ashwell, A. Morell and their coworkers (see Ashwell and Morell, 1974 for a review) at the end of the sixties. These authors showed that animal cells express sugar-binding proteins (lectins) on their surface and that these lectins very efficiently induce the endocytosis of their ligands (see Ashwell and Harford, 1982 for a review). Since, lectins have been found to be expressed both at the surface and inside a very large number of normal as well as transformed animal cells.

The present chapter describes the preparation of glycoconjugates suitable to synthesize glycosylated proteins and glycosylated polymers, starting from reducing monosaccharides or oligosaccharides. Fluorescent glycoconjugates are useful tools to study the properties of lectins, specially of cell surface lectins, but also of intracellular membrane lectins as well as of cytosolic and nuclear lectins. Glycosylated polycationic polymers are suitable to carry oligonucleotides and genes and to help them to be taken up by cells in a sugar dependent manner. Such glycopolymers very efficiently enhance the inhibitory effect of antisense oligonucleotides and the expression of genes present in plasmids used for transfection.

2. PREPARATION OF GLYCOSYNTHONS

Monosaccharide (and some disaccharide) derivatives such as *p*-nitrophenylglycosides are easy to synthesize and many of them are even commercially available; they may be easily transformed into units reacting on proteins, polymers or fluorescent tags. However, such monosaccharides are not really specific for endogenous lectins. Therefore, it is more rewarding to use complex oligosaccharides which are exquisitely specific for a given lectin expressed on or in a precise type of cells. Although it is possible to synthesize complex oligosaccharide derivatives by chemical means using an impressive series of protection, coupling, deprotection steps, it is much easier to obtain complex oligosaccharides from natural products. Such oligosaccharides usually contain at one end a reducing sugar unit; it is possible to use the property of that unit to couple directly or indirectly the oligosacharides to fluorescent tags, proteins or polymers, (for a review on the preparation of glycosynthons, see Monsigny *et al.*, 1998, and references therein).

2.1. Preparation of Unprotected Reducing Oligosaccharides

Oligosaccharides bearing a reducing sugar unit may be prepared from various biological fluids such as colostrum, milk, urine. Oligosaccharides are easily released from glycoproteins by enzymatic or chemical means, (Figures 1 and 2). Various endo-*N*-acetyl-β-glucosaminidases release oligosaccharides from *N*-glycans with a terminal reducing *N*-acetylglucosamine. The *O*-peptidyl-*N*-acetyl-α-galactosaminidases release oligosaccharides from *O*-glycans with a terminal reducing *N*-acetylgalactosamine. The *N*-glycanases cleave the GlcNAc-asparagine linkage (see for a review, Tarentino and Plummer, 1994), releasing the β carboxylic group of aspartyl residue and the oligosaccharides as glycosylamine derivatives when the hydrolysis is conducted in alkaline medium, or an amine-free moiety when the hydrolysis is conducted in neutral or acidic medium (Figure 3). Various endo-β-galactosidases have also been described. Oligosaccharides with a reducing end may also be obtained: i) from glycoprotein *N*-glycans upon hydrazinolysis followed by an acetylation step and the release of the *N*-acetyl-hydrazide moities as well as ii) from glycoprotein *O*-glycans by β-elimination in the absence of a reducing agent, using in specific conditions, either anhydrous hydrazine or aqueous hydrazine in the presence of triethylamine (Cooper *et al.*, 1994) or aqueous ethylamine.

Oligosaccharides may also be obtained from bacterial or plant polysaccharides using specific endoglycosidases. Oligosaccharides may be released from glycolipids by various enzymes including a ceramide glycanase leading to a variety of complex oligosaccharides ending with a glucose in terminal reducing position as well as by some endo-β-glycosidases such as endo-β-galactosidases.

N-linked oligosaccharides : high-mannose type

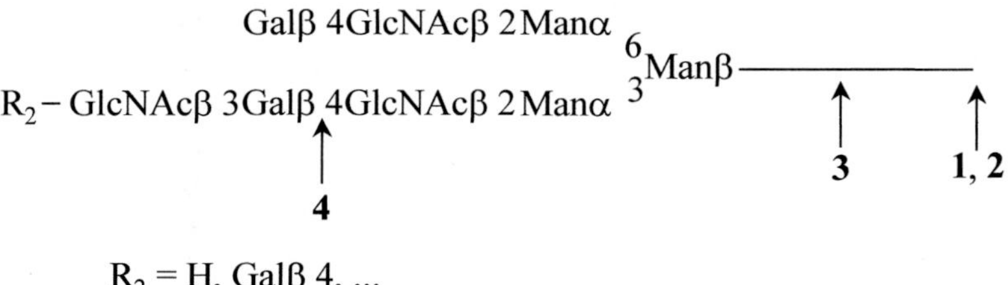

R_1 = H, Manα 2, ...

N-linked oligosaccharides : complex type

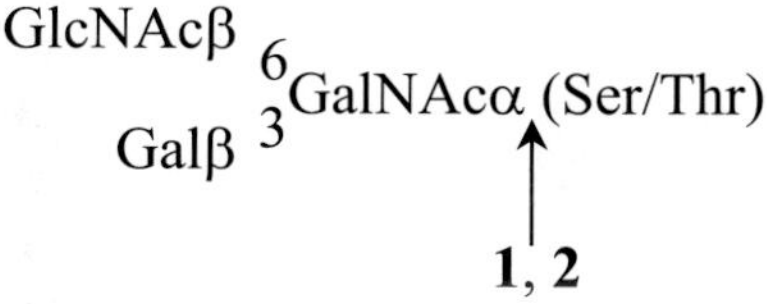

R_2 = H, Galβ 4, ...

Figure 1: Preparation of reducing oligosaccharides from *N*-linked oligosaccharides by enzymatic and chemical means.

O-linked oligosaccharides : core 2 as an example

GlcNAcβ
Galβ
GalNAcα (Ser/Thr)

1, 2

1- β-elimination
2- *O*-peptidyl *N*-acetyl α-galactosaminidase

Figure 2: Preparation of reducing oligosaccharides from *O*-linked oligosaccharides by enzymatic and chemical means.

2.2 Properties of Glycosylamines and Derivatives

2.2.1. Glycosylamines (Figure 4)

Glucose and galactose dissolved in ethanol or methanol in the presence of ammonia upon resting for few weeks give crystalline glycosylamines, glucosylamine and galactosylamine, respectively. Glycosylamines are easily prepared in the presence of

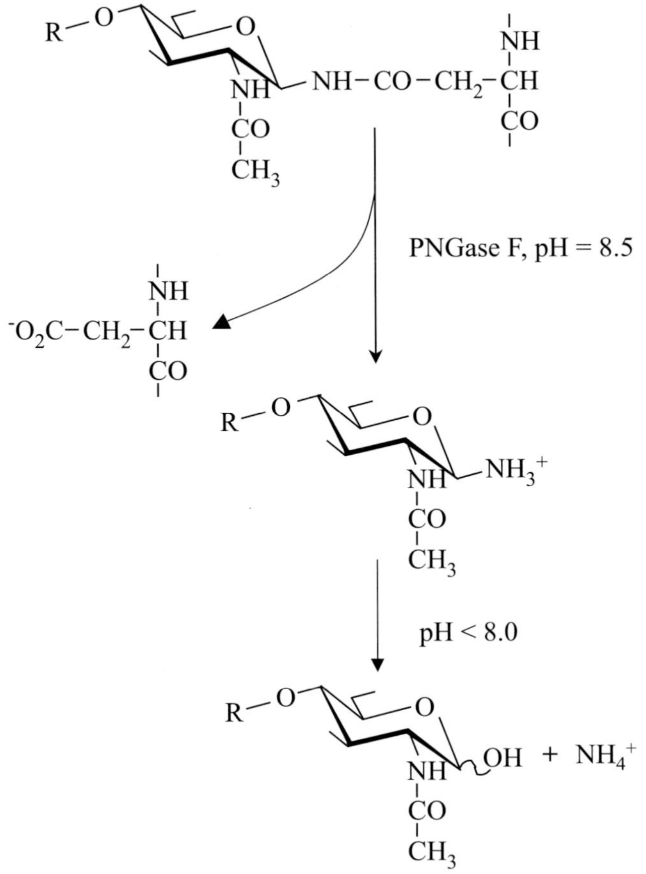

Figure 3: Products released upon *N*-glycanase (PNGase F) action on glycoproteins containing *N*-linked oligosaccharides, according to Tarentino *et al.*, 1985.

an excess of ammonia and can be isolated from such a medium, but they are not stable in neutral or slightly acidic media, however, they become stable upon acylation (see below).

2.2.2. *N-aryl and N-alkylglycosylamines* (Figure 5)

Various *N*-arylglycosamines and *N*-alkylglycosylamines were prepared by heating the sugar in the presence of an arylamine or an alkylamine, respectively; methanol being added then to induce the crystallization or being used to dissolve the amine and the sugar.

   N-alkylglycosylamines undergo mutarotation, as do unsubstituted glycosylamines and are easily hydrolyzed, specially in dilute acetic acid. The ease of hydrolysis of *N*-alkylglycosylamines parallels the basicity of the amines; the *N*-alkylglycosylamines being more labile than the *N*-arylglycosylamines. However, the *N*-acyl-*N*-alkylglycosylamides do not undergo mutarotation and they are stable in neutral or slightly acidic medium (see for reviews, Ellis and Honeyman, 1955; Isbell and Frush, 1980).

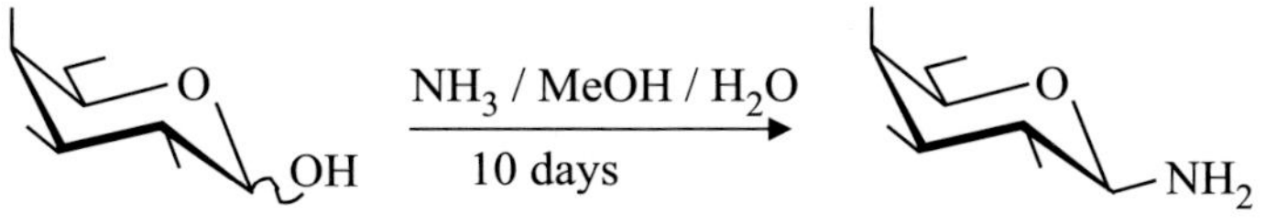

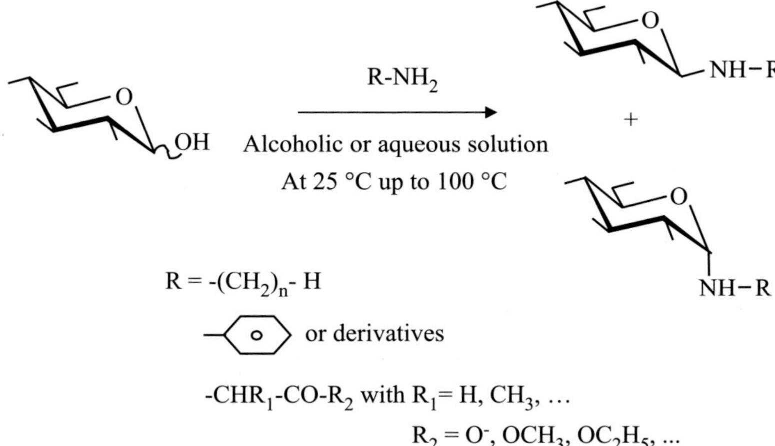

Figure 4: Galactosylamine preparation according to Lobry de Bruyn and Van Leent, 1896.

R = -(CH$_2$)$_n$- H

or derivatives

-CHR$_1$-CO-R$_2$ with R$_1$= H, CH$_3$, ...

R$_2$ = O$^-$, OCH$_3$, OC$_2$H$_5$, ...

Figure 5: Glycosylamine preparation by reaction of glucose with various primary amines.

Various *N*-glycosylamino acids were prepared according to the methods used to prepare *N*-alkylglycosylamines. These derivatives include *N*-glucosylalanine, N^α-N^ε-di-**D**-glucosyl-**L**-lysine, *N*-glucosylglycine, *N*-glucosylglycine ethyl ester, *N*-glucosylglycine sodium salt, *N*-glucosyllysine, *N*-glucosylserine (see for review Ellis and Honeyman, 1955). *N*-glycosylaminoacids were found to be easily hydrolyzed in water.

2.2.3. *Stability of glycosylamines* (Figure 6)

Glycosylamines made from **D**-sugars have usually a β-configuration, the α-configuration being less stable than the β-configuration.

Upon *N*-acylation, the *N*-acylglycosylamides are stable and it is therefore possible to isolate them. *N*-acylglycosylamides may be obtained directly by action of ketene on **D**-glycosylamine (Niemann and Hays, 1940) or, indirectly, by *per O*- and *N*-acylation with a mixture of pyridine and acetic anhydride, and then selective catalytic *O*-deacetylation in anhydrous methanol.

Glycosylamines may also undergo Amadori rearrangement. The Amadori rearrangement, a transformation of glycosylamine into 1-amino-1-deoxy-2-keto derivative, was shown to occur with *N*-arylglycosylamines when they were heated for a few hours in ethanolic solution with a weak acid as a catalyst. The Amadori rearrangement occurs specially in the presence of both a compound having an activated methylene group and a catalytic amount of a secondary amine or in the presence of glacial acetic acid (see for a review, Isbell and Frush,

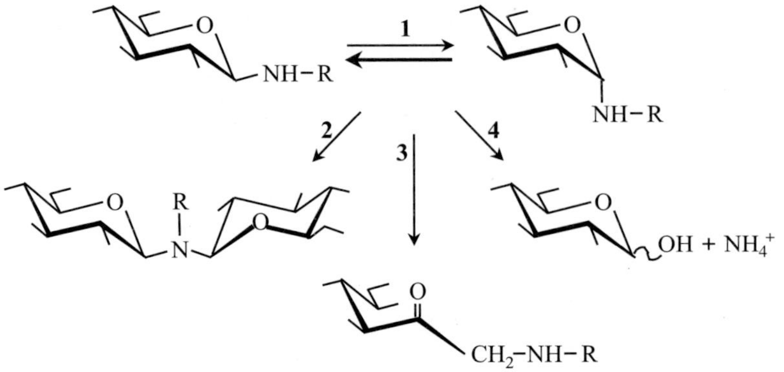

1 : epimerization
2 : transglycosylation
3 : Amadori rearrangement
4 : hydrolysis

Figure 6: Four main rearrangements of glycosylamines occuring as long as the amino group of the glycosylamine is not acylated.

1958). The mechanism involves the addition of a proton to the nitrogen atom of the glycosylamine; therefore when the amine is further substituted by acylation as in *N*-acetylglycosylamide, the protonation is inhibited and the Amadori rearrangement is limited. This rearrangement readily occurs with *N*-glycosylamino acids. The carboxylic group of neutral amino acids was shown to be efficient enough to catalyze the Amadori rearrangement.

Glycosylamines may also be transformed into diglycosylamines by a trans-glycosylation process (Figure 6). The first diglycosylamines were obtained by boiling glycosylamine in methanol (see for reviews, Ellis and Honeyman, 1955; Isbell and Frush, 1958).

2.3. Preparation of Glycosylamines from Unprotected Reducing Sugars

2.3.1. Methanolic ammonia

Various glycosylamines were obtained, a century ago, – by Lobry de Bruyn 1895 (Lobry de Bruyn and Franchimont, 1893; Lobry de Bruyn and Van Leent, 1896) – by dissolving a reducing sugar in warm water and adding anhydrous methanol saturated with ammonia. Within 10 days, the glycosylamine crystallized.

2.3.2. Liquid ammonia

A β-glycosylamine can also be obtained by dissolution of the sugar in liquid ammonia followed by evaporation of the solvent (Pigman *et al.*, 1951).

2.3.3. Ammonium hydrogen carbonate in water (Figure 7)

2-Acetamido-2-deoxy-β-**D**-glucopyranosylamine was prepared by dissolving *N*-acetylglucosamine, in saturated aqueous ammonium hydrogen carbonate; the solution was kept at

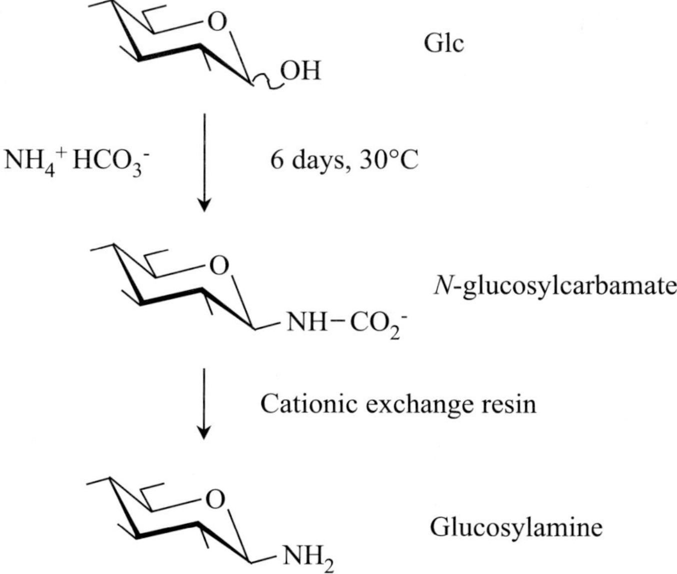

Figure 7: Preparation of glycosylamines from unprotected reducing sugars according to Likhoshertov *et al.* (1986) and Kallin *et al.* (1989).

20 °C for 45 days or at 30 °C for 6 days, with a 70 ± 10 % yield (Likhosherstov *et al.*, 1986).

Similarly, oligosaccharides with a glucose residue in the reducing terminal position were converted into glycosylamines after 7 days at room temperature with a good yield (Kallin *et al.*, 1989). The mixture was kept at room temperature up to 7 days, solid ammonium bicarbonate was added at intervals making sure that an excess of solid salt was present in the mixture. The main compound however was not the expected glycosylamine but rather a glycosylcarbamate. However when the crude product was purified by cation exchange chromatography, the expected glycosylamine was eluted with 2 M ammonia in a methanol/water mixture. This procedure was also used to prepare β-glycosylamine of oligosaccharides ended with a reducing *N*-acetylglucosamine (Manger *et al.*, 1992a).

2.3.4. Ammonium hydrogen carbonate in ammonia solution (Figure 8)

The reducing sugar and NH_4HCO_3 were dissolved in 16 M commercial aqueous ammonia, (Lubineau *et al.*, 1995). The solution was heated at 42 °C for 36 h. The yield was almost quantitative in the expected glycosylamine; the presence of glycosylcarbamate being rather low (< 8%).

2.4. Preparation of N-Amino-Acyl-Glycosylamides from Unprotected Glycosylamines (Figure 9)

As glycosylamines are not stable,- they may be easily transformed either by mutarotation, by hydrolysis, by transglycosylation or by Amadori rearrangement – they must be stabilized. An easy way to obtain a stable derivative of glycosylamines is to transform

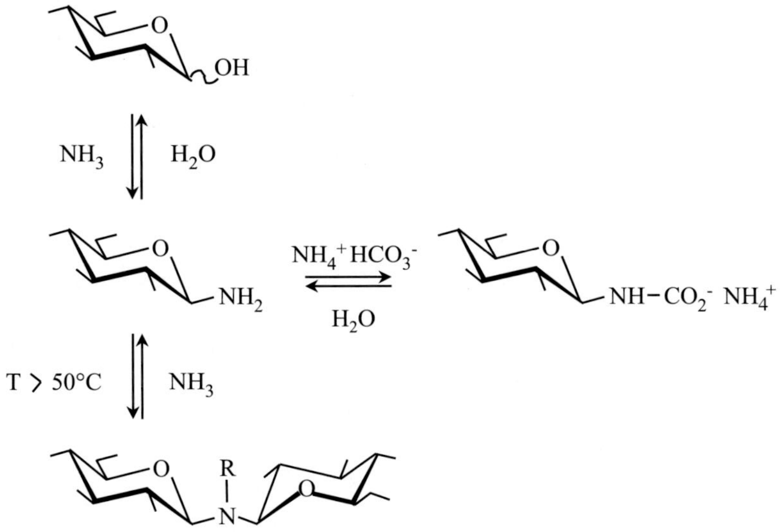

                    Michel Monsigny et al.

Figure 8: Improved synthesis of glycosylamines according to Lubineau *et al.*, (1995); the sugar is incubated in the presence of a large excess of both hydrogen ammonium carbonate and of aqueous ammonia.

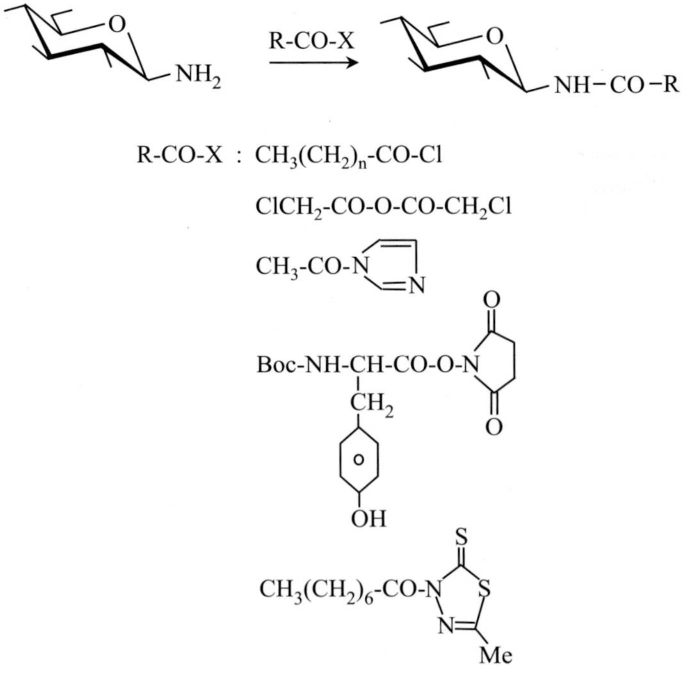

Figure 9: Main reactions leading to the obtention of stable *N*-acylglycosylamides from an unprotected glycosylamine.

them into acyl derivatives (see above). Such acyl derivatives are stable in a large range of pH at room temperature and they do not undergo mutarotation, hydrolysis, transglycosylation or Amadori rearrangement (see, Niemann and Hays, 1940; Isbell and Frush, 1951). Various amino acyl-glycosylamides were obtained by acylation of glycosylamines either with protected amino acids having one carboxylic group free or with amino acid precursors, namely halogeno acids.

2.4.1. Preparation of β-aspartylglycosylamide

The 2-acetamido-2-deoxy-1-(N′-Fmoc-β-aspartyl)-β-**D**-glucopyranosylamide was obtained by coupling Fmoc-Asp-OtBu in *N,N*-dimethylformamide (DMF) in the presence of diisopropylcarbodiimide with the β-glycosylamine of *N*-acetyl-**D**-glucosamine dissolved in water; this synthon was then used to prepare various glycopeptides (Otvos *et al.*, 1989).

Various synthons were obtained in a similar way. For instance, the β-glycosylamine of an heptasaccharide containing 5 mannoses and 2 *N*-acetylglucosamines: (Man$_5$)-GlcNAcβ 4GlcNAcβNH$_2$, was coupled with a pentapeptide Ac-Tyr-Asp-Leu-Thr-Ser-NH$_2$ in dimethylsulfoxide in the presence of 1-hydroxybenzotriazole (HOBT), *N,N*-diisopropylethylamine, and hydroxy-*O*-benzotriazol-1-yl-*N,N,N′,N′*-tetramethyluronium hexafluorophosphate (HBTU), leading to the expected glycopeptide.

2.4.2. Preparation of N-(tyrosyl)-glycosylamide

On the purpose of preparing oligosaccharide derivatives easily detected by absorption in ultraviolet light (280 nm), easily labelled by radioiodination and easily usable for further condensation onto various molecules or matrices, glycosylamines have been selectively acylated by reaction with a *N*-protected tyrosine (Tamura *et al.*, 1994). The glycosylamine derivative of oligosaccharides, isolated from glycoproteins upon enzymatic hydrolysis and treated with ammonium bicarbonate, were reacted with Boc-Tyr-OSu (*N-t*-butyloxycarbonyl-tyrosyl-succinimidyl ester) in DMF at 50 °C for 3h. Then, the conjugate was purified by gel filtration. Finally the Boc group was released by treating the dry oligosaccharide (5 μmol) with trifluoroacetic acid at room temperature for 10 min.

2.4.3. Preparation of glycosylamine derivatives of oligosaccharides released from natural products

Glycosylamine derivatives of oligosaccharides may be obtained by enzymatic hydrolysis of *N*-acetylglucosaminyl-asparagine of glyco-amino acids, glycopeptides or glycoproteins containing *N*-glycan moieties (see above). The enzyme, called PNGase F or *N*-glycanase, cleaves the linkage between the amino group of the β-glycosylamine and the β-carboxylic group of the aspartyl residue, with an optimal activity in slightly alkaline medium pH 8.5 or 9.0 (Tarentino *et al.*, 1985). The glycosylamine derivatives of the oligosaccharides released from glycopeptides are stable in slightly alkaline medium, but conversely they loose rapidly the amine group when they are kept in slightly acidic medium. On these bases, Rasmussen *et al.* (1992) showed that the glycosylamine was stable enough to be *N*-acylated with acetic anhydride or thiocarbamylated with phenylisothiocynate. Similarly, Tarentino *et al.* (1993) showed that such β-glycosylamine derivatives of oligosaccharides released from glycoproteins with PNGase F were very efficiently substituted with 2-iminothiolane at pH 8.8, leading to a derivative bearing a thiol group. Such oligosaccharides bearing a thiol group are glycosynthons suitable to specifically react with any molecule bearing a disulfide bridge or with a compound reacting with thiol, such as haloacetyl or maleimidyl groups.

2.4.4. *Preparation of N-(glycyl)-β-glycosylamides*

Glycosylamines were prepared by Manger *et al.* (1992a,b) according to the hydrogen ammonium carbonate method of Likhosherstov *et al.* (1986), with addition of solid salt to keep the solution saturated, at 30 °C for several days (up to 5). Glycosylamines were then acylated with chloroacetic anhydride in 1 M sodium bicarbonate, the pH being kept above 7. The *N*-chloroacetylglycosylamines obtained were then ammonolyzed by action of saturated hydrogen ammonium carbonate at 50 °C for 8h in a sealed tube. The *N*-(glycyl)-glycosylamide was purified by cation exchange chromatography on a carboxymethylated gel.

Alternatively, *N*-(glycyl)-glycosylamides were prepared by reaction of Fmoc-glycine in DMF with β-glycosylamine dissolved in a mixture of dimethylsulfoxide (DMSO) and DMF, *N*-ethyldiisopropylamine, HBTU and HOBT at room temperature for about 2 h. The overall yield of *N*-(glycyl)-β-glycosylamide upon deprotection was 55% from the starting sugar (Arsequell *et al.*, 1994).

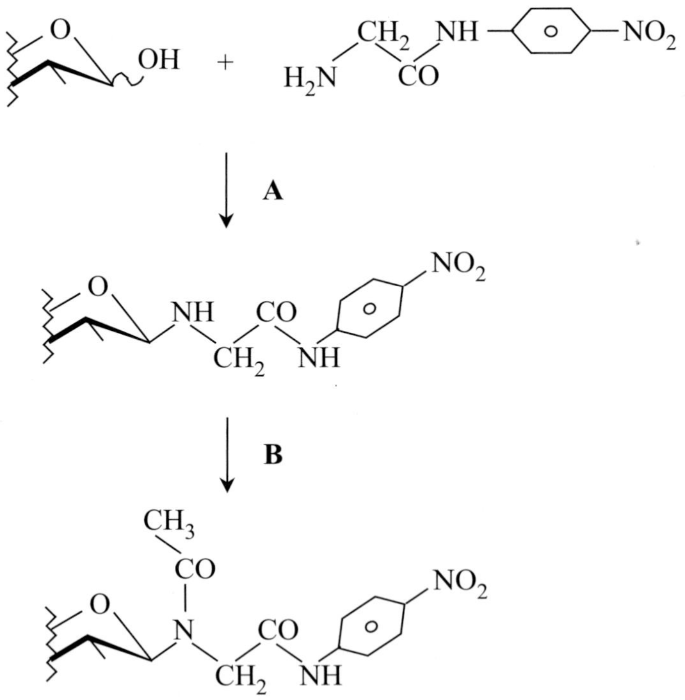

Figure 10: Preparation of an N_1-acetyl-N_1-glycosyl-glycyl derivative from an unprotected oligosaccharides containing a reducing sugar unit according to Sdiqui *et al.* (1995). A: the coupling step; B: the acylating step.

2.5. Direct Preparation of N-Glycosylamine Acid Derivatives from Unprotected Reducing Oligosaccharides

2.5.1. *Preparation of N-acetyl-N-glycosylglycine* (Figure 10)

A reducing sugar dissolved in an organic solvent such as *N,N*-dimethylformamide, 1-methyl-2-pyrrolidone or dimethylsulfoxide reacted with the amino group of glycyl-*p*-nitroanilide, at 50 °C for 5 days, leading to a *N*-(β-glycosyl)-glycyl-*p*-nitroanilide. As expected, this glycosylamino acid derivative was stable in alkaline conditions, even in aqueous solution, as in the case of glycosylamines (see above), but it was not stable in neutral or acidic conditions. Upon addition of an acylating agent, *N*-acetyl-imidazole, the conjugate was selectively *N*-acetylated within 30 min at room temperature (Sdiqui *et al.*, 1995). The N_1-acetyl-N_1-glycosyl-glycyl-*p*-nitroanilide was purified by gel filtration and was found to be quite stable in a large range of pH. The anomeric configuration of the glucose residue linked to the amino acid was found to be β on the basis of proton NMR analysis.

2.5.2. *Preparation of N-glycosylglutamyl derivatives* (Figure 11)

Another example is given with the preparation of *N*-glycosylpyroglutamyl derivatives. The incubation of a sugar with α-glutamyl-*p*-nitroanilide in the presence of imidazole led to a quantitative coupling within 10 h at 50 °C (Quétard *et al.*, 1998). In the absence of imidazole,

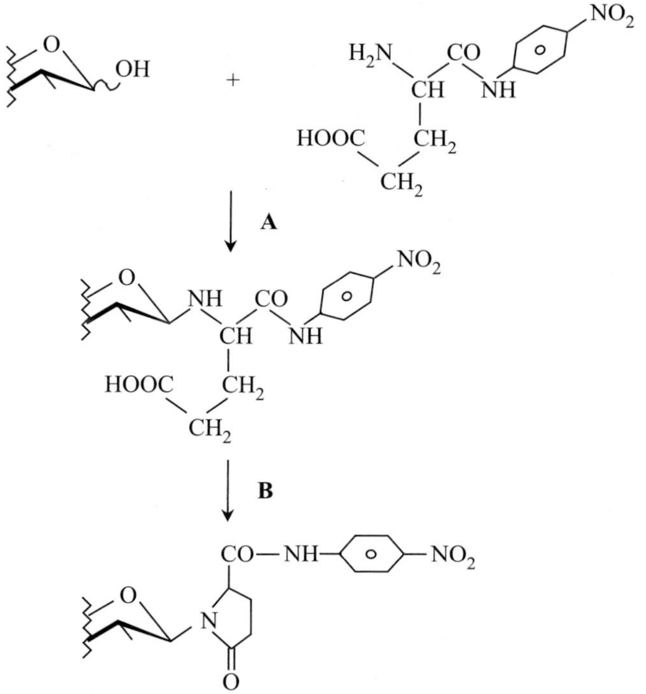

Figure 11: Preparation of an *N*-oligosaccharyl-pyroglutamyl derivative (*p*-nitroanilide) from an unprotected oligosaccharide containing a reducing sugar unit according to Quétard *et al*, (1998). A: the coupling step; B: the acylating step.

the reaction was slower and more side reactions occured. As in the case of *N*-glycosylglycyl derivatives, the *N*-glycosylglutamyl-*p*-nitroanilide were not stable in aqueous medium, except in alkaline conditions. The conjugate was readily stabilized by adding into the solution of an *N*-glycosylglutamyl-*p*-nitroanilide, benzotriazol-1-yloxy-tris (dimethylamino) phosphonium hexafluoro phosphate (BOP) and imidazole; the intramolecular acylation was complete within 30 min at room temperature. This compound was found to be stable at room temperature at any pH between 3.6 and 9, at room temperature as well as at 90 °C in either slightly acidic or neutral conditions (Quétard *et al.*, 1998). Various *N*-glycosylpyroglutamyl-*p*-nitroanilides were prepared by this procedure, starting with either monosaccharides or oligosaccharides (Quétard *et al.*, unpublished results). In all cases, the yield of the expected glycoamino acid derivatives were very high.

The coupling procedure was also efficient to prepare other *N*-glycopyroglutamyl derivatives including disulfide-containing compounds as well as fluorescent derivatives. Various *N*-glycopyroglutamyl-amido-ethyldithiopyridines were prepared (Figure 12) with a high yield (up to 94%). In all cases, the conjugates appeared to have a β-anomeric configuration with a *H*-1 resonance of the glycosylamide around 5.15 and a $J_{1,2}$ coupling constant close to 9 Hz. In the case of oligosaccharides ended with a *N*-acetylglucosamine, the

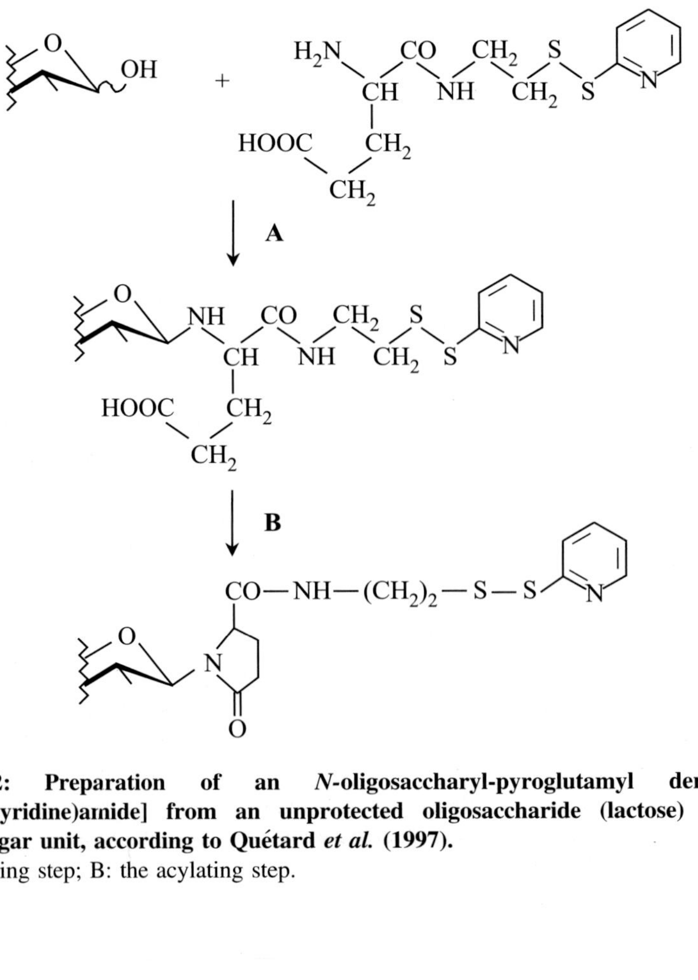

Figure 12: Preparation of an *N*-oligosaccharyl-pyroglutamyl derivative [(2-ethyldithiopyridine)amide] from an unprotected oligosaccharide (lactose) containing a reducing sugar unit, according to Quétard *et al.* (1997).
A: the coupling step; B: the acylating step.

coupling reaction was slower and the complete reaction was obtained after 24 h (Quétard *et al.*, 1997). Therefore, unprotected reducing oligosaccharides readily react with glutamyl derivatives and allow in a one-pot two-steps reaction to prepare glycosynthons ready for the preparation of more sophisticated glycoconjugates.

3. PREPARATION OF GLYCOCONJUGATES

3.1. Fluorescent Oligosaccharide Derivatives (Figure 13)

β-Oligosaccharyl-pyroglutamyl-amido-ethyl-dithio-2-pyridine was quantitatively reduced by one equivalent of tris(carboxyethyl)phosphine (TCEP), and then transformed into a fluorescent conjugate by reacting with a slight excess of iodoacetamido-fluorescein or an iodoacetamide derivative of another fluorophore. Such conjugates (β-oligosaccharyl)-pyro-glutamyl-amido-ethyl-thioacetamido-fluorescein, were readily purified by gel filtration and used to characterize endogenous lectins of various cells, such as the lung carcinoma cells A549 (Figure 14). *N*-(glycyl)-glycosylamides have also been used as glycosynthons to prepare fluorescent derivatives (Manger *et al.*, 1992a,b).

3.2. Neoglycoproteins and Glycosylated Polymers

Neoglycoproteins and glycosylated polymers must contain several tens of simple sugars to be efficiently recognized by lectins, while glycoconjugates containing a single complex oligosaccharide tightly bind on lectins. Indeed, as shown by Y.C. Lee in the case of the asialoglycoprotein receptor of parenchymal cells, the affinity of an oligosaccharide with a single galactose residue is about 10^3 L $\times$ mole^{-1}, with 2 galactose residues, about 10^6 L $\times$ mole^{-1} and with 3 or 4 galactose residues about 10^9 L $\times$ mole^{-1} (Lee *et al.*, 1983). Therefore, a single triantennary oligosaccharide will allow a protein or a polymer to be efficiently recognized by this lectin. In addition, such glycoconjugates containing a complex oligosaccharide are more specific than the glycoconjugates containing tens of simple sugars. Oligosaccharides isolated from living organisms, or released by hydrolysis from glycoconjugates may be easily transformed into glycosynthons.

Various synthetic glycoconjugates have been prepared, including neoglycoproteins, serum albumin substituted with about 25 simple sugar residues or glycosylated neutral polymers, (see for reviews, Stowell and Lee, 1980; Aplin and Wriston, 1981; Monsigny *et al.*, 1988, 1994a, b; Lee and Lee, 1994a, b; Kiessling and Pohl, 1996; Gabius and Gabius, 1997). Among those synthetic glycoconjugates, glycosylated cytochemical markers (see for a review, Schrével *et al.*, 1981) and fluorescent tag-labelled neoglycoproteins have been largely used to visualize and to study the properties and the functions of endogenous lectins (see for reviews, Monsigny *et al.*, 1988; Gabius and Gabius, 1997). Neoglycoproteins, neutral glycosylated polymers and cationic glycosylated polylysines have been prepared and shown to be quite useful to study the properties of endogenous lectins and to carry drugs, oligonucleotides and genes (see for reviews, Roche *et al.*, 1990; Molema and Meijer, 1994; Monsigny *et al.*, 1994a,b; Wadhwa and Rice, 1995); cationic glycosylated polylysines are efficient to increase the biological effect of antisense oligonucleotides (Hangeland *et al.*, 1995; Stewart *et al.*, 1996) and to allow gene targeting (Plank *et al.*, 1992; Midoux *et al.*, 1993, 1997; Wadhwa *et al.*, 1995; Perales *et al.*, 1994). Neoglycoproteins were developped independently by Lee and coworkers and

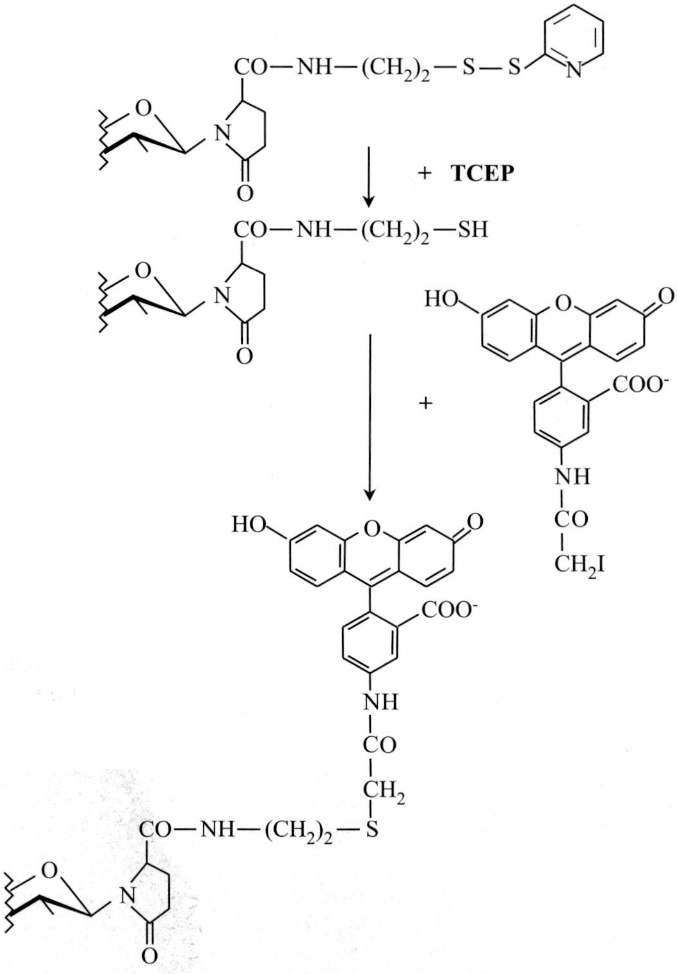

Figure 13: Preparation of a fluorescein-labelled oligosaccharide: *N*-oligosaccharylpyroglutamyl-amidoethylthio-acetamidofluorescein, from a glycosynthon containing a disulfide bridge.
TCEP, tris(carboxyethyl)phosphine quantitatively reduces the disulfide bridge, releasing a free thiol and 2-thiopyridinone, then the thiol is allowed to react with iodoacetamidofluorescein, leading to the title compound.

ourselves (Privat *et al.*, 1974) as tools to study the properties of lectins. In both cases, neoglycoproteins were made by adding activated sugars on serum albumin. About half the lysine residues of bovine serum albumin are easily substituted upon adding glycosylphenylisothiocyanate (Roche *et al.*, 1983; Monsigny *et al.*, 1984): neoglycoproteins were then made fluorescent by adding fluorescein isothiocyanate. Fluorescein labelled neoglycoproteins usually contain about 23 ± 3 sugars and 2.5 ± 0.5 fluorescein residues: for instance, when the sugar is mannose: Man_{23-}, $Flu_{2.5-}BSA$; where Man stands for α-mannopyranosylphenylthiocarbamyl and Flu for fluorescein thiocarbamyl. The apparent affinity of neoglycoproteins for lectins is much higher (more than 1000) than that of the corresponding free sugar, for instance 500 nM instead of 2 mM. Neoglycoproteins containing complex oligosaccharides were prepared in various ways by using, for instance,

i) *N*-(glycyl)-glycosylamide (Wong *et al.*, 1993): the *N*-(glycyl)-glycosylamide was converted into *N*-isothiocyanatoacetyl-β-glycosylamine:

$$Os\text{-}NH\text{-}CO\text{-}CH_2\text{-}NH_2 \ + \ CSCl_2 \ \rightarrow \ Os\text{-}NH\text{-}CO\text{-}CH_2\text{-}N{=}C{=}S,$$ (Os stands for oligosaccharyl)

and then neoglycoproteins containing about 10 oligosaccharides moieties per mole of bovine serum albumin were obtained.

ii) *N*-glycopyroglutamyl derivatives such as *N*-glycopyroglutamyl paranitroanilide (Quétard *et al.*, 1998). The nitrophenyl group was readily reduced in the presence of hydrogen and Pd/Cas catalyst and then transformed into an isothiocyanate derivative by action of thiocarbamylimidazole (Figure 15); the *N*-glycopyroglutamyl-isothiocynato-anilide readily reacted with amino groups of serum albumin leading to a neoglycoprotein.

Poly-L-lysine has been used by several authors to compact plasmid and to enhance the cell uptake of genes of interest (see for reviews: Frese *et al.*, 1994; Monsigny *et al.*, 1994a,b; Wagner *et al.*, 1994). Various glycosylated poly-L-lysine have been prepared by partially substituting the ε-amino group of lysine residues with simple sugars,

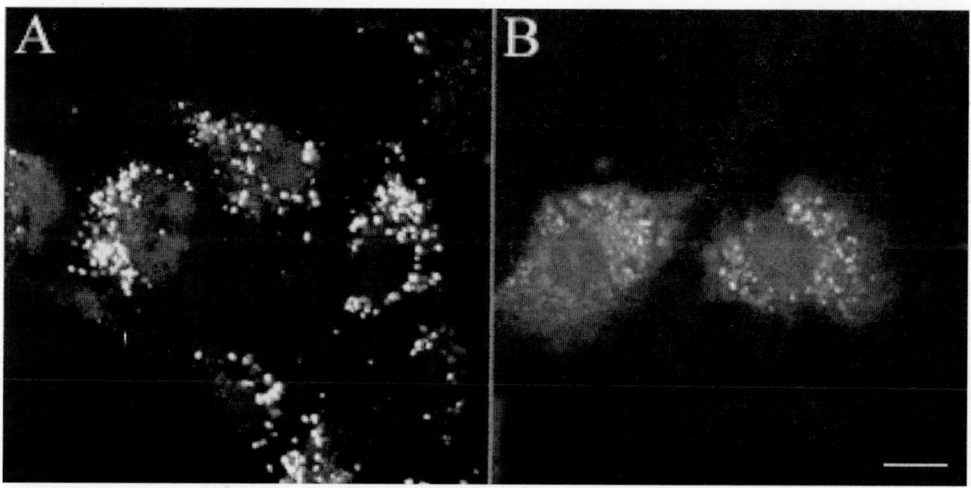

Figure 14: Fluorescein-labelled oligosaccharides, A: Galα 3Galβ 4Glcβ-R; B: Fucα 2Galβ 3GlcNAcβ 3Galβ 4Glcβ-R with R = pyroglutamyl-amidoethylthio-acetamido-fluorescein.
Human lung carcinoma cells, A549, were permeabilized by incubation at 4 °C in the presence of 1 mg/ml saponine in a (2 p 100 per volume) water solution of paraformaldehyde at 4 °C. Permeabilized cells were then incubated in the presence of the fluorescein-substituted oligosaccharide at 4 °C for 30 min. Cells were washed with a sugar-free buffer and observed with the use of a fluorescence microscope equipped with a confocal device. Scale bar = 8 μm.

oligosaccharides or glycosynthons, including the use of glycosylphenylisothiocyanate (Derrien *et al.* 1989; Midoux *et al.*, 1993) of *N*-(tyrosyl)glycosylamides (Wadhwa *et al.*, 1995) and of *N*-(glycosyl)-pyroglutamyl derivatives (Quétard *et al.*, 1998). Briefly, polylysine (as *p*-toluene sulfonic acid salt), containing about 200 or 450 residues was solubilized in DMF and substituted with a glycosyl-pyroglutamylamido phenylisothiocyanate leading to a partially glycosylated polymer containing about 60 or 100 sugar moieties, respectively (Figure 16). The smaller glycosylated polylysine in phosphate buffered saline pH 7.4, upon mixing with a plasmid in the same buffer, led to a compact complex within 20 min. The complex is then ready, without any further step, for transfection. The larger glycosylated polylysine was used to prepare oligonucleotide polymer complexes (see infra).

Similarly N-(tyrosyl)-glycosylamides were also used as starting material to prepare glycosylated carriers (Wadhwa *et al.*, 1995): the N-(N$^\alpha$-Boc-tyrosyl)-glycosylamides were

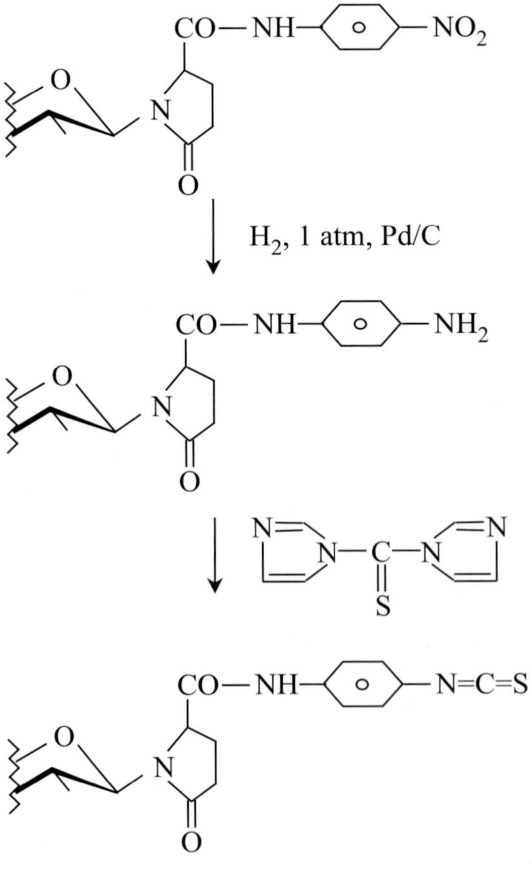

Figure 15: An *N*-oligosaccharyl-pyroglutamyl derivative (*p*-nitroanilide) was quantitatively reduced in the presence of hydrogen and palladium as a catalyst, leading to a *p*-aminoanilide derivative. The intermediary arylamine was quantitatively converted into a *p*-isothiocyanato-anilide derivative upon reaction with thiocarbonyl-bis-imidazole according to Staab and Walther (1962).

deprotected by TFA, the α-amino group was then succinylated and finally the free carboxylic group of the succinyl derivative was activated with a carbodiimide: 1-(3-dimethylaminopropyl)-3-ethyl-carbodiimide hydrochloriole, EDC. This activated glyco-amino acid was used to prepare glycosylated poly-L-lysine.

4. GLYCOTARGETING : GLYCOSYLATED POLYLYSINE AS EFFICIENT CARRIERS OF NUCLEIC ACIDS

4.1. Enhancement of the Activity of Antisense Oligonucleotides

Glycosylated polylysines of relatively large size (dp $\geq$ 455) were shown to give stable complexes with oligonucleotides containing about 20 units, (Stewart *et al.*, 1996). In addition, when the ratio of charges borne by the glycosylated polymers (polycation) and of those borne by the oligonucleotides was close to 1, the efficiency was optimal. This was clearly shown by using a phosphorothioate oligonucleotide, complementary of a segment of an intron close to the 3' end of the ICAM-1 mRNA. This antisense oligonucleotide was inefficient even at concentration as high as 5 μM when it was used alone; to become efficient, it had to be either added as a complex with lipocations (Bennett *et al.*, 1992) or as a complex with partially fucosylated polylysine in the presence of chloroquine (Stewart *et al.*, 1996). In the last case, the IC_{50} was about 0.5 μM. When the polylysine was prepared with a *N*-(oligosaccharyl)pyroglutamyl derivative, containing both the Lewis[a] and the Lewis[x] structures, the inhibition occured at still lower concentration (Figure 17). Oligonucleotides, directly coupled to synthetic glycopeptides, were shown to be taken up

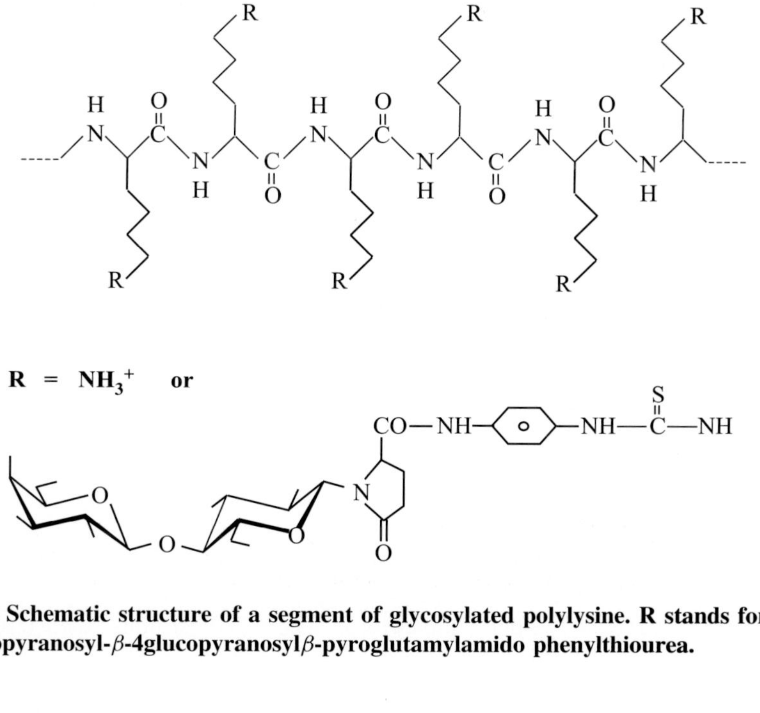

Figure 16: Schematic structure of a segment of glycosylated polylysine. R stands for NH$_3^+$ or for galactopyranosyl-β-4glucopyranosylβ-pyroglutamylamido phenylthiourea.

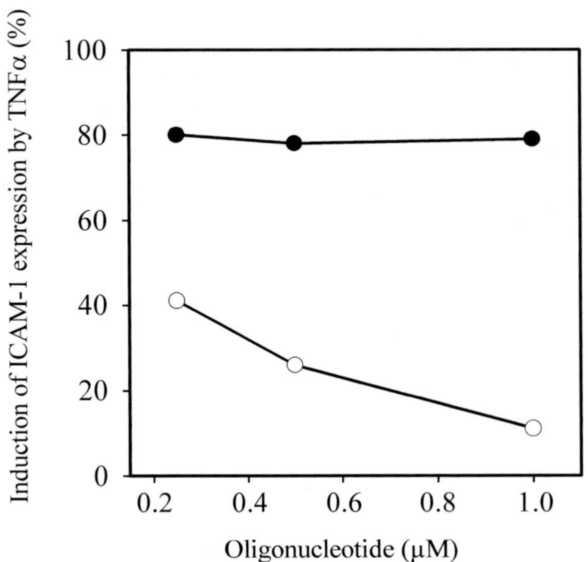

Figure 17: Inhibition of the expression of ICAM-1 induced by TNFα on cultured lung carcinoma cells, A549.

The expression of ICAM-1 was detected on the cell surface by fluorescein-labelled anti-ICAM-1 antibody using a flow cytometer. In the presence of added antisense oligonucleotides (Bennet *et al.*, 1992), the expression of ICAM-1 was slightly inhibited (-●-) when the oligonucleotide was either free or complexed on an unsubstituted polylysine (polymerization dp = 450). Conversely, (-○-) when the antisense oligonucleotide was complexed with a glycosylated polylysine (partially substituted with a fucosylated glycosynthons, Lewis[a], Lewis[x]) the expression was inhibited in an oligonucleotide concentration dependent manner.

Lewis[a]:   Fucα 4GlcNAcβ 3Galβ 4Glcβ-R
           Galβ 3
Lewis [x]:   Galβ 4GlcNAcβ 3Galβ 4Glcβ-R
           Fucα 3

where R is pyroglutamylamido phenylthiocarbamyl. Irrelevant oligonucleotides, free or complexed with unsubstituted polylysine or complexed with a glycosylated polylysine (containing the fucosylated glycosynthons) did not inhibit the expression of ICAM-1.

more efficiently than free oligonuclotides, by cells expressing a lectin with a related sugar specificity (Hangeland *et al.*, 1995).

4.2 Enhancement of the Efficiency of Gene Transfection

Polylysine as a polycation interacts with nucleic acid as a polyanion. This interaction is quite strong, the complex is stable even at high ionic strength up to 1.2 M NaCl, for instance. The formation of the complex induces an efficient compaction of this plasmid leading to toroid particles having about 50 nm in diameter for a 5kb plasmid and a polylysine containing about 200 lysine residues. Such complexes are relatively inefficient to transfer the plasmid in cultured cells. Conversely, glycosylated polylysines bearing oligosaccharides terminated either with lactosyl (Galβ 4Glcβ) units, with three or four galactose residues (triortetrantennary oligosaccaharides) were shown to efficiently target DNA into HepG2 cells (Plank *et al.*, 1992; Midoux *et al.*, 1993; Wadhwa *et al.*, 1995); this cell line is known to express a galactose specific lectin on its surface (Schwartz *et al.*, 1981).

To be efficient, the complexes, either obtained with – unsubstituted or substituted polylysines – must be used in conjunction with either chloroquine or fusogenic peptides. Glycosylated polylysines as well as glycosylated polylysines partially substituted with gluconoyl residues (Erbacher *et al.*, 1997) specifically transfect various cultured cells in a sugar dependent manner, as shown, in Table 1 and Figure 18.

Table 1. Gene transfer by glycofectins^{TM} into various cell types

Cell type		*Source*	*Sugar on pLK*
Human	Macrophage	blood monocytes	Man, Fuc
	HepG2	hepatoma	Lact
	HOS	osteosarcoma	Lact
	CF$^+$	airway epithelial line	Lact
	CF$^-$	airway epithelial line	Lact
	CT/T43	airway epithelial line	Lact
	A431	colon adenocarcinoma	Lact
	A549	lung carcinoma	Fuc, Lact
Murine	RBE4	rat brain endothelial line	Lact, Fuc
	3LL	mouse lung carcinoma	Lact, Glc
	L1210	mouse leukemia	Lact
	B 16	mouse melanoma	Fuc, Lact

Lact: Lactose

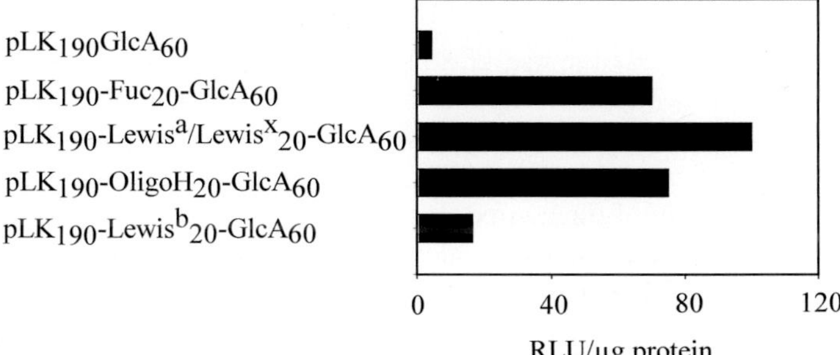

Figure 18: A549 cells were transfected with a plasmid (containing a gene encoding firefly luciferase) complexed with glycosylated polylysine in the presence of chloroquine, according to Erbacher *et al.* 1997.

Polylysine (190 amino acid residues) was partially substituted with gluconoyl residues (GlcA) and with simple sugars (fucosides) or glycosynthons (*N*-oligosaccharyl-pyroglutamylamidophenylthiocarbamyl residues). The presence of sugars enhances the transfection efficiency as monitored by luminescence due to the activity of the expressed luciferase.

Fuc, stands for fucopyranosyl-α-phenylthiocarbamyl residue. Lewis^a, Lewis^x, OligoH stand for N-oligosaccharyl-pyroglutamyl-amidophenylthiocarbamyl residue, with

Lewis^b: Fucα 2Galβ 3GlcNAcβ 3Galβ 4Glcβ-R

        Fucα 4

Oligo H: Fucα 2Galβ 3GlcNAcβ 3Galβ 4Glcβ-R

The high efficiency of glycosylated (and gluconoylated, Derrien *et al.*, 1989) polylysine on suitable cells is mainly due to two reasons: one is due to the fact that the strength of the interaction between the partially glycosylated polymer and DNA, allowing a more efficient release of the plasmid in the cell (Erbacher *et al.*, 1997), is lower than that between an unsubstituted polylysine and DNA; the second one is related to the nature of the sugar itself. Indeed, the sugar borne on the polylysine in the complex acts at the level of the cell surface as in the case of galactosides which are recognized by the lectin of HepG2 cells deriving from the liver parenchyme (Midoux *et al.*, 1993), and in addition at intracellular levels as shown by the high efficiency of glycosylated polylysine containing GlcNAc in the case of rabbit vesicular smooth muscle cells (Boutin *et al.*, 1999) and airway epithelial cells (Fajac *et al.*, 1998, and for a review see Monsigny *et al.*, 1999). Indeed, endogeous lectins are not only present on the plasma membranes, but they are also present in the intracellular membranes – including endosomes, Golgi apparatus, intermediate compartment, reticulum endoplasmic – in addition, soluble lectins have also been found in the cytosol and the nucleus (for a review, see Roche and Monsigny, 1996).

5. CONCLUDING REMARKS

Sugar binding proteins present in mammalian cells: – lectins of the plasma membrane, lectins of intracellular membranes and lectins shuttling between the cytosol and the nucleus – are efficient receptors which selectively recognized simple sugars and in a more specific way, complex oligosaccharides. Oligosaccharides are readily available from natural sources and are now quite efficiently and easily converted into glycosynthons; these glycosynthons open the way to prepare highly specific and very efficient carriers of oligonucleotides and genes.

Future developments include the use of a large, panel of complex oligosaccharides in order to master the uptake of the compounds to be delivered but also the intracellular trafficking and the precise and efficient delivery in the most appropriate intracellular compartment. In addition, new polylysine derivatives, bearing substituents such as histidyl residues, allowing a delivery of the carried nucleic acids into the cytosol without the help of any additional compound such as chloroquine or fusogenic peptides are being developed, (Midoux and Monsigny, 1999). These compounds are the basis of new glycosylated carriers (glycofectins[TM]) designed to very efficiently and quite specifically transfect selected cells.

In addition, glycosynthons made of complex oligosaccharides are being developed with the aim of a specific and efficient delivery of oligonucleotides in selected cells as glyco-oligonucleotide conjugates.

ACKNOWLEDGEMENTS

We would like to thank Philippe Bouchard, Marie-Thérèse Bousser and Philippe Marceau for excellent technical assistance.

This research was partly supported by grants from the Commission of European Communities (CEC), ANRS *(Agence Nationale de Recherches sur le SIDA,* (AIDS), ARC

(Association pour la Recherche sur le Cancer), LNC, *(Ligue Nationale contre le Cancer)*, AFLM *(Association française pour la lutte contre la mucoviscidose)*, *Biotechnocentre* et la *Région Centre*. MM is Professor of Biochemistry at the University of Orléans; CQ received a fellowship from the *Région Centre* and from *ARC;* SB received a fellowship from ANRS and is now Assistant Professor at the University of Orléans; VC received a fellowship from the Ministry of Research; ED, VA and CP are Assistant Professors at the University of Orléans; RM is Research Director at the CNRS *(Centre National de la Recherche Scientifique);* PM and ACR are Research Directors at the INSERM *(Institut National de la Santé et de la Recherche Médicale).*

REFERENCES

Aplin, J.D. and Wriston, J.C. (1981) Preparation, properties and application of carbohydrate conjugates of proteins and lipids. *Crit. Rev. Biochem.* **10**, 259–306.

Arsequell, G., Dwek, R.A. and Wong, S.Y. (1994) 9-Fluorenylmethoxycarbonyl (Fmoc)-glycine coupling of saccharide beta-glycosylamines for the fractionation of oligosaccharides and the formation of neoglycoconjugates. *Anal. Biochem.* **216**, 165–170.

Ashwell, G. and Harford, J. (1982) Carbohydrate-specific receptor of the liver. *Annu. Rev. Biochem.* **51**, 531–554.

Ashwell, G. and Morell, A.G. (1974) The role of surface carbohydrates in the hepatic recognition and transport of circulating glycoproteins. *Adv. Enzymol. Relat. Areas Mol. Biol.* **41**, 99–128.

Bennett, C.F., Chiang, M.Y., Chan, H., Shoemaker, J.E.S. and Mirabelli, C.K. (1992) Cationic lipids enhance cellular uptake and activity of phosphorothioate antisense oligonucleotides. *Molecular Pharmacol.* **41**, 1023–1033.

Boutin, V., Legrand, A., Mayer, R., Nachtigal, M., Monsigny, M. and Midoux, P. (1999) Glycofection: the ubiquitous roles of sugar bound on glycoplexes. *Drug Delivery*, **6**, 45–50.

Cooper, C.A., Packer, N.H. and Redmond, J.W. (1994) The elimination of *O*-linked glycans from glycoproteins under non-reducing conditions. *Glycoconjugate J.* **11**, 163–167.

Derrien, D., Midoux, P., Petit, C., Nègre, E., Mayer, R., Monsigny, M. and Roche, A.C. (1989) Muramyl-dipeptide bound to poly-L-lysine substituted with mannosyl and gluconoyl residues as macrophage activators. *Glycoconjugate J.* **6**, 241–255.

Ellis, G.P. and Honeyman, J. (1955) Glycosylamines. *Adv. Carbohydr. Chem.* **10**, 95–168.

Erbacher, P., Roche, A.C., Monsigny, M. and Midoux, P. (1997) The reduction of the positive charges of polylysine by partial gluconoylation increases the transfection efficiency of DNA/polylysine complexes. *Biochim. Biophys. Acta* **1324**, 27–36.

Fajac, I., Briand, P., Monsigny, M. and Midoux, P. (1999) Sugar-mediated uptake of glycosylated polylysines and gene transfer into normal and cystic fibrosis airway epithelial cells. *Hum. Gene Ther.*, **10**, 395–406.

Frese, J., Wu, C.H. and Wu, G.Y. (1994) Targeting of genes to the liver with glycoprotein carrier. *Adv. Drug Delivery Rev.* **14**, 137–152.

Gabius, H.J. and Gabius, S. (1997) Glycosciences. Status and perspectives. Chapman and Hall, Weinheim, p. 631.

Hangeland, J.J., Levis, J.T., Lee, Y.C. and Ts'o, O.P. (1995) Cell-type specific and ligand specific enhancement of cellular uptake of oligodeoxynucleoside methylphosphonates covalently linked with a neoglycopeptide, YEE(ah-GalNAc)3. *Bioconjugate Chem.* **6**, 695–701.

Isbell, H.S. and Frush, H.L. (1951) Mechanisms for the mutarotation and hydrolysis of the glycosylamines and the mutarotation of the sugars. *J. Res. Nat. Bur. Stand.* **46**, 132–144.

Isbell, H.S. and Frush, H.L. (1958) Mutarotation, hydrolysis and rearrangement reactions of glycosylamines. *J. Org. Chem.* **23**, 1309–1319.

Isbell, H.S. and Frush, H.L. (1980) Primary glycopyranosylamines. *Meth. Carbohyd. Chem.* **8**, 255–259.

Kallin, E., Lönn, H., Norberg, J. and Elofsson, M. (1989) Derivatization procedures for reducing oligosaccharides. Part 3: Preparation of oligosaccharide glycosylamines and their conversion into oligosaccharide-acrylamide polymers. *J. Carbohydr. Chem.* **8**, 597–611.

Kiessling, L.L. and Pohl, N.L. (1996) Strength in numbers: non-natural polyvalent carbohydrate derivatives. *Chem. Biol.* **3**, 71–77.

Lee, Y.C. and Lee, R.T. (1994a) Neoglycoconjugates. Part A. Synthesis. *Meth. Enzymol.* **242**., p. 238.

Lee, Y.C. and Lee, R.T. (1994b) Neoglycoconjugates. Part B. Biochemical application. *Meth. Enzymol.* **247**., p. 450.

214   *Michel Monsigny* et al.

Lee, Y.C., Townsend, R.R., Hardy, M.R., Lonngren, J., Arnarp, J., Haraldsson, M. and Lonn, H. (1983) Binding of synthetic oligosaccharides to the hepatic Gal/GalNAc lectin. Dependence on fine structural features. *J. Biol. Chem.* **258**, 199–202.

Likhosherstov, L.M., Novikova, O.S., Derevitskaja and Kochetkov, N.K. (1986) A new simple synthesis of amino sugar beta-D-glycosylamines. *Carbohydr. Res.* **146**, c1-c5.

Lobry de Bruyn, C.A. (1895) Dérivé ammoniacal de la d-glucose. *Rec. Trav. Chim., Pays-Bas* **14**, 98–105.

Lobry de Bruyn, C.A. and Franchimont, A.P.N. (1893) Dérivés ammoniacaux cristallisés d'hydrate de carbone. *Rec. Trav. Chim., Pays-Bas* **12**, 286–289.

Lobry de Bruyn, C.A. and Van Leent, F.H. (1896) Dérivés ammoniacaux de la mannose, de la sorbose et de la galactose. *Rec. Trav. Chim., Pays-Bas* **15**, 81–85.

Lubineau, A., Augé, J. and Drouillat, B. (1995) Improved synthesis of glycosylamines and a straightforward preparation of *N*-acylglycosylamines as carbohydrate-based detergents. *Carbohydr. Res.* **266**, 211–219.

Manger, I.D., Rademacher, T.W. and Dwek, R.A. (1992a) 1-*N*-(glycyl)-beta-oligosaccharide derivatives as stable intermediates for the formation of glycoconjugate probes. *Biochemistry* **31**, 10724–10732.

Manger, I.D., Wong, S.Y.C., Rademacher, T.W. and Dwek, R.A. (1992b) Synthesis of 1-*N*-(glycyl)-beta-oligosaccharide derivatives. Reactivity of *Lens culinaris* lectin with a fluorescent labeled streptavidin pseudoglycoprotein and immobilized neoglycolipid. *Biochemistry* **31**, 10733–10740.

Midoux, P., Mendes, C., Legrand, A., Raimond, J., Mayer, R., Monsigny, M. and Roche, A.C. (1993) Specific gene transfer mediated by lactosylated poly-L-lysine into hepatoma cells. *Nucleic Acids Res.* **21**, 871–878.

Midoux, P. and Monsigny, M. (1999) Efficient gene transfer by histidylated polylysine pDNA complexes. *Bioconjugate Chem.*, **10**, 406–411.

Molema, G. and Meijer, D.K. (1994) Targeting of drugs to various blood cell types using (neo-)glycoproteins, antibodies and other protein carriers. *Adv. Drug Delivery Rev.* **14**, 25–50.

Monsigny M., Midoux P., Mayer R. & Roche A.C. (1999) Glycotargeting : Influence of the sugar moieties on both the uptake and the intracellular trafficking of nucleic acid-glycosylated polymer complexes. *Bioscience Rep.* **19**, 125–132.

Monsigny, M., Quétard, C., Bourgerie, S., Delay, D., Pichon, C., Midoux, P., Mayer, R. and Roche, A.C. (1998) Glycotargeting: The preparation of glyco-aminoacids and derivatives from unprotected reducing sugars. *Biochimie* **80**, 99–108.

Monsigny, M., Roche, A.C., Kieda, C., Midoux, P. and Obrénovitch, A. (1988) Characterization and biological implications of membrane lectins in tumor, lymphoid and myeloid cells. *Biochimie* **70**, 1633–1649.

Monsigny, M., Roche, A.C. and Midoux, P. (1984) Uptake of neoglycoproteins via membrane lectin(s) of L 1210 cells evidenced by quantitative flow cytofluorometry and drug targeting. *Biol. Cell* **51**, 187–196.

Monsigny, M., Roche, A.C., Midoux, P. and Mayer, R. (1994a) Glycoconjugates as carriers for specific delivery of therapeutic drugs and genes. *Adv. Drug Delivery Rev.* **14**, 1–24.

Monsigny, M., Roche, A.C., Midoux, P. and Mayer, R. (1994b) Sugar specific delivery of drugs, oligonucleotides and genes. In Gregoriadis, G., McCormack, B. and Poste, G. (eds.), *Targeting of drugs, 4: Advances in system constructs. NATO ASI Series A: Life Sciences,* **273**, Plenum Press, New-York, 31–50.

Niemann, C. and Hays, J.T. (1940) Structure of *N*-acetyl-*d*-glucosylamine. *J. Am. Chem. Soc.* **62**, 2960–2961.

Otvos, L., Wroblewski, K., Kollat, E., Perczel, A., Hollosi, M., Fasman, G.D., Ertl, H.C.J. and Thurin, J. (1989) Coupling strategies in solid-phase synthesis of glycopeptides. *Peptide Res.* **2**, 362–366.

Perales, J.C., Ferkol, T., Beegen, H., Ratnoff, O.D. and Hanson, R.W. (1994) Gene transfer *in vivo*: sustained expression and regulation of genes introduced into liver by receptor-targeted uptake. *Proc.Nat. Acad. Sci., USA* **91**, 4086–4090.

Pigman, W., Cleveland, E.A., Couch, D.H. and Cleveland, J.H. (1951) Reactions of carbohydrates with nitrogenous substances. I. Mutarotations of some glycosylamines. *J. Am. Chem. Soc.* **73**, 1976–1979.

Plank, C., Zatloukal, K., Cotten, M., Mechtler, K. and Wagner, E. (1992) Gene transfer into hepatocytes using asialoglycoprotein receptor mediated endocytosis of DNA complexed with an artificial tetra-antennary galactose ligand. *Bioconjugate Chem.* **3**, 533–539.

Privat, J.P., Delmotte, F. and Monsigny, M. (1974) Protein-sugar interaction. Association of (beta-4) linked *N*-acetyl-**D**-glucosamine oligomer derivatives with wheat germ agglutinin (lectin). *FEBS Lett.* **46**, 224–228.

Quétard, C., Bourgerie, S., Normand-Sdiqui, N., Mayer, R., Strecker, G., Midoux, P., Roche, A.C. and Monsigny, M. (1998) Novel glycosynthons for glycoconjugate preparation: Oligosaccharyl*pyro*glutamylanilide derivatives. *Bioconjugate Chem.* **9**, 268–276.

Quétard, C., Normand-Sdiqui, N., Mayer, R., Strecker, G., Roche, A.C. and Monsigny, M. (1997) Simple synthesis of novel glycosynthons: oligosyl-pyroglutamyl derivatives. *Carbohydr. Lett.* **2**, 415–422.

Rasmussen, J.R., Davis, J., Risley, J.M. and Van Etten, R.L. (1992) Identification and derivatization of (oligosaccharyl)amines obtained by treatment of asparagine-linked glycopeptides with N-Glycanase enzyme. *J. Am. Chem. Soc.* **114**, 1124–1126.

Roche, A.C., Barzilay, M., Midoux, P., Junqua, S., Sharon, N. and Monsigny, M. (1983) Endocytosis of glycoconjugates by Lewis lung carcinoma cells. *J. Cell Biochem.* **22**, 131–140.

Roche, A.C., Midoux, P., Pimpaneau, V., Nègre, E., Mayer, R. and Monsigny, M. (1990) Endocytosis mediated by monocyte and macrophage membrane lectins-Application to antiviral drug targeting. *Res. Virol.* **141**, 243–249.

Roche, A.C. and Monsigny, M. (1996) Trafficking of endogenous glycoproteins mediated by intracellular lectins: facts and hypotheses. *Chemtracts Biochem. Mol. Biol.* **6**, 188–201.

Schrével, J., Gros, D. and Monsigny, M. (1981) Cytochemistry of cell glycoconjugates. Prog. Histochem. Cytochem. Vol 14(2). Gustav Fischer Verlag, Stuttgart, p. 269.

Schwartz, A.L., Fridovich, S.E., Knowles, B.B. and Lodish, H.F. (1981) Characterization of asialoglycoprotein receptor in a continuous hepatoma line. *J. Biol. Chem.* **256**, 8878–8881.

Sdiqui, N., Roche, A.C., Mayer, R. and Monsigny, M. (1995) New synthesis of glycopeptides. *Carbohydr. Lett.* **1**, 269–275.

Staab, H.A. and Walther, G. (1962) Reaktionen von *N,N'*-thiocarbonyl-di-imidazol. *Ann. Chem.* **657**, 98–103.

Stewart, A.J., Pichon, C., Meunier, L., Midoux, P., Monsigny, M. and Roche, A.C. (1996) Enhanced biological activity of antisense oligonucleotides complexed with glycosylated poly-L-lysine. *Mol. Pharmacol.* **50**, 1487–1494.

Stowell, C.P. and Lee, Y.C. (1980) Neoglycoproteins. The preparation and application of synthetic glycoproteins. *Adv. Carbohyd. Chem. Biochem.* **37**, 225–281.

Tamura, T., Wadhwa, M.S. and Rice, K.G. (1994) Reducing end modification of *N*-linked oligosaccharide with tyrosine. *Anal. Biochem.* **216**, 335–344.

Tarentino, A.L., Gomez, C.M. and Plummer, T.H. (1985) Deglycosylation of asparagine-linked glycans by peptide: *N*-glycosidase F. *Biochemistry* **24**, 4665–4671.

Tarentino, A.L., Phelan, A.W. and Plummer, T.H. (1993) 2-Iminothiolane: a reagent for the introduction of sulphydryl groups into oligosaccharides derived from asparagine-linked glycans. *Glycobiology* **3**, 279–285.

Tarentino, A.L. and Plummer, T.H. (1994) Enzymatic deglycosylation of asparagine-linked glycans: purification, properties, and specificity of oligosaccharide-cleaving enzymes from *Flavobacterium meningosepticum*. *Meth. Enzymol.* **230**, 44–57.

Wadhwa, M.S., Knoell, D.L., Young, A.P. and Rice, K.G. (1995) Targeted gene delivery with a low molecular weight glycopeptide carrier. *Bioconjugate Chem.* **6**, 283–291.

Wadhwa, M.S. and Rice, K.G. (1995) Receptor mediated glycotargeting. *J. Drug Target.* **3**, 111–127.

Wagner, E., Curiel, D. and Cotten, M. (1994) Delivery of drugs, proteins and genes into cells using transferrin as a ligand for receptor-mediated endocytosis. *Adv. Drug Delivery Rev.* **14**, 113–135.

Wong, S.Y.C., Guile, G.R., Rademacher, T.W. and Dwek, R.A. (1993) Synthetic glycosylation of peptides using unprotected saccharide beta-glycosylamines. *Glycoconjugate J.* **10**, 227–234.

Index